MECHANICAL SCIENCES

The Institution of Engineers (India) Textbook Series

MECHANICAL SCIENCES

G.K. Lal
Vijay Gupta
N.G.R. Iyengar
B.N. Banerjee
K. Ramesh

Narosa Publishing House

New Delhi   Chennai   Mumbai   Kolkata

G.K. Lal
Vijay Gupta
N.G.R. Iyengar
B.N. Banerjee
K. Ramesh
Indian Institute of Technology, Kanpur
Kanpur, India

Copyright © 1998 Narosa Publishing House Pvt. Ltd.
First Reprint 2008
Second Reprint 2008
Third Reprint 2009
Fourth Reprint 2010
Fifth Reprint 2011

NAROSA PUBLISHING HOUSE PVT. LTD.

22 Delhi Medical Association Road, Daryaganj, New Delhi 110 002
35-36 Greams Road, Thousand Lights, Chennai 600 006
306 Shiv Centre, Sector 17, Vashi, Navi Mumbai 400 703
2F-2G Shivam Chambers, 53 Syed Amir Ali Avenue, Kolkata 700 019

www.narosa.com

ISBN 978-81-7319-213-5

Published by N.K. Mehra for Narosa Publishing House Pvt. Ltd.,
22 Delhi Medical Association Road, Daryaganj, New Delhi 110 002

Printed in India

Dedicated

to

Professor Amitabha Bhattacharya

The Institution of Engineers (India)

(ESTABLISHED 1920, INCORPORATED BY ROYAL CHARTER 1935)

8 GOKHALE ROAD, CALCUTTA-700 020

PHONE : 248-8311/8314/8315/8316/8334/8335 □ GRAM : ENJOIND □ TELEX : 21-7885 IEIC IN □ FAX : (91) (33) 248 8345

Pro-Vice-Chancellor
Indira Gandhi National
Open University
Maidan Garhi, New Delhi 110068
Ph : (011) 6968935, (R) (011) 6493257
Tlx : 73023, IGOU IN, Gram : IGNOU
Fax : (011) 6855102
E-mail : ignou@giasd101 vsnl net in

Prof JANARDAN JHA, FIE
VICE PRESIDENT

Foreword

The primary need of professional practice in the post-industrial age is the adaptability to cope with non-formal education. Distance education has brought learning to the place of learner. The Institution of Engineers (India) is the pioneer in the field of non-formal education in India. Recently, the course structure and the syllabi of Sections A & B of Institution Examinations have been revised to commensurate with present day requirement.

A large number of our student members appear in the bi-annual examinations of the Institution. They need guidance in various forms. The Council has, therefore, decided to produce learning aids such as text-books, audio-video cassettes, etc. for the benefit of these student members.

Publication of a series of books by the Institution in collaboration with Narosa Publishing House, New Delhi, is an effort in this direction.

I thank Prof. G.K. Lal, Department of Mechanical Engineering, Indian Institute of Technology, Kanpur and his colleagues, who made this publication possible in this form.

I am confident that this textbook on "Mechanical Sciences" will help the students in a big way as very few such books are available in the market.

(Prof. Janardan Jha)
Vice-President
The Institution of Engineers (India)

Preface

An engineer is confronted with the broadest spectrum of activities which range from research to design, applications and sales. Because of this, an engineer not only should have a good understanding of the basic sciences, engineering sciences and mathematics, but must also be knowledgeable about social sciences. A broad based training with basic science, engineering science, mathematics and social science along with design and applications oriented courses is therefore essential for an engineer.

Engineering science courses are designed to emphasise the principles of science in such a way that they are useful for the solution of engineering problems. Their requirements may be different for different disciplines, but engineering mechanics and thermodynamics background is considered essential for all engineers. This book has been written as an introductory book on mechanical sciences and covers the fundamentals of solids mechanics, fluid mechanics and thermodynamics. The mechanics of solids (Chapter 2-13) primarily emphasises on statics, dynamics and strength of materials. Fluid mechanics (Chapters 14-19) deals with elementary aspects of fluid at rest as well as in motion. Chapters 20 to 23 discuss the basic thermodynamic principles.

For better understanding, almost every topic has been clarified by means of solved examples and numerical problems added at the end of each chapter. Subjectwise bibliography at the end will benefit those readers who would like to have further clarification and go beyond the scope of this book.

Written as a textbook, it is primarily meant for students preparing for undergraduate engineering examinations at university or professional society levels. The Institution of Engineers (India) which introduced a compulsory course on mechanical sciences for all disciplines, funded the preparation cost of this book. This assistance is gratefully acknowledged. In fact, the book has been prepared keeping the Institution of Engineers (India) students in mind.

The book would not have been completed without the help and assistance of our friends and colleagues, and encouragement

and support of our wives and children. Our special thanks are due
to Mrs. Sandhya Agnihotri for her painstaking efforts in typing
the manuscript in LaTex, and to Dr. N. Venkata Reddy, Dr.
Philip Koshy and Mr. O.P. Bajaj for their assistance in finalizing
the manuscript.

Authors

Contents

1. INTRODUCTION

Engineering has often been defined as the art of effective utilization of natural resources for the survival and benefit of mankind. Engineering, therefore, demands creative imagination and innovative application of natural phenomena. It continually seeks newer, better and cheaper means of harnessing the rich natural resources of materials and energy. It has been an integral and perhaps indispensible part of the development process. The success of engineering activities in the developed countries in eliminating poverty is obvious. These transformed societies are not only totally dependent on their engineering personnel but have been demanding newer working ideas and solutions from them.

The word 'engineer' perhaps originated from the group of people who were originally involved in the development of war materials and called 'ingenia'. These 'ingenia' were subsequently provided scientific training and towards the end of 17th Century became known as 'ingeniers'. With increase in developmental activities they were divided and trained as military and civil engineers. With the passage of time, the engineering activities became more complex and further division took place. While civil engineers continued to deal with static structures like buildings, bridges, dams, roads etc., a new discipline of mechanical engineering concentrated on dynamic structures such as machines and engines. Many other disciplines got evolved over the years with electrical and chemical engineers concentrating on practical applications of electricity and chemistry, mining and metallurgical engineers applying themselves to excavation of ore bodies and extraction of ores, and so on.

The general conception of an engineer gradually became that

of a mediator between the philosopher (creator) and the working mechanic (fabricator), and of one who understands and interprets the language of both. The engineer, therefore, must possess both practical and theoretical knowledge.

Three basic human activities which have been the understanding of nature and natural phenomena, the generation of ideas and application of these for the welfare of mankind are depicted in Fig. 1.1. The generation of ideas, in today's termi-

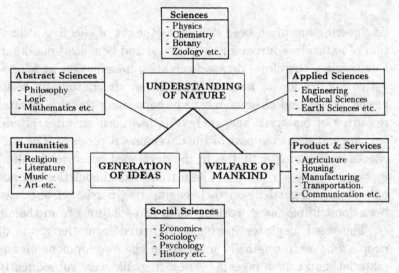

Fig. 1.1 Human activities.

nology, falls under the 'Humanities' activities which deals with religion, literature, music, art etc. Similarly, understanding of nature and natural phenomena are the activities of 'Sciences' such as physics, chemistry, biology etc. Welfare of mankind is essentially through products and services which include agriculture, housing, manufacturing, transportation, communication and several others. A number of activities, however, fall in-between. For example, the activities in the areas of economics, psychology, sociology etc. which are known as 'Social Sciences' come in-between second and the third activity, while 'Abstract Sciences' such as philosophy, logic, mathematics etc. fall between first and second activity. The 'Applied Science' area obviously comes between the second and the third activity.

Persons in the engineering profession work very closely with scientists for the application of scientific principles – both new and old for development of products and services. Scientists are mainly concerned with the nature and behaviour of natural systems, while engineers deal with realities and are good at inventing new ways of using natural phenomena. Invariably engineering problems have more than one solution and the engineer must obtain the best solution within the resources at his disposal. A best solution is rarely obtained in the first attempt and the proposed design usually requires several attempts at refinement over a long period of time before a near optimum solution is obtained.

The range of activities of an engineer is very wide. At times he thinks and acts like a scientist (research), at other times as designer of a new device or a system, while on other occasions he acts like a sales or applications engineer and deals with the customers. A spectrum of engineering activities indicated in Fig. 1.2 shows how his activities vary from 'Science' to 'Product and Services' In a scientific-technical society the relationship of engi-

Science

* Research
* Analysis & Design
* Development
* Design and Manufacture
* Production Planning
* Production
* Control of Quality
* Maintenance
* Marketing
* Customer Service & Feedback

Fig. 1.2 Engineering activities.

Product & Service

neering to science on one hand and to production on the other cannot be ignored. Engineers now must know, understand and utilize the basic physical and mathematical sciences and at the same time relate their work to suit the production conditions. After all a product that cannot be manufactured is not useful. In addition to the science-production relationship, the functional-aesthetic relationship cannot be ignored. A product which is not reasonably attractive is seldom acceptable. Similarly, all engi-

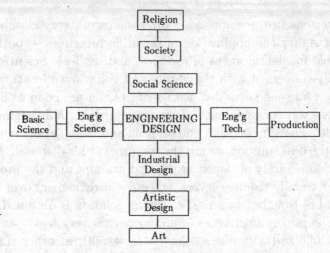

Fig. 1.3 Engineering Design.

neering activities have a link with social sciences. For example, economic resources must be available and the engineers must be able to comprehend not only the economies of his department and the organization but also of the society and the country. Since an engineer is a human being, one cannot reject the influence of religion, philosophy and society on his expression. Fig. 1.3 shows these relationships where engineering design is the centre-piece as this is the ultimate epitome of an engineer's expression. Clearly that an engineer is confronted with the broadest spectrum of subject essentially because of the broad range of activities with which he is involved. He not only should have a good understanding of the basic sciences, engineering sciences and mathematics but must be knowledgeable in social sciences. The training of an engineer should, therefore, be sufficiently broad-based and should include courses in basic sciences, mathematics, engineering sciences and social sciences in addition to other design and applications oriented courses.

Engineering science courses have been evolved to emphasise the principles of science in a form which is well suited for the solution of engineering problems. It may be mentioned that a very few engineering problems can be solved in a completely rational way. Invariably it is found that the scientific theories available are too complex for an analytical solution or that all the facts are

not known. Engineers, therefore, most often resort to approximations, often drastic ones, in order to arrive at a solution and then
use their experience or experimental/practical results in the form
of empirical rule to make appropriate corrections in the solution.
Often more than one approach is considered, and it is only after
several attempts that a meaningful solution is obtained.

The requirements for engineering science courses vary from
discipline to discipline but for potential mechanical engineers the
areas that have been considered important are engineering mechanics and thermodynamics.

Engineering mechanics is based on the knowledge concerning
the state of rest or motion of material bodies under the influence
of forces. Engineering mechanics has become one of the most fundamental subject of study for engineers since static or dynamic
phenomena are involved in just about every engineering problem.

Mechanics is perhaps the oldest physical science dating prior
to the time of ancient Egyptian builders (prior to 2000 BC) and
Archimedes (287-212 BC). The basic laws of mechanics - the laws
of motion were, however, formulated by Newton (1642-1727 AD)
which were later extended and modified by Eüler, (1707-1783
AD), d'Alembert (1717-1783 AD), Lagrange (1736-1813 AD),
Hamilton (1805-1865 AD), and others.

Engineering mechanics is usually divided into two parts: mechanics of solids, and mechanics of fluids. Solid mechanics is
further divided as mechanics of rigid bodies and mechanics of
deformable bodies. Mechanics of rigid bodies leads to statics
and dynamics, while mechanics of deformable bodies is further
studied as theories of elasticity and plasticity, and as strength of
materials (Fig. 1.4). The basic laws and principles on which the
entire science of mechanics is based could be summarized as
1. Force laws:
   Newton first law of motion (inertia)
   Newton second law of motion
   Newton third law of motion (action & reaction)
   D'Alembert principle
2. Energy principles:
   Principle of virtual work
   Principle of potential energy

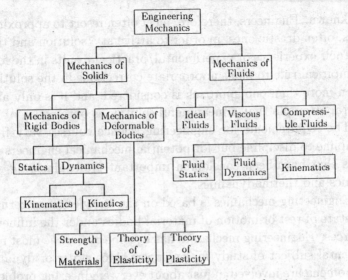

Fig. 1.4 Scope of engineering mechanics.

Principle of work and energy

Principle of conservation of energy

3. Other laws:

Newton law of gravitational attraction

The parallelogramme law

The law of superposition and transmissibility

Statics deals with the equilibrium of bodies under the action of forces, while dynamics with the motion of bodies — kinematically or kinetically, under kinematics the motion is investigated without any regard to the forces and deals mainly with position, displacement, velocity, acceleration and time. The relationship of forces and the resulting motion of bodies are investigated under kinetics.

In modern-day factories, thousands of automatic operations are involved in handling and movements of jobs including electronic devices which require mechanical parts to perform the assigned tasks. Such tasks are accomplished with the help of machines and mechanisms. While machine is a device which uses power to accomplish a physical effect, mechanism is a combination of machine elements arranged to produce a specified motion. In other words, a machine performs work, while a mechanism has a characteristic motion which helps machine perform work.

Consider a simple slider-crank mechanism shown in Fig. 1.5. In actual internal-combustion engines, the reciprocating piston sucks in a air-fuel mixture and then compresses it. At a prede-

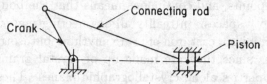

Fig. 1.5 The slider-crank mechanism.

termined compression ratio defined in terms of the position of the piston, spark-ignition takes place and the resulting explosion causes the piston to move downwards, which in turn rotates the crank. In this set-up, the piston-connecting rod-crank combination is a mechanism. In designing an arrangement of this kind, an engineer must understand and analyse several aspects of the problem. He must know the relationship between the motions of the piston and the crank; the effects of varying the crank radius or the length of the connecting rod and the magnitude and direction of acceleration on the piston during a complete cycle. Answer to these and several other questions are obtained through kinematic analysis of the mechanism.

The science of 'strength of materials' deals with the relations between internal forces, external loads, stability and deformation of various machine elements and structures subjected to straining action. Here the member is assumed to be in equilibrium and the relationship between the external forces acting on the body and the internal forces resisting the action of external forces is obtained from equations of static equilibrium. The internal resisting forces are generally expressed in terms of stresses acting over the area under consideration and its value is evaluated as the integral of stress times the elemental area over which it acts. For evaluating this integral, the stress distribution over the area has to be known. This is obtained by measuring or evaluating the strain distribution on the plane under consideration, since stress is related to strain for small deformations. Strain expression obtained is finally substituted in the stress-equilibrium equations for evaluating loads on the member.

In all strength-of-materials problems, it is assumed that the body undergoing deformation is isotropic, homogeneous and continuous. Isotropic means that the properties of the material are same in all directions, homogeneous means that identical properties at all points, and continuous means that the body has no voids or empty spaces. In reality all engineering materials such as steel, aluminium, cast iron etc. are anything but isotropic and homogeneous since they are made up of crystal grains having different properties along crystallographic planes. These grains are, however, very small and on a macroscopic scale they are statistically isotropic and homogeneous.

The theory of elasticity gives much greater considerations to the stresses and strains in a loaded member than in the strength-of-materials analysis. For arriving at a solution, the strength-of-material analyses are often made simpler by making simplifying assumptions, particularly with respect to the strain distribution in the loaded member. Such assumptions usually satisfy the physical situation but may not be mathematically rigorous. In elasticity theory, no simplifying assumptions are made concerning the distribution of strain in the loaded member. Here the equations of equilibrium have to be satisfied and at the same time the deformation of each element must be such that elastic continuity is preserved. This means that no cavity or pile-up can occur in a deforming material. This is ensured through use of compatibility equations which define strains in terms of displacements. Most of the complications in the theory of elasticity arise out of the necessity of satisfying the requirements of continuity in elastic deformation.

The theory of plasticity deals with the behaviour of materials when they are deformed beyond the elastic range where Hooke law no longer holds. Mathematical description of plastic deformation is much more complicated than elastic deformation since there is no simple relationship between stress and strain in the plastic region. It gets further complicated since the plastic strain depends not only upon the final load but also on the loading path. In elastic deformation the strain is not influenced by the loading path and depends only on the initial and the final state of stress. The analysis of plastic strains is required for analysis of metal forming processes and other problems involving plastic deformations.

Mechanics of liquids and gases are based on the same basic laws and principles that are employed in the mechanics of solids. The scientific study of fluid flow started with an imaginary ideal fluid which was considered to be frictionless. This approach of the mathematical physicists developed into a theoretical science known as classical hydrodynamics. The engineers claimed that results of such idealized theories where properties of real fluids are not considered have no practical value. Consequently, they turned to experiments and proposed empirical and formulae for applications in specific fields. This applied science later became known as hydraulics. Hydraulics was mainly confined to water but with developments in the petroleum and chemical industries and later in aeronautics need for broader treatment became inevitable. Physicists and engineers then started combining the basic concepts of fluid motion of classical hydrodynamics with experimental techniques of hydraulics which led to the development of mechanics of fluids. The experimental data is now used for verifying the theories or for providing information which supplement theoretical analyses. The end result of this effort is a set of basic laws and principles of fluid mechanics which can be applied to obtain solutions of the fluid-flow problems encountered by engineers.

Engineering applications of fluid mechanics can be broadly classified into two categories. The first one involves the study of forces which cause or result from the motion of fluids. In second, the effect of fluid motions on heat and mass transfer through fluid body is examined. The forces in fluids have diverse applications ranging from motion of a sail boat and driving of a wind mill to designing of automobiles, submarines and aircrafts. Similarly, by appropriately controlling the motion of the fluid, the heat and mass transfer rates can be modified in cooling towers, chimneys, heart-and-lung machines and spacecrafts.

Like solid mechanics, fluid mechanics can also be divided into three areas: fluid statics, fluid dynamics and kinematics. As the name imply, fluid statics deal with the mechanics of fluids at rest, while fluid dynamics is concerned with the relationship between the velocities and accelerations, and the forces during fluid motion. Kinematics, on the other hand, deals with velocities and streamlines without considering forces or energy.

The prosperity of a nation, in general, is a direct function of the per capita energy consumption. Thus energy and its transformation is most important for man's societal needs, and thermodynamics essentially deals with this aspect. Although all types of energy fall within the scope of thermodynamic analysis, most studies of thermodynamics are concerned primarily with two forms of energy — heat and work.

As indicated earlier, engineering mechanics deals with force, mass, distance and time and is developed through the application of basic laws and principles. The behaviour of a mechanical system is described in terms of its interaction with its surroundings through the application of force. Just as mechanical system is analysed using the free-body technique, the thermodynamic system is analysed by identifying a certain quantity of matter and a study is made of its interaction with the surroundings. In mechanics we study the forces which act on the free-body, while in thermodynamics we study the effect of various forms of energy as they interact with a system and its surroundings. Here the emphasis is on energy interchanges — energies that enter or leave the system.

The analytical approach in thermodynamics could be macroscopic or microscopic. In macroscopic studies, the quantity of matter is considered without any considerations to the events at the molecular level, while in microscopic the behaviour is described by summing up the behaviour of each molecule. Thus, macroscopic analyses are completely independent of the assumptions regarding the nature of matter, but the results can be derived from microscopic considerations.

Thermodynamics is based on observations which have been formulated into thermodynamic laws which govern the principles of energy conversion. These thermodynamic laws and principles have found wide engineering applications in power plants, internal combustion engines and gas turbines, air conditioning and refrigeration, chemical plants, jet propulsion etc.

2. FORCE SYSTEMS

2.1 Introduction

The concept of force is of fundamental importance in mechanics as well as in all fields of physics. This chapter deals with composition and resolution of forces, various force systems, equivalence and resultants of force systems and other related concepts. The concepts discussed in this chapter are common to both statics and dynamics. Towards the end of the chapter the stage wherein statics and dynamics branch out into different disciplines is pointed out.

2.2 Concept of a Rigid Body

Idealisation of system characteristics for the purpose of analysis is one of the very crucial steps in engineering analysis. In general, the actual behaviour of physical systems are very complex. From the analysis point of view, considering all the features of a system would be very difficult or even impossible to solve in most cases. The trick is to set up a model of the physical system which is simple enough to analyse and yet it exhibits the phenomena under consideration. Such simplified models can be called mathematically ideal models, or simply mathematical models or ideal models. Idealisation and use of such models are valid if the analytical solution checks well with the results of experimentation or observation.

One of the important and basic idealisations used in mechanics is to consider a body as *rigid*. A body is said to be rigid if

11

it does not deform under the action of forces, i.e., *a rigid body is one in which all particles remain at fixed distances from each other irrespective of the forces that act on the body.* All bodies deform under the action of forces, but if the deformations are negligible, the body can be considered to be rigid for the purpose of analysis. However, one must be careful not to apply this concept indiscriminately. Some of the early studies involving the determination of how beams resist loads were completely erroneous because the pioneering investigators overlooked the effects of elastic deformation.

Idealisation of the given body as a rigid body is used in the study of both — *statics* and *dynamics*. In these studies one is interested in finding the external effects of the force on the body and not on how the internal particles of the body resist the external loads. However, when one has to analyse how the body resists the external loads, i.e., in the study of *Strength of Materials/Mechanics of Solids* one has to take into account the deformation behaviour of the body. Even while doing so, one generally has to idealise the deformation behaviour. This would be discussed later in this book.

In addition to idealizing *physical systems*, one may also idealise *physical actions.* One of the important idealisations used is the concept of point force or concentrated force. A detailed description on how a distributed force can be idealised to a concentrated force is discussed in the later part of this chapter. It is important to note that the concept of a concentrated force depends on the idealisation of the body as rigid.

2.3 Concept of a Force

A force is the action of one body on another body which changes or tends to change the motion of the body acted on. Force is characterised by its point of application, its magnitude, and its direction. Hence, force is represented by a vector.

According to Newton's third law of motion, because of the inertia of all material bodies, the body acted upon will react to or oppose the force acting on it with a force of equal magnitude and

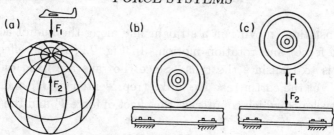

Fig. 2.1 Force interactions

opposite sense. Thus, force always exists in pairs. For example, in Fig. 2.1(a), there is an attraction between the aeroplane and the centre of the earth indicated by the pair of vectors F_1 and F_2. In Fig. 2.1(c), there is a force interaction between the rail road wheel and the rail indicated by the pair of vectors F_1 and F_2. The vectors F_1 and F_2 are equal in magnitude but with opposite sense having the same line of action. It has become customary to apply the term force indiscriminately to either the pair F_1, F_2 or to the single vectors F_1 or F_2 separately.

Looking from a different point of view, *force* is a useful and simple concept for describing a very complex physical interaction between "bodies". Force interactions may occur when there is direct contact between systems, (Fig. 2.1(b)) or between systems which are physically separated (Fig. 2.1(a)).

If an isolated system such as an aeroplane (Fig. 2.1(a)) is taken up for analysis, then the interaction with the earth can be represented by the vector F_1, where F_1 is the force exerted by the earth upon the aeroplane. Similarly, the interaction of rail road wheel with the rails can be represented by F_2 (Fig. 2.1(b)).

As mentioned earlier, the effect of a force is to change the motion of the body. In Fig. 2.2(a), the attraction of the earth has a tendency to alter the motion of the aeroplane. In general, the effect of a force on a rigid body is to produce a combination

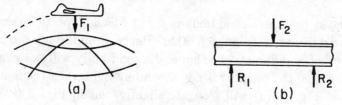

Fig. 2.2 Effect of force.

of translation (moving in a straight line along the line of action of the force) and rotation motion. In Fig. 2.2(b), the effect of force is to translate the rails. Since the rails are fixed on the surface of the earth, reactive forces (represented by R_1 and R_2) are developed which counteract the effect of force F_2 and the rails are under rest.

In International System of Units (SI system), the unit of force is Newton. A Newton is defined as that force which gives an acceleration of 1 m/s^2 to a mass of 1 kg.

2.4   Transmissibility of a Force

In dealing with the mechanics of rigid bodies, the principle of transmissibility states that *the external effects of a force are independent of the point of application of the force along its line of action.* Only the external effects on the body remain unchanged, the internal effects, however, may vary greatly as the force is moved along its line of action. For example, when two equal, opposite, and collinear forces F_1 and F_2 are acting on the body at points A and B, respectively (Fig. 2.3(a)), the body is in tension. If force F_1 is moved along its line of action to B, and F_2 is moved to A, (Fig. 2.3(b)), then the body is in compression. The body is in equilibrium before and after the forces have been moved along their line of action but the internal effect of the force on the body is changed.

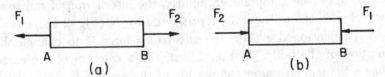

Fig. 2.3 Transmissibility of a force.

By definition, a rigid body is one in which all particles remain at fixed distances from each other. Hence, even though the internal effect is different, the shape of a rigid body will not change. However, if the member is not absolutely rigid, then the member AB in Fig. 2.3(a) will elongate slightly and in Fig. 2.3(b) the member will contract slightly.

The principle of transmissibility, therefore, must be avoided or at least used with care, in determining the internal forces and deformations. It may, however, be used freely to determine the conditions of motion or equilibrium of rigid bodies and to compute the external forces acting on these bodies.

As another example, consider the situtation shown in Fig. 2.4. As long as the line of action and the direction is the same, pushing or pulling a body has the same effect. The choice of pushing or pulling is just a matter of convenience. The forces acting on a

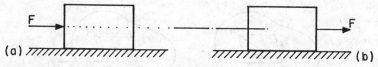

Fig. 2.4 Pushing or pulling effect of a rigid body.

rigid body can be represented by *sliding vectors*, since these vectors may be allowed to slide along their line of action in view of the principle of transmissibility.

2.5   Classification of Force Systems

Any number of forces, considered as a group, constitute a force system. Force systems can be classified as *concurrent*, *coplanar*, *collinear* etc.. It is said to be *concurrent*, if the action lines of all the forces intersect at a common point. If they do not intersect at a common point it is a *non-concurrent* force system. When all the forces lie in the same plane, the force system is *coplanar*, and when the forces do not lie in the same plane, it is *non-coplanar*. *Collinear* is one in which all the forces have a common action line. When the action lines of the forces are parallel, the system is said to be *parallel*.

2.6   Composition and Resolution of Forces

The process of reducing a force system to an equivalent and simplest force system, which has the same external effect on the body as the original force system, is called *composition*. Determination of the resultant of concurrent forces, is one of the simplest examples of composition. The resultant in such a case can be determined using the parallelogram law of vector addition.

In Fig. 2.5(a), the resultant of forces F_1 and F_2 is determined using the parallelogram law, as R which passes through the intersection of the two forces. If the two concurrent forces are applied at different points and their line of action intersect at P, then using the principle of transmissibility the forces can be moved to the point P and the resultant is again obtained as R (Fig. 2.5(b)).

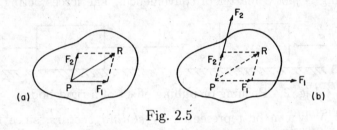

Fig. 2.5

The resultant of the two forces (Fig. 2.5) has the same external effect on the body as that of the original force system. The reverse of this process, namely, finding two components of a force which will have the same external effect on a body as the force itself is very useful in the study of mechanics. This process is known as the *resolution of a force*.

While dealing with cartesian co-ordinates, the most common two-dimensional resolution of a force F, is the resolution into rectangular components F_x and F_y (Fig. 2.6). If the unit vectors along x and y directions are designated as i and j respectively, then

$$F = F_x + F_y = F_x\, i + F_y\, j \qquad (2.1)$$

where F_x and F_y are the magnitudes of the components along x and y directions. From Fig. 2.6 one can obtain the magnitudes of the components as

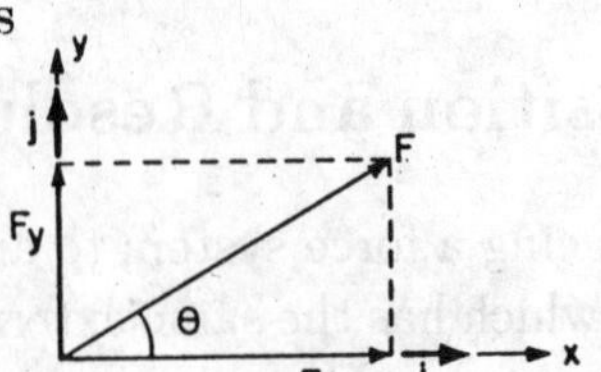

Fig. 2.6

$$F_x = F \cos \theta; \quad F_y = F \sin \theta;$$

$$\tan \theta = F_y / F_x \qquad (2.2)$$

and the magnitude of the force as

$$F = \sqrt{(F_x^2 + F_y^2)} \qquad (2.3)$$

The direction cosines of this force are

$$\ell = F_x / F, \quad m = F_y / F, \quad n = 0 \qquad (2.4)$$

and the unit vector along this force is

$$e_F = \ell i + m j + n k \qquad (2.5)$$

Example 2.1

Determine the magnitude and direction of the resultant of the two forces shown in Fig. 2.7.

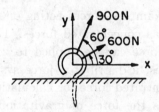

Fig. 2.7

Solution

Resolving 600 N and 900 N forces along x and y directions gives

$$600 \cos 30° \, i + 600 \sin 30° \, j = 519.6 \, i + 300 \, j$$

$$900 \cos 60° \, i + 900 \sin 60° \, j = 450 \, i + 779.4 \, j$$

and the resultant

$$\begin{aligned}
R &= (519.6 + 450)i + (300 + 779.4)j \\
&= 969.6i + 1079.4j
\end{aligned}$$

The magnitude of R is $\sqrt{(969.6)^2 + (1079.4)^2} = 1450.94$ N and its direction, $\theta = \tan^{-1}(1079.4/969.6) = 48°$

2.7   Body and Surface Forces

Forces acting on a body can be classified into body forces and surface forces. Body forces act on each volume element of the body and surface forces act on each surface element of the body. Body forces are exerted when the body is kept in a gravitational, magnetic, electric or centrifugal field. The force of the earth on an object at or near the surface (the object in the gravitational field of the earth) is called the weight of the object. In the SI system, the acceleration due to gravity near the earth's surface is approximately 9.81 m/s^2. A mass of 1 kg on the earth's surface will experience a gravitational force of 9.81 N, i.e., weight of 1 kg mass is 9.81 N. Surface force is exerted on the body by direct mechanical contact (Fig. 2.1(b)).

2.8   Distributed and Concentrated Forces

In the preceeding examples, forces acting on a body have been represented simply as concentrated forces. However, in reality only a distributed force can be applied to a body. Using the composition of forces, a distributed force system is reduced to an equivalent concentrated force for calculation purposes. The weight of a body is the force of gravitational attraction distributed over its volume and may be taken as a concentrated force acting through the centre of gravity. Since, surface forces are applied by direct mechanical contact, force is applied over a finite line/area and is, therefore, a distributed force. A distributed force could be reduced to a single equivalent force which would have the same overall effect at points away from the region of application. Thus, surface forces can also be reduced to an equivalent concentrated force. Representation of distributed forces by means of a concentrated force acting at specific points is actually an abstraction, justified only because it greatly simplifies the analysis.

Figure 2.8 shows the body acted upon by a surface force at C. To develop physical understanding, it is often desirable to visualize how a force can be applied. Further, it also indirectly helps in modelling physical situtations for obtaining engineering

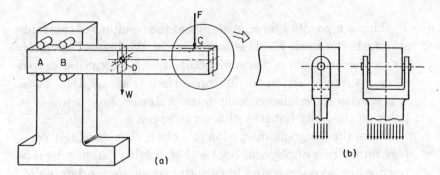

Fig. 2.8 Method for applying a concentrated force.

solutions. In Fig. 2.8, the vector F indicates that the force is
applied at point C and that the action is concentrated along the
line of the vector. One possible way of applying the force F is
shown in Fig. 2.8(b).

The force is actually applied to the beam by the pressure of
the pin on the lower surface of the hole as shown in Fig. 2.9(c).
At any cross section, the interaction between the pin and the hole
(Fig. 2.9a) can be represented by a uniformly distributed force
system (Fig. 2.9(b)). The force system is co-planar. Further, in
view of the geometry of the interacting bodies, the force system
is concurrent and passes through the centre of the pin. Using
the principle of composition, the force system can be replaced
by a concentrated force f_2 passing through the centre of the pin.
Since the pin has a line contact through the thickness of the
beam, another distributed system is obtained (Fig. 2.9(c)).

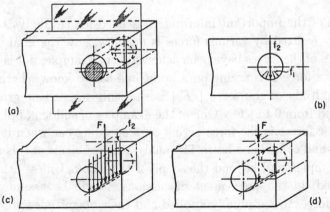

Fig. 2.9 Concentrated force is an abstraction. In general forces
are distributed.

This is a *parallel force system* and the resultant of this force system is the force *F*. This force acts at the base of the hole. Using the principle of transmissiblity, this force can be moved along its line of action to the top surface (Fig. 2.9(d)). Thus, it is easy to show schematically a concentrated force acting at a point on the body but it is difficult to apply.

From the foregoing discussion it is clear, that seemingly complex interaction of physical bodies (Fig. 2.8(b)) can just be represented by a concentrated force acting at an appropriate point. The discussion also reinforces the point that all surface forces in reality are distributed forces.

Apart from the surface force at C (Fig. 2.8(a)), the body force due to gravitational field of earth also acts on the body. The force due to gravitational field is nothing but the weight of the member ABC and it will act through its *centre of gravity* located at D. When the magnitude of weight *W* is very much smaller than force *F*, one can neglect the effect of the body force. This is done to simplify the calculation and thus the solution obtained is only approximate. Neglecting the self weight of the member in the analysis is a very useful approximation practiced by engineers in solving complex real life problems.

2.9   Rate of Application of Forces

One of the important information needed for an analysis of the body exerted by various forces is to know, how the final magnitudes of the forces have been achieved. For example, if it is shown that a force *F* is acting on a body, one has to know whether the magnitude of the force $|\ F\ |$ is suddenly applied or gradually raised from 0 to $|\ F\ |$. One of the examples of sudden application of the forces is the force acting on a speeding car when it hits a tree and comes to a halt. The analysis of problems of this nature is complex and beyond the scope of this book. Unless otherwise stated, in the subsequent discussions, it will be assumed that the forces are gradually applied and their magnitudes are slowly raised from zero to their final values.

2.10   Moment of a Force

As stated earlier, a force acting on a body has a tendency to produce in general both translation (moving the body along the line of action) and rotation motions. *Moment of a force is the measure of the tendency of the force to rotate the body about the point of interest.*

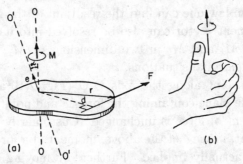

Fig. 2.10 (a) Moment of a force; (b) the right-hand rule.

Figure 2.10(a) shows a two-dimensional body acted upon by a force F in its plane. The tendency of the force to rotate the body about the point P is a function of both the magnitude of the force and the moment arm d, which is the perpendicular distance from point P to the line of action of the force. The magnitude of the moment is defined as

$$|M| = Fd \tag{2.6}$$

While defining the moment of a force, not only the magnitude but also the sense of its rotation is to be specified. If the force tends to produce a counter-clockwise rotation about the point of interest then the moment of force is +ve otherwise it is −ve. In the present example, the moment of the force about the point P is +ve and has a magnitude of Fd. The basic units of moment in SI units are Newton-meters (Nm).

Moment of a force is a vector quantity. Vectorial representation of Eq. (2.6) is

$$M = r \times F \tag{2.7}$$

where r is the position vector of the force from point P. By virtue of Eq. (2.6) the magnitude of M is independent of the operation of sliding the force along its line of action. The direction is fixed by the right-hand rule. When the fingers of the right hand curl in the direction that F tends to turn about P, the right thumb points in the direction of the moment vector (Fig. 2.10(b)).

It is customary to call *moment of a force* simply as *moment*. Since moment is a vector similar to a force, it also follows the laws of composition and resolution discussed in Sec. 2.6. Using parallelogram law one can find the resultant of the moment vectors. A moment vector can also be resolved into its rectangular components M_x and M_y in two dimensions or M_x, M_y and M_z in three-dimensional situations.

For any point along the axis O-O (Fig. 2.10(a)), which is normal to the plane containing the force F and point P, the value of the moment of force is unchanged. Customarily the moment about a point, in view of the above, moment of a force about the line O-O is actually implied. Further, by moving the moment vector M acting at P along its line of action O-O, the external effect on the body is not altered. Thus, similar to a force, moment of a force about a point is also a *sliding vector*.

If moment about the line O'-O' is to be determined, it is nothing but the component of $r \times F$ along line O'-O'. The magnitude of this component along the line O'-O' is the projection of the vector M along O'-O' and is given by the dot product of M and a unit vector e in the direction of O'-O'.

$$| M |_{O'-O'} = e.r \times F \qquad (2.8)$$

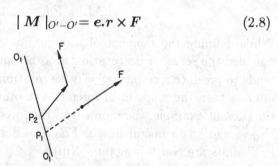

Fig. 2.11 Moment of a force about a line.

Moment of a force about an axis vanishes when

(i) the line of action of force cuts the axis or

(ii) the line of action of force is parallel to the axis (Fig. 2.11).

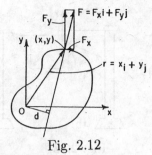

Fig. 2.12

From Fig. 2.12, the moment of the force F about the point O is

$$M = r \times F = dF k \qquad (2.9)$$

where k is the unit vector in the direction perpendicular to the plane of x and y. Writing the vectors r and F in component form gives

$$
\begin{aligned}
M &= (x i + y j) \times (F_x i + F_y j) \\
&= (x F_y - y F_x) k
\end{aligned}
\qquad (2.10)
$$

From Eq. (2.10) it is seen that the magnitude of moment is given by the algebraic sum of the magnitudes of the moments of the components about O. The result just obtained, namely, the moment of a force about any point is equal to the sum of the moments of the components of the force about the same point, is known as *Varignon's theorem*. To get a physical insight into the problem, it is advisable to work with the moments of the components. In such a case, simple scalar algebra is sufficient. In complicated problems, particularly those involving three-dimensional situations one may have to resort to vector algebra.

From the foregoing discussions it is clear that the concept of 'moment' as it is used in mechanics, is the scientific formulation of what is everybody's daily experience of the *turning effect* of a force.

Example 2.2

Determine the moment of force F about the base point A shown in Fig. 2.13(a).

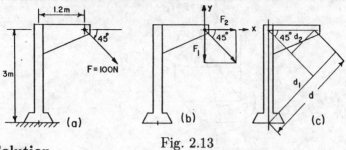

Fig. 2.13

Solution

Method 1

Resolving the force along x and y directions,

$$F_1 = 700\cos 45° = 494.97 \text{ N}$$
$$F_2 = 700\sin 45° = 494.97 \text{ N}$$

By Varignon's theorem, the moment becomes

$$M = -494.97(1.2) - 494.97(3) = -2078.87 \text{ Nm}$$

The −ve sign indicates a clockwise moment.

Method 2

Position vector of point B,

$$r = 1.2i + 3j$$

The force vector

$$F = 494.97i - 494.97j$$

and the moment

$$\begin{aligned}
M_A &= r \times F = (1.2i + 3j) \times (494.97i - 494.97j) \\
&= -494.97 \times 1.2k - 494.97 \times 3k \\
&= -2078.87k \text{ Nm}
\end{aligned}$$

Method 3

Referring to Fig. 2.13(c),

$$\begin{aligned}
d_1 &= 3\cos 45° \\
d_2 &= 1.2\sin 45° \\
d &= d_1 + d_2 = 3\cos 45° + 1.2\sin 45° \\
&= \frac{1}{\sqrt{2}}(4.2) = 2.97 \text{ m}
\end{aligned}$$

The moment about A is $700 \times 2.97 = 2079$ Nm (clockwise).

2.11   Moment of a Couple

The force system consisting of two equal and parallel but non-collinear forces with opposite sense is called a couple. Referring to Fig. 2.14, the sum of the moments of forces F_1 and F_2 about P is denoted as M and is given as

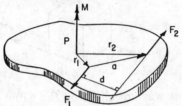

Fig. 2.14 Moment of a couple.

$$M = r_1 \times F_1 + r_2 \times F_2 \tag{2.11}$$

In the above equation r_2 can be replaced as $r_1 + a$ which gives

$$\begin{aligned} M &= r_1 \times F_1 + (r_1 + a) \times F_2 \\ &= r_1 \times (F_1 + F_2) + a \times F_2 \end{aligned} \tag{2.12}$$

Since F_1 and F_2 are equal and opposite their sum is zero. Hence,

$$M = a \times F \tag{2.13}$$

Point P is taken arbitrarily and the above equation is independent of the location of point P. Thus, the moment of a couple is the same about all points in space. This result indicates that the rotatory effect of the couple force system is the same about all points in space. Since the two forces forming the couple are equal and opposite, the net translatory effect on the body is zero. Using the principle of composition, couple force system can be just represented by a moment vector M. Since the rotatory effect is the same about all points in space, the moment vector M has no specific point of application, in contrast to the moment of a force. Thus, *moment of a couple* or simply a *couple* is a *free vector.* Similar to a moment, a couple also has Newton-meter (Nm) as the basic unit.

A couple is unchanged, by changing the values of F and d as long as their product remains the same. Further, a couple is not altered when the forces are allowed to act in any one of the parallel planes. Fig. 2.15 shows three different configurations of the same couple M.

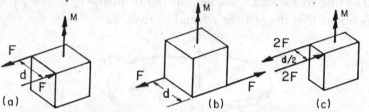

Fig. 2.15 Different ways of effecting the same couple vector.

Moments and couples are vectors and they do follow the laws of vector addition. It is often convenient to distinguish between vectors representing moment and couple to that of a force vector by using some notational device. Fig. 2.16 shows two possible representations of moment and couple vectors. As shown

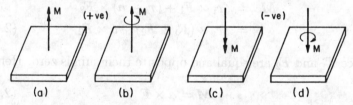

Fig. 2.16 Represenations of moment vector.

in Fig. 2.16(a), one can use two headed arrows or a single arrow with an encircling arrow (Fig. 2.16(b)) indicating the direction of rotatory effect. Both, these conventions will be used in this text. To indicate a moment vector, the point of its application also is to be specified, whereas couple being a free vector the representation shown in Fig. 2.16 is sufficient. When sketching a plane figure acted on by a couple whose axis is perpendicular to that plane, the notation indicated in Fig. 2.16(b) is convenient to use as shown in Fig. 2.17. On the other hand, while finding the resultant of couples/moments using vector algebra, the notation of Fig. 2.16(a) is simpler to use. The concepts of composition and

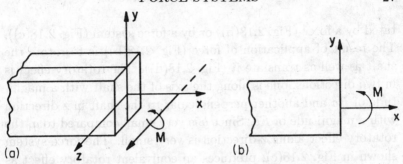

Fig. 2.17 Representation of a couple in a plane sketch.

resolution of a force discussed in Sec. 2.6 are equally applicable to moment and couple. While using those concepts it is to be remembered that force and moment are *sliding vectors* whereas a couple is a *free vector*. The difference between the effect of a

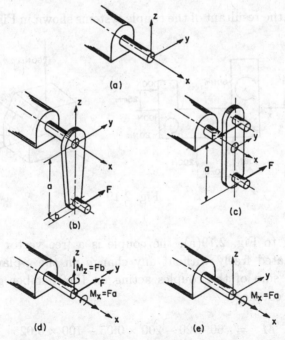

Fig. 2.18 Distinction between moment of a force and a couple.

force and a couple can be understood from the following example. The shaft shown in Fig. 2.18(a) needs to be rotated. This can be physically achieved in two different ways. The shaft can be ro-

tated by a force (Fig. 2.18(b)) or by a force system (Fig. 2.18(c)).
The result of application of force (Fig. 2.18(b)) is to rotate the
shaft as well as translate it (Fig. 2.18(d)). The rotatory effect is
in two directions, one is along the axis of the shaft with a magnitude of Fa and another perpendicular to the shaft in z direction
with a magnitude of Fb. Since b is very small compared to a, the
rotatory effect along z-direction is very small. The force system
shown in Fig. 2.18(c), produces an equivalent rotatory effect as
in the previous case (along the axis of the shaft) with no translation. This example clearly shows that a force in general tends
to translate the body along its line of action and also rotate it.
On the other hand a couple tends to only rotate the body.

Example 2.3

Find out the resultant of the couple systems shown in Fig. 2.19(a)
and (b).

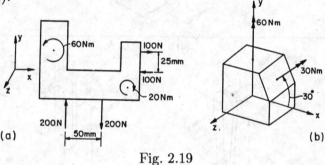

Fig. 2.19

Solution

Referring to Fig. 2.19(a), the couple is a free vector and can
be translated freely without any change on the z-plane. The
algebraic sum of the couples acting on the member gives the
resultant couple

$$M = 60 - 20 - 200 \times 0.05 - 100 \times 0.025$$
$$= 27.5 \text{ Nm}$$

Since couple is a free vector, it has no specific point of application.

Referring to Fig. 2.19(b), the resultant couple can be obtained
using the parallelogram law.

Resolving couple 30 Nm along x and y axes gives

$$30\cos 30° \, i + 30\sin 30° j = 25.98 i + 15 j$$

The vectorial representation of 60 Nm couple is $60j$ and the resultant couple is

$$25.98 i + 15 j + 60 j = 25.98 i + 75 j$$

The magnitude of couple is $\sqrt{25.98^2 + 75^2} = 79.37$ Nm and its direction is $\tan^{-1}\left(\frac{75}{25.98}\right) = 70.89°$ from the x-axis.

Note that couples can be summed up just like forces using the parallelogram law.

2.12 Resolution of a Force into a Force and a Couple

In several problems of mechanics, as an intermediate step, it is necessary to change the point of application of force to a different point other than along its line of action without changing the external effect on the body. This can be achieved by resolving a force into a force and a couple.

It is desired that the point of application of force F is to be changed from point P_1 to P_2 in Fig. 2.20 without changing the external effect on the body. As a first step, two equal and

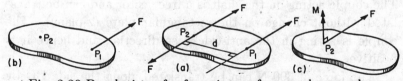

Fig. 2.20 Resolution of a force into a force and a couple.

opposite forces of magnitude $|\,F\,|$ can be added at point P_2 (Fig. 2.20(b)). Introduction of these forces does not in any way alter the external effect on the body produced by the force acting at point P_1. Fig. 2.20(b) shows that one of the forces acting at point P_2 and the force at P_1 form a couple. Thus, the point of application of force F can be changed from P_1 to P_2 by resolving the force into an equivalent force and a couple (Fig. 2.20(c)). The magnitude of the couple is given by Fd. Reversing the above procedure, it is also possible to combine a force and a couple to an equivalent force acting at a different point.

The resolution of a force into an equivalent force and a couple is a step that finds repeated applications in the study of mechanics and should be thoroughly mastered.

Example 2.4

A shaft is welded to a rectangular plate as shown in Fig. 2.21(a). The shaft is subjected to a torque (couple) of 54 Nm and the plate to a 300 N force. If the couple and the force are to be replaced by a single equivalent force at E, determine the distance y.

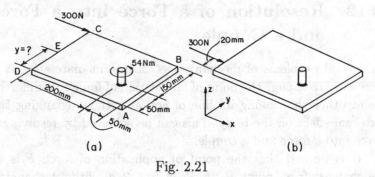

Fig. 2.21

Solution

The couple acting on the shaft is a free vector and can be translated without change on the rectangular plate (z-plane). The couple is acting in the anti-clockwise direction and hence it is positive.

The force of 300 N can be translated along the side CD by an equivalent force and a couple. If the force is moved towards D, the force produces a clockwise couple. At some point along CD, the couple produced by the force is balanced by the couple acting on the shaft.

When the point E is at $(54/300)$ m from C both the couples are balanced. This means $y = 0.02$ m.

The effect of the force and the couple acting on the rectangular plate can therefore, be replaced by a force of 300 N acting at E along the x-direction.

2.13   Resultants of Force Systems

Resolution of a force into a force and a couple helps to move the point of application of the force to a different point. Thus, if several forces are acting on a body with different points of applications, it is always possible to move these forces to a point of interest by resolving these forces into appropriate force-couple combinations to the selected point. The new force system thus obtained will have the same external effect on the body as that of the original force system. In Fig. 2.22(a), the moment arms $d_1, d_2, d_3,$ and d_4 for forces F_1, F_2, F_3 and F_4 are about the selected point P. Fig. 2.22(b), shows these forces translated to point P, and the resultant magnitude the associated couples is obtained as

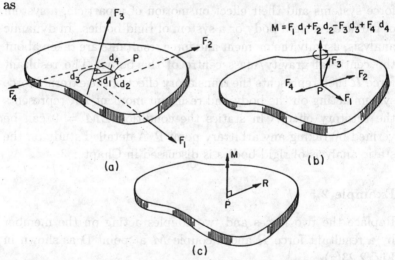

Fig. 2.22 Resultant of a general co-planar force system.

$$| M | = F_1 d_1 + F_2 d_2 - F_3 d_3 + F_4 d_4 \qquad (2.14)$$

A simple algebraic summation shown in Eq. (2.14) is possible since the force system is co-planar. If the force system is non-coplanar, then the resultant couple vector M could be found by the vectorial addition of the couples M_1, M_2, M_3 and M_4 using the parallelogram law of vector addition. The force system acting at point P is concurrent and their resultant **R** can be found using the parallelogram law of vector addition as discussed in sec. 2.6. Thus, the force system is reduced to a single resultant force and a single resultant couple acting at point P (Fig. 2.22(c)). Point P is selected arbitrarily and the magnitude and direction of **R** are the same no matter which point is selected. However, the magnitude and the direction of **M** will depend on point P.

Analysis of forces and determination of resultants of forces is common to both *statics* and *dynamics*. In statics one deals with zero resultants of force systems such that

$$\Sigma F = 0 \quad \text{and} \quad \Sigma M = 0 \tag{2.15}$$

The body is said to be in equilibrium if the above condition is satisfied. In *dynamics* one deals with non-zero resultants of force systems and their effect on motion of a particle, a system of particles, a rigid body or a system of rigid bodies. In dynamic analysis, usually the moment and force resultants are taken about the centre of gravity/mass centre of the body. The resultant force **R** then represents the translatory effect of the external force system acting on the body and resultant moment **M**, represents the rotatory effect. In statics the condition $\Sigma M = 0$ can be verified by taking any arbitrary point P. A detailed study on the static analysis of rigid bodies is discussed in Chapter 3.

Example 2.5

Replace the two forces and two couples acting on the member by a resultant force **R** and a couple **M** at point D as shown in Fig. 2.23(a).

Solution

The couples acting at A and C can be translated to point D without any change since the couple is a free vector.

Contribution of couples at A and C to D is

$$M_1 = 60 - 25 = 35 \text{ Nm}$$

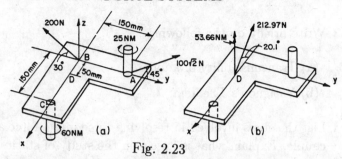

Fig. 2.23

The forces at A and B can be easily translated to D after resolving them into components.

Force at A is $(-100i + 100j)$. Force at B is $(-200\sin 30°\, i - 200\cos 30°\, j) = (-100i - 173.2j)$.

Contribution to couple at D while translating force at A to D is

$$M_2 = 100 \times 0.15 - 100 \times 0.05 = 10 \text{ Nm}$$

Contribution to couple at D while translating force at B to D is

$$M_3 = 100 \times 0 + 173.2 \times 0.05 = 8.66 \text{ Nm}$$

Resultant couple at D is $M = M_1 + M_2 + M_3 = 53.66$ Nm.

The resultant force at D is

$$
\begin{aligned}
R &= -(100 + 100)i + (100 - 173.2)j \\
&= -200i - 73.2j
\end{aligned}
$$

Magnitude of R is $\sqrt{200^2 + 73.2^2} = 212.97$ N and its direction is $-20.1°$ from the x-axis.

The resultant force R and moment M are shown in Fig. 2.23(b).

Problems

2.1 (a) State and explain the most important and fundamental approximation used in representing physical systems in the study of statics.

(b) Explain the notion of a concentrated force. Under what circumstances such an approximation is valid?

2.2 Compare the vectorial properties of a force, moment of a force and moment of a couple. Also comment on the composition and resolution of these.

2.3 Write briefly on the following :

 (a) Transmissibility of a force,

 (b) Varignon's Theorem.

2.4 List the steps involved in resolving force into a force and a couple. Explain what is its use in the study of statics.

2.5 A cable exerts a force F on the bracket of the structural member as shown in Fig. 2.24. If the magnitude of the y- component of the force F is 600 N, calculate the x-component and the magnitude of F.

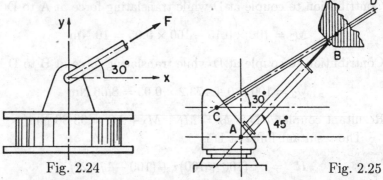

Fig. 2.24   Fig. 2.25

2.6 Figure 2.25 shows a portion of a mechanism forming an excavator. The hydraulic cylinder AB exerts a force P directed along AB, on the member CBD. The hydraulic cylinder must exert a force of 1000 N perpendicular to CBD. What is the magnitude of force P and its component along the member CBD?

2.7 A bolt is tightened by a pipe wrench as shown in Fig. 2.26. Calculate the moment of the 300 N force about the bolt centre.

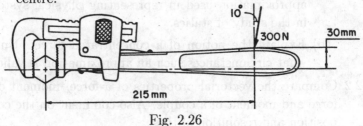

Fig. 2.26

2.8 Figure 2.27 shows a hook carrying a load of 500 N. Determine the force and couple acting at point B of the hook.

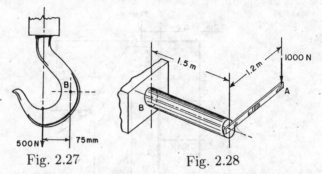

Fig. 2.27   Fig. 2.28

2.9 Determine the effect of 1000 N force at point B in Fig. 2.28.

2.10 Replace the couple and force (Fig. 2.29) by a single force at point A. Locate point A along the length of the arm.

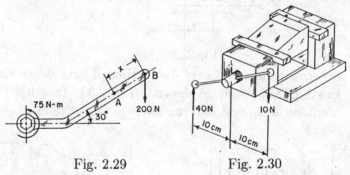

Fig. 2.29   Fig. 2.30

2.11 Replace the two forces on the slide handle of the bench-vise of Fig. 2.30 by a single force.

2.12 In an assembly operation of a wheel, five nuts are tightened simultaneously by the same torque M as shown in Fig. 2.31. Calculate the net moment M_o acting on the wheel.

Fig. 2.31

2.13 Replace the force system shown in Fig. 2.32 by a force and a couple at point A.

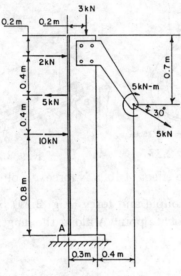

Fig. 2.32

2.14 Determine the resultant of the distributed force shown along length CD of the beam shown in Fig. 2.33. Translate this resultant force, and also the forces acting at points B and E to F.

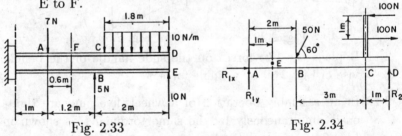

Fig. 2.33   Fig. 2.34

2.15 A force and a couple are acting on the beam as shown in Fig. 2.34.

(a) Translate the force/force system acting at points B and C to point E.

(b) Translate the force/force system acting at points A,B,C and D to E. Comment on the results obtained.

3. EQUILIBRIUM OF RIGID BODIES

3.1 Introduction

'Statics' deals primarily with the descriptions of forces that are both − necessary and sufficient − to maintain the state of equilibrium of engineering structures. Newton's first law of motion states that if the resultant force acting on a particle is zero, the particle will remain at rest or keep moving with a constant velocity along a straight line. This law provides the basis for the equations of equilibrium which are developed in this chapter.

Newton formulated the fundamental laws of mechanics, the laws of motion of particles, basing them on evidence he obtained from a large number of experiments. For problems dealing with small-scale phenomena such as sub-atomic particles, and large-scale phenomena such as astronomical bodies with velocities approaching the speed of light, Newton's formulations have been found to be inexact. In such cases, quantum mechanics and the theories of relativity should be used. However, for solving nearly all engineering problems, Newtonian mechanics is still valid and remains the basis of today's engineering mechanics.

It is the ingenuity of Eüler which has helped us to extend the laws of Newtonian mechanics to a collection of particles and eventually to rigid bodies. The rigid body is a special case of a system of infinitely many particles with fixed distance between them.

The necessary and sufficient condition for a particle to be in equilibrium is that the resultant force acting on the particle should be zero. Necessary conditions for a system of particles to be in equilibrium are that the resultant force acting on the particles and the resultant moment of these forces about any point should vanish. These conditions become sufficient if every conceivable sub-system satisfies the above condition. In case of rigid bodies, since the distance between the particles remain fixed, the conditions that resultant force and moment should vanish are both necessary and sufficient.

This chapter discusses the analysis of equilibrium of rigid bodies using both − the 'Force Method' and the 'Method of Virtual Work'. The necessary and sufficient conditions of equilibrium for interconnected rigid bodies are also discussed. Advantages of identifying two- and three- force members in analysing structures are also brought out. In contrast to evaluating the forces to keep the body in equilibrium, the role of method of virtual work in determining the *position of equilibrium* is highlighted.

3.2   Necessary and Sufficient Conditions of Equilibrium of Rigid Bodies

In the preceeding chapter we have seen that the external forces acting on a rigid body can be reduced to a resultant force and a resultant couple at some arbitrary point P. When the force and the couple are both equal to zero, the external forces form a system, equivalent to zero force system, and the rigid body is said to be in *equilibrium*. According to Newton's second law of motion, the body has no acceleration if the resultant force acting on it is zero. It is to be remembered, that zero acceleration implies only *constant* velocity. Hence, a rigid body in equilibrium can remain at rest or can move with a constant velocity along a straight line.

In general for several civil engineering structures like bridges, roof trusses etc., equilibrium also implies a state of rest (i.e., the velocity is zero). However,in the analysis of mechanical engineering components such as shafts, gear pairs in rotation etc., equilibrium only implies that acceleration is zero. The bodies in general can have a constant non-zero velocity and in reality a shaft or a gear pair in equilibrium rotate at a constant speed.

To summarise, the conditions for equilibrium of a rigid body are

$$\sum F = 0 \quad \text{and} \quad \sum M = 0 \qquad (3.1)$$

Writing it in the component form in a three-dimensional orthogonal co-ordinate system Eq. (3.1) reduces to

$$\sum F_x = 0 \qquad \qquad \sum M_x = 0$$
$$\sum F_y = 0 \quad \text{and} \quad \sum M_y = 0$$
$$\sum F_z = 0 \qquad \qquad \sum M_z = 0$$

The conditions mentioned in Eq. (3.1) are necessary and sufficient for a single rigid body to be in equilibrium. However, for interconnected rigid bodies, Eq. (3.1) should not only be valid for the overall system but also be satisfied for every conceivable subsystem.

Analysis of a rigid body in equilibrium, under the action of several forces, can begin by asking "How does a body satisfy the requirements of Eqs. (3.1)?" That is, to determine the unknown forces applied to the rigid body or unknown reactions exerted on it by its supports, one can use Eqs. (3.1).

3.3 Simplified Use of Equilibrium Equations for Special Situations

A rigid body cannot remain in equilibrium under the action of a single force. At least two forces are required to keep the body in equilibrium. A member acted upon by two forces is termed as a *two force member*. If three forces are acting on a member then it is a *three force member*. In several structures of engineering importance, the loading can be simplified to an extent, that one can identify the presence of two force and three force members in the overall structure. The application of equilibrium conditions becomes very simple to two and three force members and these are discussed in the following sections. Further, in the case of a body under the action of a co-planar system, the number of equilibrium conditions in component form reduce to three.

3.3.1 Conditions for Equilibrium of a Two Force Member

In Fig. 3.1, the body is acted upon by two forces. For the case shown in Fig. 3.1(a), the effect of the two forces can be reduced to a single force R using parallelogram law of vector addition. In Fig. 3.1(b), the effect of the external forces can be reduced to a couple of magnitude Fd. In view of these, the cases shown in Figs. 3.1 (a and b) are not in equilibrium. The only way resultant R can be made zero is to have the same magnitudes for F_1 and F_2 but with opposite sense. If their lines of action are different as shown in Fig. 3.1(b), the body will not be under equilibrium. For the couple to be zero, the lines of action of these forces must be identical as shown in Fig. 3.1(c) and in this case the body is in equilibrium.

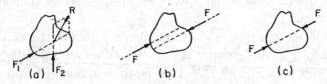

Fig. 3.1 Equilibrium conditions for a two force member.

The above discussion clearly brings out that for a two force member to be in equilibrium, the two forces must be of equal magnitude and opposite sense having the same line of action.

Two force members often occur in analysis when the weight of a member, i.e., the body force due to gravitational attraction, is negligible in comparison to external forces acting on the member.

3.3.2 Conditions for Equilibrium of a Three Force Member

Some examples of a force system consisting of three forces acting on a body are shown in Fig. 3.2. The force system shown in Fig. 3.2(a) has a non-zero resultant R. Hence, condition $\sum F = 0$ is not satisfied and the body is not in equilibrium. The force system in Fig. 3.2(b) has a zero-force resultant, but if moments are summed about any point P, the moment resultant is found to be non-zero. Thus, the body is not in equilibrium under the

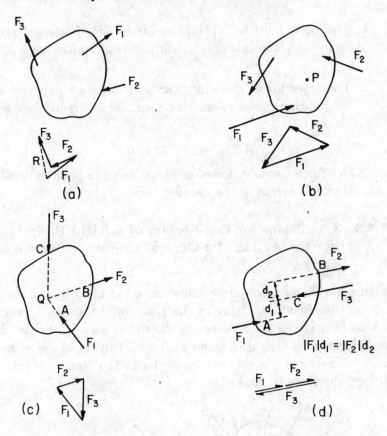

Fig. 3.2 Equilibrium conditions for a three force member.

action of force system shown in Fig. 3.2(b). The force system of
Fig. 3.2(c) is one of the two possible configurations under which
the body is in equilibrium. The magnitude and the directions of
the forces are such that their resultant is zero. Further, the line
of action of these forces meet at a common point Q, this ensures
that the moment resultant is zero. This result (that the three
forces must intersect at a common point) is a useful one to keep
in mind. A limiting case occurs when the forces F_1, F_2 and F_3
become parallel as shown in Fig. 3.2(d). It is to be noted that
in Fig. 3.2 (c and d), A, B and C are the point of application of
forces F_1, F_2 and F_3 and these forces lie in the plane containing
the points A, B and C. The conditions for equilibrium for a three
force member can be summarized as follows:

1. The forces must be coplanar and their line of actions must lie in the plane containing their points of application.

2. The three forces must intersect at a common point or in the limiting case they should be .parallel satisfying the conditions

$$\Sigma F = 0 \text{ and } \Sigma M = 0$$

Three-force members are ,often encountered in practical analysis, when the weight of the member is not negligible.

3.3.3   Conditions for Equilibrium of a Rigid Body Under the Action of a General Co-planar System of Forces

Figure 3.3 shows a body in equilibrium acted upon by a system of coplanar forces, all lying in the plane of the sketch. There are no force components perpendicular to the plane of the sketch and if moments are taken about any point P lying in the plane, only moment components perpendicular to the plane will exist. Hence, the three equations

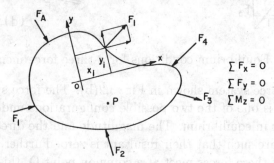

$$\Sigma F_x = 0$$
$$\Sigma F_y = 0$$
$$\Sigma M_z = 0$$

Fig. 3.3 Equilibrium conditions for a co-planar force system.

$$\sum F_z = 0, \ \sum M_x = 0 \text{ and } \sum M_y = 0 \qquad (3.2)$$

are automatically satisfied.

The force system shown should satisfy the following equations to keep the body in equilibrium:

$$\sum F_x = 0$$

$$\sum F_y = 0 \qquad (3.3)$$
$$\sum M_z = 0$$

Since summation of only one moment component, namely M_z, is to be zero, it is customary to leave the subscript z and denote it just as M.

As moment of a force depends on the selection of the point, the point about which moment of force is taken is denoted as the subscript. Thus, for the case shown in Fig. 3.3, $\sum M_P = 0$

3.4 Free-body Diagram (FBD)

Selection of an appropriate system is the first step in any analysis. The system may consist of a body, a part of a body or a collection of interconnected bodies, depending on the problem on hand.

A *free-body diagram is a sketch of the selected system consisting of a body, a portion of a body, or a collection of interconnected bodies completely isolated or free from all other bodies, showing the interaction of all other bodies by forces on the one being considered.* A free-body diagram has three essential characteristics; (i) it is a diagram or sketch of the system selected, (ii) the system is shown completely separated (isolated/cut free) from all other bodies including foundations, supports etc., and (iii) the interaction on the system by each body removed in the isolating process is shown as a force or forces on the diagram.

A set of exact rules for drawing free-body diagrams cannot be given to cover all problems. Nevertheless, certain general principles that are common to all free-body diagrams can be stated. One of the key steps is to decide what should constitute a system that need to be isolated.

In simple cases, it is obvious what system should be isolated. The problem on hand may be simple enough that a single isolation is sufficient to solve the problem. In complex analysis many different isolations may be required. This leads to analysis of interconnected systems. An intricate pattern of partial results, obtained from each such isolation, may have to be assembled before the problem can be completely solved. Identification of an appropriate system/systems to be isolated depends on the specific problem and a proper selection is possible only through experience.

Once the system is identified, it has to be properly isolated and one must ensure that all external forces are accounted for. The best way to perform an isolation is to draw a reasonably careful sketch of the periphery of the isolated system and then to show all external forces acting. While showing the forces one has to keep in mind that all surface forces are applied through direct mechanical contact but the body forces are applied through remote action. Hence, special attention need to be paid whether the body forces due to gravitational, magnetic or electric fields are properly accounted for. Once this is ensured, then one has to go carefully around the entire periphery and indicate all the forces which make direct contact with the system.

Both known and unknown forces must be properly indicated. Known external forces generally include the weight of the body (body force due to gravitational field) and forces applied for a given purpose. Unknown external forces usually consist of the constraining forces, known as reactive forces, through which the ground and other bodies oppose a possible motion (both translation and rotation) of the free body and thus constrain it to remain at rest or move at a constant velocity.

While indicating the known external forces, both magnitude and direction of the forces should be carefully indicated. Force interaction at the supports are usually not known and their direction can be arbitrarily taken for showing on the FBD. However, consistency in their directions should be maintained if the complete analysis involves several isolations from the original system. If at the end of the analysis, the force magnitude is +ve then the initial assumed direction of the force is correct. However,if the force magnitude is −ve then the force direction has to be reversed. In dealing with forces due to friction one cannot assume any arbitrary direction but actual direction must be indicated. This is discussed in detail in Chapter 5.

The free-body diagram should also include essential dimensions that are needed in the computation of moments of forces.

However, too much of detailed dimensions should be avoided.

Success of the analysis of the problem on hand solely depends on the correct drawing of the FBD. Insufficient information or the incorrect understanding of the force interaction at the supports can lead to errors in drawing FBD. This would eventually affect the analysis of the problem. Hence, the student is advised to adopt the habit of drawing clear and complete FBDs for every mechanics problem which he undertakes to solve. Section 3.5 gives the idealised force interaction most commonly used at supports.

3.5   Reactions at Supports for a Two-dimensional Structure

The construction of free-body diagrams requires an understanding of what kinds of forces act at various contiguous surfaces particularly at the various supports and connections.

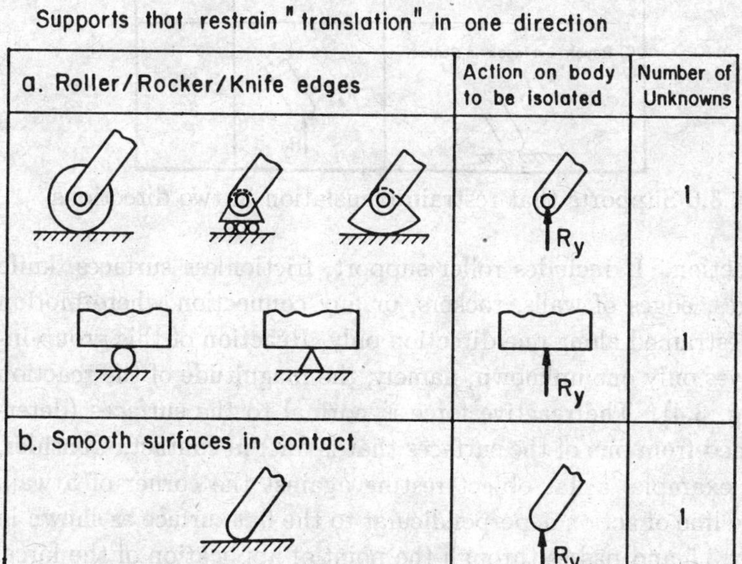

Supports that restrain "translation" in one direction

a. Roller/Rocker/Knife edges	Action on body to be isolated	Number of Unknowns
	R_y	1
	R_y	1
b. Smooth surfaces in contact	R_y	1

Fig. 3.4 Supports that restran translation in one direction

There are three basic classes of support conditions in two-dimensions. The first provides translation restraint only in one

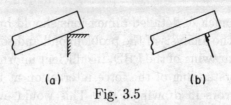

(a)   (b)

Fig. 3.5

Supports that restrain "translation" in two directions

Type of Support / Contact	Action on body to be isolated	Number of unknowns
a. Frictionless Pin Connections	R_x, R_y	2
	R, R	2
b. Frictionless thin bearings 	 R	2
c. Rough Surfaces in Contact 	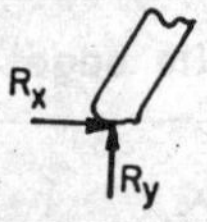 R_x, R_y	2

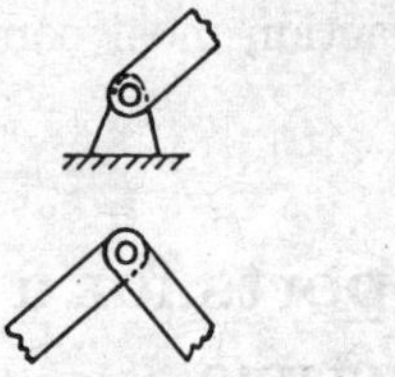

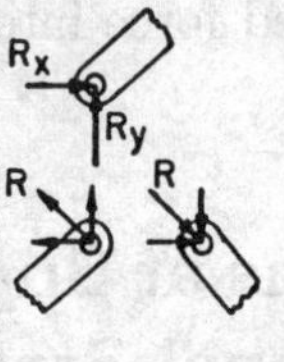

Fig. 3.6 Supports that restrain translation in two directions

direction. It includes roller support, frictionless surfaces, knife edges, edges of walls, rockers, or any connection where motion is restrained along one direction only. Reaction of this group involves only one unknown, namely, the magnitude of the reaction (Fig. 3.4). The reactive force is normal to the surfaces (determined from one of the surfaces that is flat) in contact. Consider, for example, a flat object resting against the corner of a wall. The line of action is perpendicular to the flat surface as shown in Fig. 3.5 and passes through the point of application of the force, i.e., the point at which the flat object rests on the corner

The second class is one in which translation is restrained but rotation is allowed. This includes frictionless pin connections, hinges, frictionless thin bearings and rough surfaces. Both, mag-

Supports that restrain both translation and rotation

Type of Support/Contact	Action on body to be isolated	Number of Unknowns
Fixed support, weld	R_x M R_y	3

Fig. 3.7 Supports that restrain both translation and rotation

nitude and direction of the force at the support, are not known. Hence, reactions of this group involve two unknowns and are usually represented by their x and y components (Fig. 3.6).

Third class of support is one in which both rotation and translation are totally restrained, e.g. fixed supports, welded ends, and even two closely spaced pin joints (Fig. 3.7). Forces at the supports form a force system that could be reduced to a force and a couple. In case of fixed supports, the actual distribution of force at the wall is very difficult to determine but a possible distribution is shown in Fig. 3.8. This force system could be reduced to a force and a couple. Welded ends also behave in a similar way. In case of two closely placed pin joints, the force system is as shown in Fig. 3.9 and can be reduced to a force and a couple which arrests both translation and rotation (in the plane of the sketch) at the support. Reaction of this group involves three unknowns consisting of the x and y components of the force and the magnitude of the couple.

A flexible cable/string finds wide application for supporting loads. A weightless (it is an idealization) flexible string or cable can support only a tensile force along its length (Fig. 3.10).

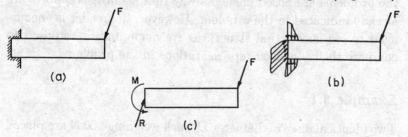

Fig. 3.8 Force distribution at the built-in end.

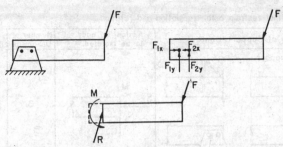

Fig. 3.9 Force system for two closely spaced pin-joints

No compressive forces are supported by it. If the cable /string passes over a frictionless peg or pulley,the direction of the force in the cable/string is altered but its magnitude remains constant (Fig. 3.10). Flexible cable is also widely used to support tall slender structures such as electrical poles, flag poles, radio transmitters etc., to support them against wind forces. In such applications flexible cable is named as *guy wire*.

A spring can support both tensile and compressive forces (Fig. 3.10). Spring is a deformable member and spring force is tensile if the spring is stretched and compressive if compressed from its neutral position. For a linearly elastic spring, the spring stiffness k denotes the force required to deform the spring by a unit distance.

The supports discussed above are the idealized supports. In real life structures, the actual supporting elements may be quite involved and they could be reduced to any of the above mentioned idealized supports. A practising engineer should develop the acumen to simplify the real supporting elements in practical structures to idealized supports for the purpose of analysis. An illustration of the connections in a truss is explained in Chapter 4. Fortunately, in a first course in Engineering Mechanics the problems are posed in such a way that idealized supports are already indicated in the problem. However, the reader is encouraged to see how actual structures are made and are advised to compare their idealized representations in the problems.

Example 3.1

Two identical discs of diameter D, each weighing 100 N are placed in a box as shown in the Fig. 3.11(a). Determine the reactions

Commonly used other types of connections/Supports

Type of Support/Connections	Action on body to be isolated	Number of Unknowns
a. Weightless flexible string or cable 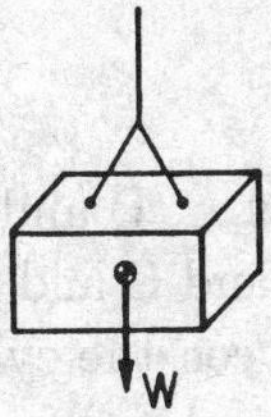	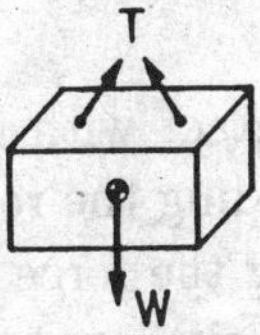	1
b. Cable over a frictionless peg or pulley 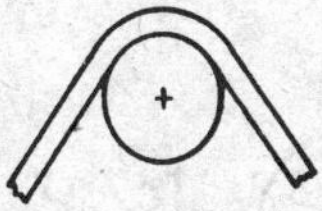	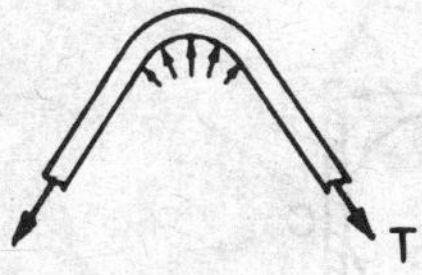	1
c. Action of spring 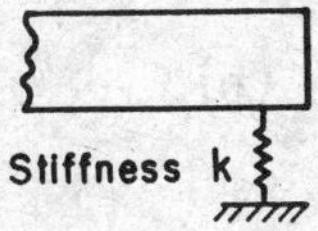	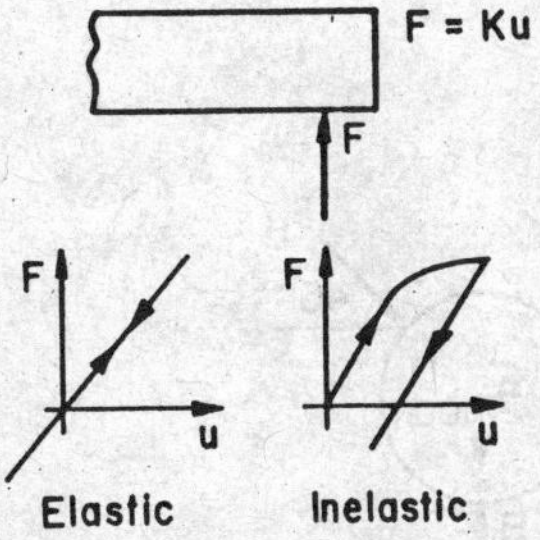	1

Fig. 3.10 Commonly used other types of connections/supports.

at A,B and C neglecting friction between the cylinders and the box. (This example is chosen to illustrate the identification of an appropriate system for analysis and also to illustrate the concept of necessary and sufficient conditions of equilibrium.)

Solution

The discs make contact with the box at points A, B and C. We are interested in finding the reactions at A, B and C and are not interested in finding the force interaction between the cylinders. Hence, a system consisting of two discs can be isolated for analysis. As it is given in the problem, the contacts are frictionless and the force interaction, at points A, B and C arrests translation in one direction only and can be taken as shown in Fig. 3.11(b).

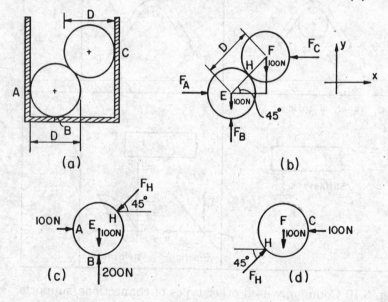

Fig. 3.11

Body force due to gravitational attraction is the weight of each disc and is given as 100 N. Weight of the discs can be represented as a concentrated force passing through the centre of gravity of the disc. For circular discs, the centre of the disc itself is its centre of gravity. Hence, through points E and F two concentrated forces of magnitude 100 N are shown in the figure.

From the geometry of the problem, distance EF is equal to D, the diameter of the discs. Reference axes are also shown in the figure. Thus Fig. 3.11(b). represents the complete FBD of the problem taken up for analysis.

For the system to be in equilibrium, the following conditions are to be satisfied:

$$\sum F_x = 0$$
$$\sum F_y = 0 \qquad (3.4)$$
$$\sum M_z = 0$$

Application of condition $\sum F_x = 0$ yields,

$$F_A = F_C \qquad (3.5)$$

and condition $\sum F_y = 0$ yields

$$F_B = 200 \text{ N} \qquad (3.6)$$

For applying the condition $\sum M_z = 0$ one has to select an appropriate point. For the present analysis, point B is a convenient point to take moment of forces.

From the geometry of the problem,

$$EG = FG = D/\sqrt{2} \qquad (3.7)$$

Since forces F_A and F_C are equal and opposite they form a couple. Application of condition $\sum M_B = 0$ yields

$$F_c.D/\sqrt{2} - 100 \times D/\sqrt{2} = 0$$

or

$$F_c = 100 \text{ N} \qquad (3.8)$$

From Eq. (3.6)

$$F_A = 100 \text{ N} \qquad (3.9)$$

Satisfaction of equilibrium condition (Eq. 3.5) is only a necessary condition but not sufficient. For the system to be in equilibrium, the equilibirium conditions (Eq. 3.5) should be satisfied for every conceivable sub-system. The sub-systems that are possible in this case are the two individual discs.

Note the consistent representation of F_H in the above FBDs (Figs. 3.11 (c and d)). Using the FBD of Fig. 3.11(c), $\sum F_x = 0$ yields $100 - F_H \cos 45° = 0$, or

$$F_H = 100\sqrt{2} \text{ N} \qquad (3.10)$$

This value of F_H satisfies equations $\sum F_y = 0$ and $\sum M_B = 0$. Hence, disc E (one of the sub-system) is in equilibrium.

Now consider the FBD of disc F.

Here too the value of $F_A = 100\sqrt{2}$ N satisfies the conditions $\sum F_x = 0., \sum F_y = 0$ and $\sum M_z = 0$. Hence, disc F (another sub-system) is also in equilibrium.

Since the system as a whole and also the sub-systems are in equilibrium, the reaction forces evaluated are correct. They are

$$F_A = 100 \text{ N}; \quad F_B = 200 \text{ N}; \quad F_C = 100 \text{ N} \qquad (3.11)$$

Problem 3.1 is simple and the checking for sufficient conditions of equilibrium appeared to be redundant. The force interaction at points A, B and C are simple and obvious. While analyzing a complicated system, lack of proper idealization of supports can lead to errors and the system as a whole may be in equilibrium but the subsystem may not be.

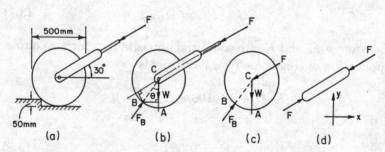

Fig. 3.12

Example 3.2

A circular disc of weight 200 N is pushed by a force F to overcome a step of height 50 mm as shown in Fig. 3.12a. Find the force F. (This example is to illustrate the judicious choice of equations to solve a problem.)

Solution

For the disc to be moved up, the contact with the ground at point A has to be severed and thus there is no force interaction at point A (Fig. 3.12(b)). Weight of the disc acts through point C. The system is acted upon by only three forces, namely, F, W and F_B. The line of action of W and F intersects at C. Hence, line of action of F_B also must pass through C.

The line of action of force F is known and we need to find only its magnitude. Hence, only one equation is sufficient for evaluation.

$\sum F_x = 0$; and $\sum F_y = 0$ involves also the other force F_B. Although, for the system to be in equilibrium any point can be taken to check the condition $\sum M_z = 0$, if point B is selected for taking moments, the contribution of F_B is zero. Thus, a judicious choice of point such as B greatly simplifies the number of computations.

Application of $\sum M_B = 0$ yields,

$$W\,25\sin\theta = F\,25\sin(60-\theta) \qquad (3.12)$$

From geometry $\cos\theta = 20/25$ or $\theta = 36.87°$.
Solution of Eq. (3.13) gives F = 305.5 N.

It is easy to see from Fig. 3.12(c) that subsystems also satisfy the equilibrium conditions.

Example 3.3

Figure 3.13(a) shows a crimping tool is used to crimp (make curly/wavy) terminals onto electrical wires. When a force of 140 N is applied at the end of the handle, find out the magnitude of the crimping forces which will be exerted on the terminal. (This example is to illustrate that identification of two and three force members in a system greatly simplifies calculations.)

Solution

Figure 3.13(a) shows a photograph of the crimping tool and Fig. 3.13(b) shows a dimensional sketch of the tool. Figure 3.13(a) also shows the close-up view of the cutter portion for

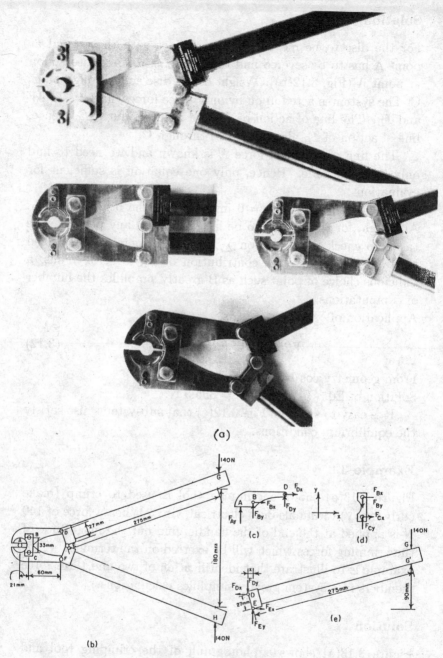

Fig. 3.13

various orientation of the handle. These figures clearly bring out the fact that the joints B, C, D, E and F can be idealized to be frictionless pin joints. This idealization is very important to proceed further in analyzing the problem. The student is advised to experiment on a cutting plier (which is commonly available) to unerstand the above mentioned idealization for a crimping tool. The commonly available cutting pliers have only one pin joint whereas the crimping tool has five pin joints. A very high magnification of the applied force is possible with the crimping tool.

Referring to Fig. 3.13(d), $\sum M_c = 0$ yields

$$F_{Bx} = 0 \tag{3.13}$$

and $\sum F_x = 0$ gives

$$F_{cx} = F_{bx} = 0 \tag{3.14}$$

This shows that member BC is a two force member. Referring to Fig. 3.13(c), the member ABD is a three force member and since forces at A and B are vertical the force at D must also be vertical and hence $F_{Dx} = 0$.

From Fig. 3.13(c), $\sum M_B = 0$ yields,

$$F_{Ay} = F_{Dy}60/21 \tag{3.15}$$

Referring to Fig. 3.13(e), the member DEG is again a three force member. The forces at D, E and G are parallel and $F_{Dx} = F_{Ex} = 0$ and $\sum M_E = 0$ gives

$$F_{Dy}x = 140(275^2 - 90^2)^{1/2} \tag{3.16}$$

From property of triangle,

$$\frac{x}{27} = \frac{90}{275}; \quad x = 8.836 \text{ mm} \tag{3.17}$$

Therefore,

$$F_{Dy} = 4117.23 \text{ N}; \quad F_{Ay} = 11763.5 \text{ N} \tag{3.18}$$

The ratio of the output force to the input force is $\frac{11763.5}{140} \simeq 84$. This ratio is known as *mechanical advantage* and is quite high for a crimping tool. (If you come across a crimping tool never make an attempt to put your finger between the cutters, even for curiosity sake—lest you might injure yourself.)

Example 3.4

A small lathe is driven by a motor that supplies 10 kW power and turns at a constant speed of 1725 rpm. The shaft is supported by two thin bearings and the belt tension on one side is found to be 420 N. The pulley weighs 100 N. Determine the reactions at the bearings and also the belt tension (Fig. 3.14). Note: In

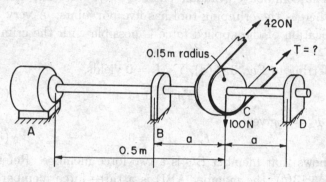

Fig. 3.14

case of belt drives, due to friction between the belt and pulley the tension on both sides of the belt are different. (This example is chosen to illustrate the point that a system in equilibrium does not always mean that the system is at rest. It can have constant velocity.)

Solution

The torque delivered by the motor is $(M_A) = \frac{(10,000)(60)}{(2\pi)(1725)} = 55.4$ Nm and $\sum M_z = 0$ gives

$$-(0.15)(420) + 0.15T + 55.4 = 0$$

or

$$T = 50.7 \text{ N} \tag{3.19}$$

$\sum F_x = 0$ gives

$$R_{Bx} + R_{Dx} = 100 \tag{3.20}$$

$\sum F_y = 0$ gives

$$R_{BY} + R_{DY} = 420 + 50.7 \tag{3.21}$$

$\sum M_x = 0$ and $\sum M_y = 0$ give

$$R_{Bx} = R_{Dx} = 50 \text{ N}; \quad R_{BY} = R_{DY} = 235.35 \text{ N} \qquad (3.22)$$

The system is in equilibrium since it satisfies all the conditions of equilibrium.

3.6 Work and the Principle of Virtual Work

3.6.1 Work

The term work, as used in mechanics, should not be confused with such common terms as housework, yard work etc. The work done by a force is the product of the force and the corresponding displacement. This simple definition has to be carefully interpreted and used. The term *corresponding displacement* signifies that the displacement is in the direction of the force. Consider a force F that acts on a particle P. The work dW that the force performs when the particle undergoes an infinitesimal displacement ds is defined by the equation,

$$dW = F.ds = F_t \, ds \qquad (3.23)$$

where F_t is the projection of the force F along the directed line in which the displacement occurs. Consequently, if particle P describes a curve segment C, (Fig. 3.15) the work that force F performs is

$$W = \int_c F_t ds = \int_c F \cos \theta ds \qquad (3.24)$$

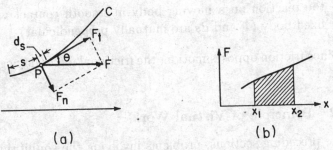

(a)   (b)

Fig. 3.15 (a) Work absorbing component of force; (b) Graphical representation of work.

where θ is the angle between the vector F and the tangent to the curve. In other words, the work done by a force is the integral sum of the product of the displacement of its point of action and the component of the force along the line of displacement.

Work is a scalar quantity and is obtained as a dot product of two vectors; force and linear displacement. It can be either positive or negative, depending on whether the force and the in-line displacement are in the same direction or opposite to each other. Work is measured in Newton-meters (Nm). The work done is zero if $F = 0$; $ds = 0$ or F is perpendicular to ds.

In the above discussion F is represented by a concentrated force and ds the corresponding linear displacement. It is possible to extend the term force to include not only a concentrated force but also a bending moment or a torque. Similarly, the term 'displacement' may mean linear or angular displacement. Such a definition is known as generalized forces and displacements. The work done by a moment M is given as

$$dW = |M| \, d\theta \qquad (3.25)$$

or

$$W = \int M \, d\theta \qquad (3.26)$$

where $d\theta$ is the corresponding angular displacement.

A number of forces frequently encountered in statics do no work. For example,

(i) the reaction on a fixed hinge ($ds = 0$)

(ii) the reaction on a moving body in smooth contact with a fixed body (F and ds are mutually perpendicular)

Since friction opposes motion, the frictional forces do negative work.

3.6.2 Principle of Virtual Work

In the preceding sections, problems involving the equilibrium of rigid bodies were solved by expressing that the external forces acting on the bodies were balanced. Principle of virtual work is an alternate way of investigating/analyzing the equilibrium of a given system. The approach can be likened to a situation wherein one shakes a ladder to see whether it is steady before he starts to climb. Unless one is convinced that the ladder is stable he would not attempt to climb. Rather than performing an actual experiment as mentioned above, in applying this principle, one has to place oneself in the imaginary role of an experimenter. The system being investigated is supposed to be given arbitrary infinitesimal and consistent displacements and the sign of the resulting work of the forces that act on the system is investigated to verify the condition of equilibrium. Since these displacements are not necessarily realized in an actual movement of the system, they are called *virtual displacements*, and the work of the forces that act during virtual displacements is called *virtual work*.

The principle of virtual work states that a *"mechanical system is in equilibrium if the virtual work is negative or zero for every small virtual displacement"*. The virtual displacement may be any arbitrary infinitesimal displacement *consistent with the constraints of the system*. One need not restrict oneself to infinitesimal displacements. One can work with virtual displacements of finite dimensions, however, this would take us beyond the scope of this book.

As a simple application of the principle of virtual work, consider a brick that rests on the floor. The brick is obviously in equilibrium. Let us now give *consistent* virtual displacements to the brick. If one lifts the brick, the force of gravity does negative work since the direction of force and displacement are opposite to each other. Consider now that the brick is given a horizontal displacement. If the floor is frictionless, the virtual work is zero since the weight of the brick acts in a direction normal to the displacement. It is important to note, that to give a displacement one has to apply a force. However, *while calculating virtual work, the force needed to provide virtual displacement is not considered for calculations*. If the floor has friction, then the frictional forces oppose the motion and hence does negative work. Thus, for any consistent virtual displacement, the virtual work is zero or negative and hence the brick is in equilibrium.

The method of virtual work for a particle or a single rigid

body usually provides no advantage over using the equations of equilibrium. However, in the determination of equilibrium configuration for a system of rigid bodies, the method proves to be advantageous. Unlike in the force method, it is not necessary to dismember the interconnected rigid bodies and analyse the force system acting on each body separately. This is because the forces acting at mechanical connections are equal but opposite in sense. If the mechanical connections are assumed to be frictionless, then one of the forces acting at the joint produce positive virtual work and the other produces negative virtual work and the overall effect is zero.

For an interconnected mechanical system, one can identify three types of forces that act. They are,

(a) *Active forces*: These are the external forces capable of doing virtual work during possible virtual displacements.

(b) *Reactive forces*: These are the forces which act at position of fixed support where no virtual displacement in the direction of the forces takes place. Hence, the contribution of virtual work by reactive forces is zero.

(c) *Internal forces*: These are the forces that act in the connections between members. As mentioned before the net effect of these forces to virtual work is zero.

Since the internal forces and the reactive forces do not contribute to virtual work, the relations between the active forces can be directly determined without evaluating the reactive forces. This advantage makes the method of virtual work particularly useful in determining the *position of equilibrium* of a system under known loads. This type of problem is in contrast with the problem of determining the forces acting on a body whose equilibrium position is fixed or specified.

The method of virtual work can also be used to determine the reactive forces. In order to do this, the restraint/support is replaced by a force. The body is then given a virtual displacement and the virtual work done by the reaction and all other forces acting on the system is computed. If several forces are to be determined, the system can be given a series of separate virtual displacements in which only one of the unknown forces does virtual work during each displacement.

The imaginary work done by a force F during a virtual displacement can be expressed as

$$\delta W = F \delta r \tag{3.27}$$

Similarly, for the virtual work of a moment M

$$\delta W = M \delta \theta \tag{3.28}$$

where $\delta \theta$ is the virtual angular displacement of infinitesimal magnitude. The symbol δ is used instead of d in the above equations to signify that we deal with *virtual* quantities.

It is usually assumed that both forces, couples, and moments remain constant during the virtual displacements. This assumption leads to the simplification that in scalar form, the work is equal to the product of force and the virtual displacement. In reality, the force/couples do not remain constant and they change slightly and this effect is neglected.

Example 3.5

Using the principle of virtual work, find the reactions at A and B for the system shown in Fig. 3.16(a).

Solution

The support at A is a pin joint and it can allow rotation. Let us give a virtual angular displacement of $\delta \theta$ about A which is consistent with the constraints of the system. Figure 3.16(c) shows the displaced configuration in dotted lines. For infinitesimal angular displacement, one can consider that the point of application of force R_{By} moves by $b \, d\theta$. Virtual work done by force R_{By} is $W_1 = R_{By} \, b \, \delta\theta$.

For a virtual displacement of $\delta \theta$ about A, point D has vertical displacement of $a\delta\theta$ and negligible horizontal displacement. Only the vertical component of force F does work. The virtual displacement and the force direction are opposite and hence the virtual work is negative, i.e.

$$W_2 = -F\sin 30° a\, \delta\theta \tag{3.29}$$

For the system to be in equilibrium, the total virtual work is zero. Applying this condition i.e., $W_1 + W_2 = 0$, gives

$$R_{By} b\delta\theta - F\sin 30° a\delta\theta = 0$$

(Note: Interestingly, if $\delta\theta$ is omitted from above equation, it rep-

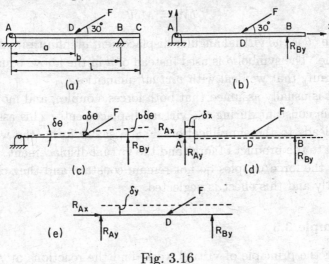

Fig. 3.16

resents the moment equilibrium equation about point A. Thus, principle of virtual work is an alternative way of expressing the equilibrium conditions). Or

$$R_{By} = \frac{1}{2}\frac{a}{b}F \tag{3.30}$$

We are also interested in determining the reactive forces at A. Now, replace the support by the forces R_{Ax} and R_{Ay} as shown in Fig. 3.16(d). Let a virtual displacement of δx be given as shown in Fig. 3.16(d). The *reactive force* R_{Ax} and the horizontal component of active force F contributes to the virtual work. From the principle of virtual work, for equilibrium to exist the virtual work has to be zero. This gives

$$R_{Ax}\delta x - F\cos 30° \delta x = 0$$

or

$$R_{Ax} = \frac{\sqrt{3}}{2}F \qquad (3.31)$$

Let a virtual displacement of δy be given as shown in Fig. 3.16(e). The reactive forces R_{Ay}, R_{By} and the vertical component of active force F contributes to virtual work. Therefore,

$$R_{Ay}\delta y + R_{By}\delta y - F\sin 30°\delta y = 0$$

or

$$R_{Ay} = \frac{F}{2}(1 - \frac{a}{b}) \qquad (3.32)$$

It is to be noted that the virtual displacements δx and δy were possible and were consistent with the constraints of the system only because the supports at A and B were replaced by their respective reactive forces.

The above example has brought out the methodology of virtual work. Since only a single rigid body is considered the method is not really attractive. Nevertheless, the general procedure need to be followed has been brought out by this example.

Example 3.6

Figure 3.17(a) shows two uniform hinged bars having weights W and length ℓ that are loaded by a force P. Determine the angle θ for equilibrium.

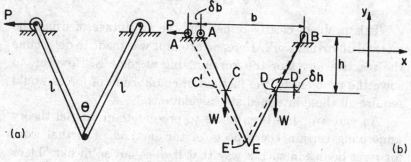

Fig. 3.17

Solution

Figure 3.17(b) shows the active force diagram and the dotted lines show the configuration for a virtual displacement of δb. The work done by the applied force is $W_1 = P\delta b$. As the point A moves horizontally, the mass centers C and D of the bars moves up by δh and the virtual work done by the weight of the bars is

$$W_2 = -2W\delta h \tag{3.33}$$

Using the principle of virtual work, i.e. $W_1 + W_2 = 0$,

$$P\delta b - 2W\delta h = 0 \tag{3.34}$$

From geometry,

$$h = -\frac{\ell}{2}\cos\frac{\theta}{2}$$

$$\delta h = +\frac{\ell}{4}\sin\frac{\theta}{2}\delta\theta \tag{3.35}$$

$$b = 2\ell\sin\frac{\theta}{2}$$

$$\delta b = \ell\cos\frac{\theta}{2}\delta\theta$$

On substitution

$$P\ell\cos\frac{\theta}{2}\delta\theta - 2W\frac{\ell}{4}\sin\frac{\theta}{2}\delta\theta = 0$$

or

$$\theta = 2\tan^{-1}\left(\frac{2P}{W}\right). \tag{3.36}$$

This problem clearly brings out the advantage of using the method of virtual work. Here, no attempt was made to determine the reactive forces or the forces acting at joint E. However, to solve the problem by the method of equilibrium of forces would require all these intermediate calculations.

An endeavor has been made to present the essential theory underlying certain key aspects of the method of virtual work for rigid bodies in such a way that it does not seem like "black magic", as students in the past have put it. The method of virtual work can be extended for deformable bodies.

The principle of virtual work has dominated the development of mechanics for the last four Centuries and continues to be useful. Although the approach is less physically obvious and more mathematical in nature, the resulting computational abilities makes for tremendous simplification for certain problems. The method was variously known as *principle of virtual velocities, principles of virtual displacement* or *principle of virtual work* in the literature. The principle of virtual work is closely related to the *principle of minimum potential energy*. These methods are often used to prove theorems and to derive equations used in advanced mechanics.

Problems

3.1 State the necessary and sufficient condititons for the equilibrium of a (a) particle, (b) collection of particles, (c) rigid body, and (d) system of interconnected rigid bodies.

3.2 List the conditions to be satisfied by a *two force* and *three force* member for equilibrium.

3.3 What is meant by a free body diagram? List the general principles to be followed in drawing a FBD.

3.4 (a) Classify and list the idealised supports. (b) A flat object is resting on a knife edge. Indicate how would you find the direction of reaction. (c) Explain how a welded end or two closely spaced pin joints can be considered as a fixed support for the purpose of analysis.

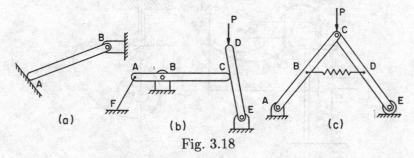

Fig. 3.18

3.5 Draw complete free-body diagrams for the cases shown in Fig. 3.18 assuming all surfaces to be smooth.

(a) Member AB

(b) (i) Overall system, (ii) Member ABC, (iii) Member DCE;

(c) (i) System ABCDE, (ii) Member ABC, (iii) Member CDE.

3.6 A cylinder of 60 cm diameter weighing 100 N is resting between two smooth surfaces, (Fig. 3.19). Determine the reaction at the points of contact.

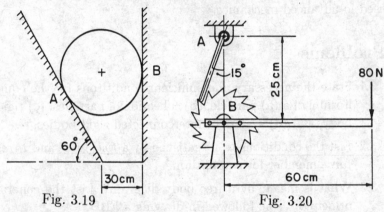

Fig. 3.19   Fig. 3.20

3.7 Evaluate the force acting on the pawl shown in Fig. 3.20.

3.8 Determine the force acting at point A of the hand pump shown in Fig. 3.21.

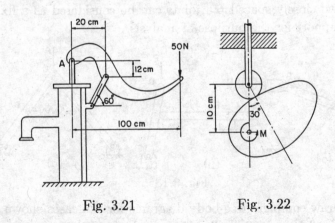

Fig. 3.21   Fig. 3.22

3.9 Determine the torque M on the camshaft required to hold the follower weighing 3 kg in the position shown in Fig. 3.22.

3.10 Consider, a two stroke engine of a scooter shown in Fig. 3.23. A force of magnitude 3 kN is applied on the piston as shown in the figure. Determine the torque M required to hold the system in equilibrium.

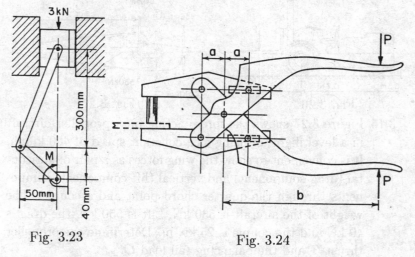

Fig. 3.23   Fig. 3.24

3.11 Figure 3.24 shows a paper puncher. For the force P applied at the handle, what is the punching force?

3.12 Using the equations of equilibrium, determine the reactions at the supports A and B of the example problem 3.5.

3.13 Figure 3.25 shows a collapsable table and a man rests his elbow on it which exerts a downward force of 250 N on the table top. Determine the magnitude of the shear force on

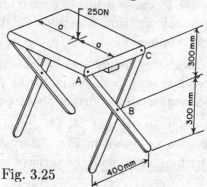

Fig. 3.25

the pin B and the reactions at the legs resting on the floor.
Assume that all contacts are frictionless.

3.14 A fork-lift truck (Fig. 3.26) is used for loading and stacking
materials. The truck weighs 30 kN. Determine the reactive
forces developed at the front and rear wheels when the fork-
lift truck is lifting a 22 kN crate.

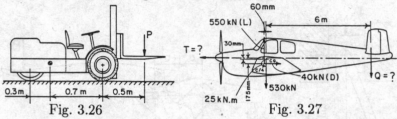

Fig. 3.26   Fig. 3.27

3.15 Figure 3.27 shows the forces acting on a propeller aircraft
in a level flight, cruising at a constant speed of 400 kmph.
It is convenient to show the wing forces as a pair of horizontal (drag component) and vertical (lift component) components through the quarter chord point and a couple. The
weight of the aircraft is 530 kN. Lift is 550 kN, the drag is
40 kN and the couple is 25 kN-m. Determine the propeller
thrust T and the balancing tail load Q.

3.16 (a) Define "work" as used in mechanics.

(b) What is meant by generalised forces and generalised
displacements.

3.17 (a) Define the principle of virtual work for rigid bodies.

(b) Explain the condition that the virtual displacement be
consistent with the constraints of the system.

(c) In general to cause a displacement one requires an
appropriate force. How does a virtual displacement is
effected? Comment.

(d) While calculating virtual work, what is the assumption
made with regard to the value of the forces, couples
and moments during a virtual displacement?

3.18 Two blocks weighing W_1 and W_2 kN are supported by an
inextensible wire as shown in Fig. 3.28. Assuming that
there is no friction between all the surfaces in contact what

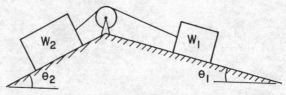

Fig. 3.28

is the relationship between W_1, W_2, θ_1 and θ_2 for the blocks to be in equilibrium. Solve it by both the force method and the method of virtual work.

3.19 Figure 3.29 shows a frame made of two uniform links each weighing W/N and the length of them is ℓ. The frame is in equilibrium under the action of the force P. Express the angle in terms of P.

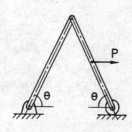

Fig. 3.29

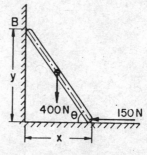

Fig. 3.30

3.20 A 8m long ladder weighing 400 N is resting against a wall. A horizontal force of 150 N is applied at the bottom as shown in Fig. 3.30. Assume that all the contacting surfaces are smooth.

(a) Using the principle of virtual work determine the distance x for which the ladder will be in equilibrium.

(b) For the configuration obtained, determine the reactions at the supporting points A and B by both the force method and the method of virtual work.

Fig. 338

4. TRUSSES AND CABLES

4.1 Introduction

In a wide range of engineering applications, trusses and cables are employed to support transverse loads. Members forming the truss and cable are predominantly subjected to axial loads though the external loading is in the transverse direction. This aspect leads to economic solutions for the specific application. This chapter first discusses the way trusses are built, their idealization for the purpose of analysis and the evaluation of forces in the members followed by cables subjected to concentrated and distributed loads. The method of finding the tension in the cable, sag and length of the cable for parabolic and catenary cables are also discussed in detail.

4.2 Trusses

A framework composed of straight members joined at their ends to form a structure to bear loads is called a truss. It is one of the major structures commonly used in engineering applications. It provides practical and economical solution to many engineering situations. Trusses may be seen in steel bridges, supporting structures of large roofs, electrical transmission towers, roller coasters and giant wheels of amusement parks, and in innumerable other constructions.

Actually most structures are made of several trusses joined

Fig. 4.1 Some examples of trusses.

together to form a space framework. Each truss is designed to carry those loads which act in its plane and thus may be treated as a two dimensional structure and can be considered as a plane truss for purposes of analysis. Figure 4.1(a, b and c) shows a typical bridge truss a roof truss and a space framework to support a water tank, respectively. The loading of truss is designed in such a way that all external forces are applied at the joints. If loads are applied elsewhere, it will introduce bending loads and a truss cannot resist bending loads. In bridge trusses, the deck is usually laid on cross beams that transfer the loads to the joints (Fig. 4.2(a)). In the case of roof trusses, the rafters or purlins are supported only at the joints (Fig. 4.2(b)). The rafters are beams in bending which carry the wind load and weight of the roof, more or less perpendicular to each other, and transmit this load by bearing reactions to the joints of the truss underneath. Thus the purlins are bent, but the members forming the truss are not.

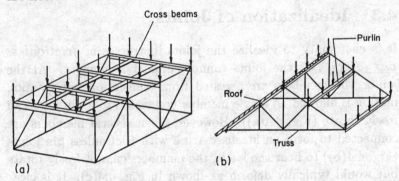

Fig. 4.2 Loading of bridge and roof trusses.

A sketch of a typical joint in a truss (Fig. 4.3). Clearly shows that no member is continuous through a joint. In the example chosen, the members are angles joined by riveted connections. In general, the cross-sections of members can be of I-section, channels, angles, circular rods, or any other special shape required for specific application. The joints can be made by welding, riveting, bolting etc. For making these connections, the members are individually joined to a plate, known as a *Gusset plate*, such that the centroidal axes of these members intersect at a common

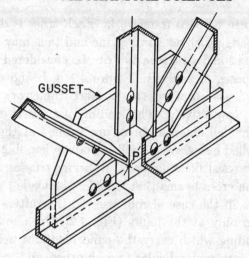

Fig. 4.3 A typical joint in a truss.

point (Fig. 4.3).

4.3 Idealization of Joints

It is customary to idealize the joints in trusses as *frictionless pin joints*, i.e., the joints cannot support a moment. At the joints, the members are arrested from translation but rotation motion is allowed. A single member connected to a pin joint can freely rotate (Fig. 4.4(a)). However, when several members are connected to form an idealised truss with frictionless pin joints (Fig. 4.4(b)) to bear the loads, the members cannot freely rotate but would typically deform as shown in Fig. 4.4(c). It is clear from the figure that the members rotate slightly about the pin axis and the angle α between two members changes after the

Rotation

(a)

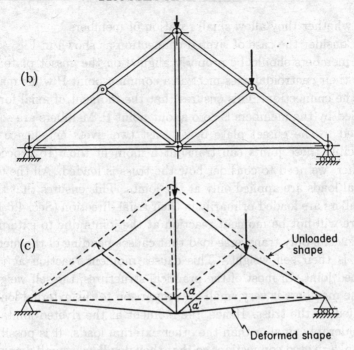

Fig. 4.4 Magnified deformation of a pin-jointed truss.

loads are applied.

In case of fixed end conditions the angle α is unaltered after the application of loads. The deformed truss (Fig. 4.4(c)) is highly magnified for the purpose of illustration. In reality, the deformation is very much smaller and consequently, the rotation of the members about the pin joint is also very small. Since the amount of rotation that is to be allowed is very small any other type of joint which allows small rotation about the joint can be idealised as a pin joint.

Though, for the purpose of engineering analysis one makes approximations, the success of the analysis depends on whether such approximations are valid and to what extent these approximations affect the final result. Very few trusses in practice are constructed with pin joints as shown in Fig. 4.4(b). The present practice is to resort to welding, or riveted or bolted connections. Hence, one has to verify whether idealisation of such joints as frictionless pin joints is a valid approximation. The key points to see are whether the joints transfer significant amount of moment and whether they allow small rotation of members.

Consider the case of riveted connection as shown in Fig. 4.3. The members should be properly aligned on the gusset plate so that their centroidal axes meet at a common point P while making the connection. This ensures that the moment of axial force carried by the members is zero about point P. Members are connected to the gusset plate by at least two rivets and hence in principle these joints can transmit a moment too. To proceed further, we need to consider how the truss is loaded. All the external loads are applied only at the joints. This ensures that the members are loaded primarily in their axial direction (Sec. 4.6.1). There will not be moment reaction at the joints due to external loads. The only transverse load that causes bending of the members is their self weight. This causes moment reaction at the riveted joint. In most of the practical structures, the self weight of the members is very small in comparison to the external loads applied on the truss. Hence, the moment at the riveted joints is considerably smaller than the other external loads. It is possible to make riveted connections so that they do allow a small amount of rotation of members at the joints. In view of this, for the purpose of analysis, the joint can be considered as a frictionless pin joint at P. This assumption is valid as long as the centroidal axes of members are concurrent at the joint and all external loads are applied only at the joints.

Following the same line of argument, both bolted connections and welded connections can also be idealised as frictionless pin joints. Bolted connections also allow small amount of rotation at the joints. Welded connections are more rigid and do not allow rotation at the joints. Nevertheless, welded connections are very widely used in less critical structures because of the ease of manufacturing process. Riveted connections comparatively require precision in manufacturing. In critical structures such as bridges, only riveted connections are preferred. In roof trusses one finds both welded and riveted connections.

A word of caution is necessary here. One should not conclude that wherever riveted, bolted or welded joints are used, they can be idealized as pin joints. This idealization is relevant and meaningful only in the case when the structure could be classified as a truss. This is because of the special way the trusses are constructed and loaded as pointed out earlier.

4.4 Common Types of Trusses

The basic element of a plane truss is a pin-jointed triangular structure (Fig. 4.5(a)). It is convenient to use line diagrams of the trusses for both illustration and analysis purposes. The joints are marked by small circles to indicate that no member is continuous through a joint. The reader should visualise that in reality the joints are quite complicated and with members made of various structural sections (angles, channels, I-sections etc).

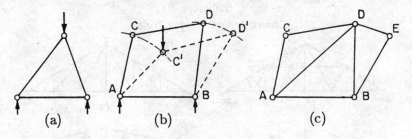

Fig. 4.5

If loads are applied as shown in Fig. 4.5(a), the structure will not collapse and can withstand the external loads. On the other hand, a structure made of four or more pin-jointed members to form a polygon will collapse under the action of external loads as shown in Fig. 4.5(b). One can make the structure shown in Fig. 4.5(b), non-collapsable by adding a diagonal bar joining A and D or B and C and thereby forming two triangles. The structure may be extended by adding two members starting from two previous joints as shown in Fig. 4.5(c). In this manner one can arrive at a non-collapsable structure. Although most trusses in practice are made in this manner, it is not the only way in which a rigid plane truss can be made.

Plane trusses are generally supported at the two ends. Provision for expansion and contraction due to temperature changes and for deformations resulting from applied loads is usually made at one of the supports. A roller support is commonly used for such purposes. Several examples of commonly used trusses that can be analysed as plane trusses are shown in Fig. 4.6.

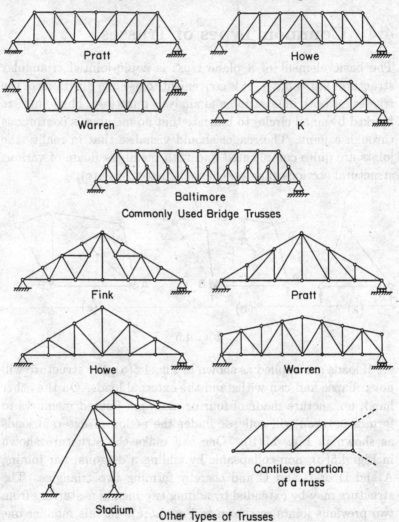

Fig. 4.6 Some typical trusses.

4.5 Classification of Statically Determinate and Indeterminate Trusses

The way a truss is constructed we can deduce a simple relation between the number of members m, joints j and reactions r. For

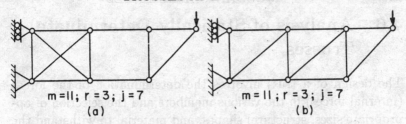

$$m = 11; \quad r = 3; \quad j = 7 \qquad\qquad m = 11; \quad r = 3; \quad j = 7$$
$$\text{(a)} \qquad\qquad\qquad\qquad\qquad \text{(b)}$$

Fig. 4.7 $m + r = 2j$ is satisfied (a) only for the overall structure (b) for every conceivable sub-systems.

each joint, idealized as a pin joint, two independent equations can be written. The total number of unknowns are the forces in the members and the number of reactions. If the condition, is satisfied then the truss is known as a *statically determinate* truss because the forces in the members and the reactions at the supports can be completely determined using the principle of statics alone.

$$m + r = 2j \qquad\qquad (4.1)$$

If $m + r > 2j$ then the unknowns are more than the number of equations that could be written using the principle of statics. Then it is called *statically indeterminate* truss. If $m + r < 2j$ then the framework can have rigid body motions and is known as a *mechanism*. A common example of a simple mechanism made of pin jointed straight members is the mini-drafter used for engineering drawing.

The abovementioned conditions are necessary but *not sufficient* to classify a structure as statically determinate or statically indeterminate truss. For example, both Figs. 4.7(a) and (b) satisfy Eq. (4.1), but the structure shown in Fig. 4.7(b) is a statically determinate truss and Fig. 4.7(a) is not a structure to bear loads and is collapsable. The way the individual members are connected is important to make the structure non-collapsable. The conditions become sufficient, if they are applied to each sub-system of the structure to verify whether they are collapsable or not.

4.6   Analysis of Statically Determinate Trusses

The design of a truss involves the determination of the forces (internal forces) in the various members and the selection of appropriate sizes, structural shapes, and material to withstand the forces. Determination of the forces in the various members of the truss will be discussed in this chapter. This information is necessary to select the size, structural shape and material whose determination is under the purview of the Strength of Materials/Mechanics of Solids.

In order to calculate forces in individual members, it may be desirable to determine the reactions at the supports. In most analysis, weight of the individual members are so small compared to the load applied that it may be neglected. However, if the small effect of weight is to be accounted for, weight W_m of member m may be assumed to be replaced by two forces, each $W_m/2$, acting at each end of the member. These forces, in effect, are treated as loads externally applied at the joints for further analysis. If the weight of members taken into account, is reactions at the supports have to be calculated appropriately. Having determined the reactions, we can choose between *the method of joints* and the *method of sections* for obtaining forces in the members.

Method of joints is useful in finding forces in all the members of the truss while *method of sections* leads to a quicker result in determining the forces in a few selected members of the truss.

In both these methods the accounting for the weight of the members is only an approximation. The result obtained does not account for the effect of bending of the member as it is usually small. However, lumping of the weight of the members does improve the estimation of average axial forces (tension/compression) acting in the members.

4.6.1   Method of Joints

It is already clarified in Sec. 4.2 that a truss can be idealized to have frictionless pin joints though the joints may be made by welding, rivetting or bolting. In this approach only an idealised truss is analysed. Necessary and sufficient conditions of equilibrium states that for the truss to be in equilibrium under the action of external loads, every conceivable subsystem must also be in equilibrium, i.e. every joint must be in equilibrium. In this method, equilibrium of each joint is considered separately and consecutively. An imaginary section is passed to isolate a single joint of the truss. The force system acting on the isolated joint is *concurrent* and *coplanar*. Such a force system has to satisfy only two independent equations of equilibrium. Therefore, the solution of a truss problem should be started at a joint where only two unknown forces act. The following example illustrates the use of method of joints.

Example 4.1

Determine the forces in all the members of the truss shown in Fig. 4.8(a).

Solution

First, determine the reactions at the supports. Support at A is a pin joint and at D is a roller. Figure 4.8(b) shows the FBD of the whole truss and gives

$$\sum F_x = 0; \qquad R_{Ax} = 0$$
$$\sum F_y = 0; \qquad R_{Ay} + R_{Dy} = 4 \text{ kN}$$
$$\sum M_A = 0; \qquad -2a - 4a + 3aR_{Dy} = 0$$

Therefore,

$$R_{Dy} = 2 \text{ kN and } R_{Ay} = 2 \text{ kN}$$

Figure 4.8(c) shows the FBD of the truss with the joints labled. Since the joints are labled with letters, the force in each member can be designated by two letters defining the ends of the member. Members are two force members and can carry either tension or compression indicated by a positive force or negative force, respectively.

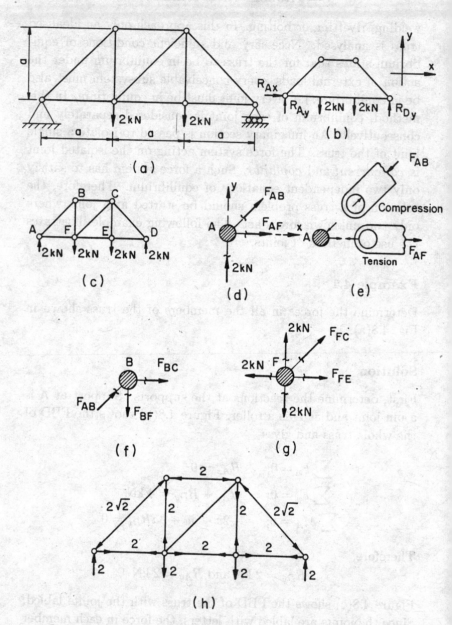

Fig. 4.8

In principle, one can assume any direction for the unknown force acting at a joint. The arithmetic will help in determining the correct direction. At the end of the calculations, if the magnitude of the force is obtained as negative then the originally assumed force direction has to be reversed. It is convenient to associate the positive and negative signs of the arithmetic calculations with tension and compression forces of the members. This can be achieved by assuming that the unknown force in the members at a joint is in tension. This force is indicated by an arrow along the member *acting away from the joint*. This convention will become clear as we proceed further.

At joints A and D the unknown forces are two. Analysis can begin either at A or D. For this example let us begin from joint A.

From the geometry, all inclined members are at 45°.

Equilibrium of joint A

Equilibrium equations give

$$\sum F_x = 0; F_{AF} + F_{AB} \cos 45° = 0$$
$$\sum F_y = 0; 2 \text{ kN} + F_{AB} \sin 45° = 0$$

Therefore,

$$F_{AB} = -2\sqrt{2} \text{ kN and } F_{AF} = 2 \text{ kN}.$$

i.e. the assumed direction of F_{AB} is incorrect and needs to be reversed.

The FBDs of members AB and AF are shown to indicate clearly the mechanism of the action and reaction. Note in the diagram, the assumed force direction of F is reversed to take into account our recent results. The members AB and AF are two force members. Member AB is in compression and member AF is in tension. It is to be observed that from the Fig. 4.8(e), the *tension* (such as AF) will always be indicated by an arrow away *from the joint*, and *compression* (such as AB) will always be indicated by an arrow *towards the joint*.

Equilibrium of joint B

The direction of force F_{AB} is known from the analysis of joint A. Member AB is in compression and the force is shown with an arrow towards the joint. The force in members BF and BC are assumed to be in tension and hence, arrows must be drawn away from the joint.

From equilibrium equations,

$$\sum F_x = 0 \quad ; \quad F_{BC} + F_{AB}\cos 45° = 0; F_{BC} = -2 \text{ kN}$$

$$\sum F_y = 0 \quad ; \quad -F_{BF} + F_{AB}\sin 45° = 0; F_{BF} = +2 \text{ kN}.$$

The result indicates that the assumed directions of forces in member BC is to be reversed. Member BC is in compression and member BF is in tension.

Joint F has only two unknown forces. Hence, let us analyse joint F.

Equilibrium of joint F

Indicate the correct directions of the forces F_{BF} and F_{AF} from the analyses of joints A and B. Equilibrium equations give

$$\sum F_x = 0; \quad F_{FE} + F_{FC}\cos 45° - 2 = 0; \quad F_{FE} = 2 \text{ kN}$$

$$\sum F_y = 0; \quad F_{FC}\sin 45° + 2 - 2 = 0; \quad F_{FC} = 0$$

Proceeding in this way, the forces in the members CE, CD and DE can be determined. This is left as an exercise to the reader.

The forces in the members are tabulated in Table 4.1. Tension and compression forces of various members are conveniently shown diagrametically (Fig. 4.8(h)).

Table 4.1

Member	Nature of force	Magnitude (kN)
AB	C	$2\sqrt{2}$
BC	C	2
CD	C	$2\sqrt{2}$
DE	T	2
EF	T	2
AF	T	2
BF	T	2
FC	-	0
CE	T	2

T: Tension ; C: Compression.

4.6.2 Method of Sections

Generally, the forces in all the members of a truss during its design are not determined. Rarely a few members need to be analysed, for example, checking the critical members in a truss. In such cases, the *method of section* is an ideal choice for quickly obtaining the results.

The necessary and sufficient conditions of equilibrium states that for the truss to be in equilibrium, every conceivable subsystem must also be in equilibrium. Rather than taking the subsystems as the various joints, a section of the truss is considered as a subsystem for the analysis. Depending on the problem on hand, suitable imaginary section is passed through the truss to separate it into two sub-systems. In principle, one can consider the equilibrium of either of these sub-systems for analysis. In general the force system acting on each of these sub-systems is *nonconcurrent* and one can write three independent equations. In view of this, while choosing a section of the truss not more than three members whose forces are unknown are cut. The imaginary section need not be a straight line and can be any arbitrary curve. Occassionally, sections may be cut with more than three unknowns. In such cases additional free bodies of sections or joints may be necessary for the solution of additional unknowns.

The moment equilibrium condition is used to a great advantage in the method of sections. Through proper choice of moment centres, either on or off the section, each of three unknowns can often be determined by a single equation, thus avoiding the solution of solving simultaneous equations.

It is not always possible to assign an unknown force in the proper sense when the free-body diagram of a section is initially drawn. It is convenient to assign all unknown forces arbitrarily as positive by showing arrows pointing away from the section (similar to the method of joints). This convention indicates that at the end of the calculations if the algebraic sign of the force is positive, the member is subjected to tension and if negative it is compression. The following example illustrates the use of method of sections.

Example 4.2

Determine the force in member BC of Fig. 4.8(c).

Solution

Let a section 1-1 be taken as shown in Fig. 4.9(a). FBD of the left portion of the truss is shown in Fig. 4.9(b).

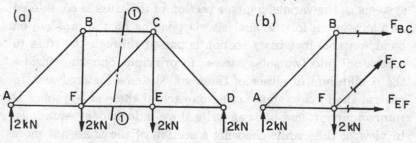

Fig. 4.9

Selecting the moment centre at F eliminates the determination of forces F_{FC} and F_{FE}. Checking for moment equilibrium about F gives,

$$\sum M_F = 0; \quad 2 \times a + F_{BC} \times a = 0$$

or, $\qquad\qquad\qquad F_{BC} = -2 \text{ kN}$

Hence, the force in F_{BC} is 2 kN compression.

Example 4.3

Determine the forces in members BC, BJ and BK of the truss shown in Fig. 4.10(a).

Solution

Normally we first find the reactive forces at the supports. But in this case we can find the forces in members BC, BJ and BK without the evaluation of reactive forces if the equilibrium of only the top portion of the truss is considered.

Consider the equilibrium of the top portion of the truss using section 1-1 (Fig. 4.10). If moment equilibrium is considered about point G, the contribution of forces F_{JG}, F_{CJ} and F_{GH} is zero since their line of action pass through G. The only unknown is F_{BC}. Now,

$$\sum M_G = 0; \quad F_{BC} \times 3 + 10 \times 3 - 10 \times 1.5 = 0$$

Therefore
$$F_{BC} = -5 \text{ kN}$$

F_{BC} is 5 kN compression.

Now consider the equilibrium of top portion of the truss obtained by section 2-2 (Fig. 4.10). If the moment about point H is considered, the only unknown to be evaluated is F_{BJ}.

From geometry all the inclined members are at 45°. Therefore,

$$\sum M_H = 0$$

and

$$F_{BJ} \cos 45° \times 1.5 + F_{BJ} \sin 45° \times 1.5 - 5 \times 3$$

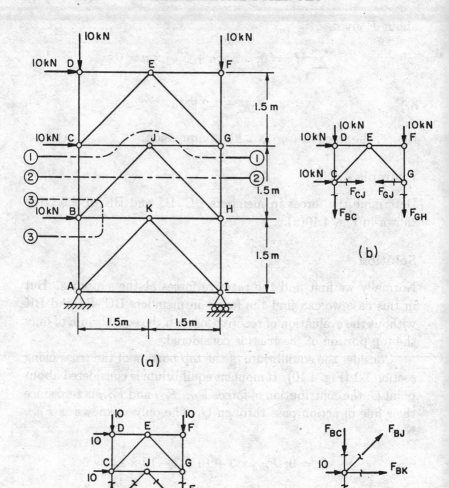

Fig. 4.10

$$-10 \times 1.5 + 10 \times 3 - 10 \times 3 = 0$$

or
$$F_{BJ} = 14.14 \text{ kN (tension)}$$

To evaluate the force in member BK consider the equlibrium of joint B, section 3-3 (Fig. 4.10). This gives

$$\sum F_x = 0; \; 10 + 14.14 \cos 45° + F_{BK} = 0$$

$$F_{BK} = -20 \text{ kN};$$

or
$$F_{BK} = 20 \text{ kN (compression)}$$

This example is solved easily by judiciously using the method of sections and method of joints. The moment centre G used in conjuction with section 1-1 is on the section and the moment centre H used in conjection with section 2-2 is off the section. Thus, by selecting the moment centres either 'on' or 'off' the sections can directly evaluate the member forces without solving simulataneous equations.

4.6.3 Discussion of Method of Joints and Method of Sections

Example 4.3 illustrated that isolations of the truss using both — method of sections and the method of joints — expedited the computation of member forces. With experience in such computations the student will learn how to combine these two methods most effectively. It is the purpose of this particular section to summarize and clarify the important points concerning these methods.

Method of joints can be used to determine the unknown forces acting on an isolated joint, provided there are not more than two unknown forces and they have *different lines of actions*. If there are more than two unknown forces acting on an isolated joint, it is usually not possible to obtain a solution for any of the unknowns using the equilibrium condition of that joint alone.

In one important case there are more than two unknown member forces acting on an isolated joint but are arranged in such a

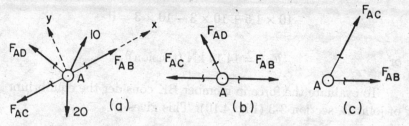

Fig. 4.11 Some cases of forces acting at a pin joint.

way that it is possible to obtain the value of one of them. If the unknown forces except one have the same line of action, the force in that one particular member can be determined by appropriately choosing the co-ordinate axis while analysing the joint. If the x-axis is taken as shown in Fig. 4.11(a), then using $\sum F_y = 0$ we can find force F_{AD}.

One such case is shown in Fig. 4.11(b), where the joint is acted upon by three unknown forces. $\sum F_y = 0$ directly indicates that $F_{AD} = 0$. It is also of interest to consider the case shown in Fig. 4.11(c), where the joint is acted upon by only two forces which do not have the same line of action. In order to satisfy $\sum F_x = 0$ and $\sum F_y = 0$ for such a joint, both F_{AB} and F_{AC} must be zero.

When applying the method of sections, if the isolated portion of the truss is acted upon by three unknown forces that are *neither parallel nor concurrent*, then the three unknown forces can be determined from the three equilibrium equations available for the isolated portion. It is presumed, that the reactions, if any, acting on any such portion have been determined previously.

It is sometimes possible to find some of the unknown forces by the method of sections even when there are more than three unknowns acting on the isolated portion. One case is when all the unknown member forces except one are *concurrent*. The force in the remaining member can be found, by considering the moment equilibrium about the point of concurrency. For example, the section 1-1 used in example problem 4.3, cuts four members and the forces acting on them are not known. Of the four forces, three are concurrent (F_{JG}, F_{CJ} and F_{GH}). Using the point of concurrency (point G), the force in the fourth member F_{BC} was found out.

Another case would be when all the unknown member forces except one are *parallel*. The force acting on this remaining member can be determined by summing up all the force components that are perpendicular to the direction of the other unknown member forces. In each of the above cases, the remaining two equilibrium conditions for the isolated portion will involve more unknowns than there are equations, and hence no solution for these remaining unknowns is possible using the selected portion alone.

In applying either the method of joints or the method of sections, it is important to realize that it makes no difference how many members have been cut in which the member forces are *known*. Only the number of *unknown* member forces is important.

In both the cases, the unknown member forces are assumed to be in tension. However, as soon as a particular member force is evaluated by considering the equilibrium of an isolated joint or a section, only its correct direction must be used consistantly in all the subsequent calculations involving this member.

4.7 Analysis of Simple Statically Indeterminate Trusses

Statically indeterminate trusses can be of three types. It can have more external supports than are necessary to ensure a stable equilibrium configuration or more members than are necessary to prevent collapse, or a combination of both. The first case is known as *external redundancy* and the second case is known as *internal redundancy*.

As stated earlier, a provision is usually made for free expansion and contraction of the truss by having one of the supports as a roller support. If both supports are pin jointed then the truss has external redundancy and is statically indeterminate. Problems of this nature can be solved if the external loading is a simple one. For example, if all the loads are vertical (Fig. 4.12), then at the supports the reactions also have to be vertical and there would be no horizontal component of reaction. This leads to a solution using the principle of statics. However, if the external loads are inclined then the solution is beyond the scope of this book.

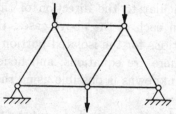

Fig. 4.12 The truss having external redundancy.

Certain classes of internal redundancy problems can also be solved. Truss panels are frequently cross-braced (Fig. 4.13(a)). Such a panel is statically indeterminate if each diagonal member is capable of supporting both tension and compression. However, when the diagonal members are flexible cables incapable of supporting compression, then only the member that carries tension has to be considered as a part of the truss and the truss becomes statically determinate. In structures that have to bear wind loads, which change its direction, cross-braced trusses are used. It is usually evident from the asymmetry of the loading how the panel will deflect. If the deflection is as indicated in Fig. 4.13(b), then member AB should be disregarded and member CD should be retained. When this choice cannot be made by inspection, one can make an arbitrary selection of the member to be retained. If at the final calculations, the member carries tension then the choice is correct. However, if the result shows compression then this member has to be disregarded and other member is to be retained for recalculation.

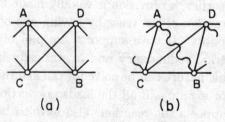

Fig. 4.13 Portion of truss having internal redundancy.

Example 4.4

Figure 4.14(a) shows a simplified model of a truss supporting a water tank. Force in the horizontal direction indicate the idealisation of the force due to wind. The cross braced members can support only tension and incapable of supporting compression. Determine the forces in members CE and DF.

In the truss as shown, both the supports are hinged and the truss has cross-braced members. Here

$$m = 16; \ r = 4; \ j = 8$$

and
$$m + r = 20$$

which is greater than $2j = 16$.

The truss is statically indeterminate. It has internal redundancy as well as external redundancy. The degree of indeterminancy is 4 in the present case.

It is given in the problem statement that the cross-braced members cannot support compression. In view of the wind load (Fig. 4.14(a)), it is reasonable to assume that the truss would deflect (Fig. 4.14(b)). In such a case, the presence of member CE can be neglected. This assumption helps in proceeding with the problem using the equilibrium of joint E. Now,

$$F_x = 0; \ -F_{DE} - 10 = 0$$

or
$$F_{DE} = -10 \text{ kN}; \ F_{DE} = 10 \text{ kN (compression)}$$

and
$$F_y = 0; \ -30 - F_{EF} = 0$$

or
$$F_{EF} = -30 \text{ kN}; \ F_{EF} = 30 \text{ kN (compression)}$$

Equilibrium of joint D (Fig. 4.14(d))
From geometry, the inclined members are at $45°$. Hence,

$$\sum F_x = 0; \ F_{DF} \cos 45° - 10 = 0$$

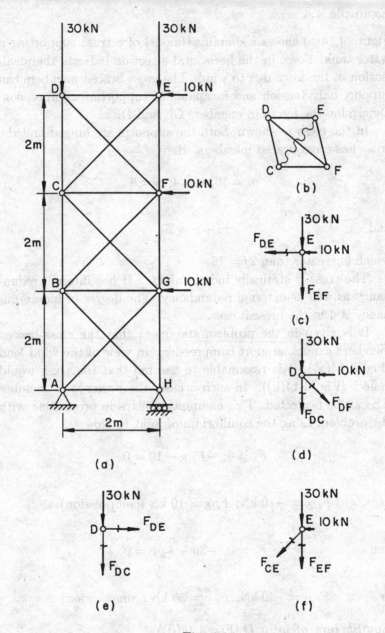

Fig. 4.14

or $$F_{DF} = 10\sqrt{2} \text{ kN} \quad \text{(tension)}$$

The result tallies with the deflection pattern assumed but the deflection pattern shown in Fig. 4.14(b) has to be verified. Assume that the truss deflects in the opposite sense and the member DF is in compression and hence this can be removed for the purpose of analysis.

Equilibrilum of joint (Fig. 4.14(e))

Here $$\sum F_x = 0; \quad F_{DE} = 0$$

Equilibrium of joint (Fig. 4.14(f))

This gives $$\sum F_x = 0; \quad -F_{CE} \cos 45° - 10 = 0$$

or $$F_{CE} = 10\sqrt{2} \text{ kN (compression)}$$

This can never be since the cross-braced members cannot support compression. Thus, the defelction pattern assumed in Fig. 4.14(b) is correct and the member forces are $F_{DF} = 10\sqrt{2}$ (tension) and $F_{CE} = 0$.

4.8 Role of Computers in Truss Analysis

Most of the practical structures like transmission towers, cranes etc., have hundreds of members forming the truss. In such cases, using the method of joints with hand calculations will take months to solve the problem. With the advent of digital computers, the procedure for evaluation of not only member forces but also the stresses and deflections can be computerised. The formulation of the problem for such an approach and its solution procedure is beyond the scope of this book. Nevertheless a cursory look at the available computer oriented methodology is appropriate.

One of the earlist approaches in this direction was expressing the force displacement relations for each member of the truss in a matrix form. Subsequently, individual matrices of members are assembled appropriately to obtain a global matrix equation. Solving this equation leads to the evaluation of displacements at the joints. Using this, the member forces and stresses can be computed. The entire methodology is amenable for computerisation. Several classical books are available on Matrix methods in structural analysis.

When there is a digital computer to assist the analysis procedure, one can improve the idealisations so that the mathematical model is closer to the real life situations. Idealisation of riveted and bolted connections as pin joints is to some extent acceptable. However, a welded joint is more closer to fixed end conditions. Nevertheless, it was also assumed as a pin joint to simplify the calculations.

A further discussion on the role of idealised pin connection and idealised fixed supports may be appropriate now. In reality, every pin connection in the actual structure can take some amount of moment due to friction and every fixed connection cannot completely restrict the rotary motion. So, the idealised representation of the supports actually serves as two extereme situations of actual support conditions. If the real structure is idealised to have pin connections then one has modelled it to be a less rigid structure for analysis purpose. The analysis would then lead to conservative estimates of the load carrying capacity of the structure which is permissible from the engineering point of view. Further, idealising the connections as pin joints removes the evaluation of one unknown per joint. Hence, the analysis would be very simple to perform. On the other hand, idealising the connections as rigid connections leads to more complications in performing the analysis. Thus, idealisation of joints or connections has to be done with care. Since the behaviour of the actual joint or connection lies between the two idealised connections, in special situations one may like to perform two separate analysis to know the bounds of the load carrying capacity.

Before the use of computers became widespread in design offices, trusses were normally assumed to be pin jointed to simplify the analysis. Today computers allow advantage to be taken of the additional stiffness that comes from riveted, bolted or welded connections in the analysis. Pin-jointed trusses, therefore, are no longer of service in the design office. They do, however, still have a role to play in engineering education.

4.9 Flexible Cables

Cables are used in many engineering applications. They are used to support axial loading (tension) such as guy wires in high towers, long chimneys and cooling towers, and for lifting loads in cranes and cable-stayed bridges. Vidyasagar Sethu (completed in 1992) the new bridge across the Ganges in Calcutta is an example of a cable-stayed bridge. Cable-stay technology is fast developing and a detailed analyais must also include the self weight of the cable which introduces transverse loading on the cable. Cables are also used to support transverse loading in applications such as aerial tramways, suspension bridges and transmission lines.

In this chapter, cables supporting transverse loading will be discussed. Cables are assumed to be flexible and any resistance offered to bending is negligible. The transverse loading is supported by axial tension in the cable. Transverse loading of the cable can be divided into two categories, according to their loading : (i) cable supporting concentrated loads, (ii) cable supporting distributed loads. In the design of cables for such applications, the tension, sag and length of the cables have to be determined. The determination of tension in the cable is required to select the appropriate material and the cross-sectional area of the cable.

4.10 Flexible Cable Supporting Concentr- ated Loads in the Transverse Direction

Cable is flexible and consequently, the cable is in axial tension between neighbouring points where the loads are applied. The unknown forces in the analysis of cables are the reactive forces at the supports and the tensions in the individual segments of the cable. The solution for these unknowns is straight forward if the loads and the geometry of the cable are known. It must be remembered that cable is not a rigid body and hence, changes in loading would change its shape.

Example 4.5

A flexible cable supports three loads (Fig. 4.15(a)). Under the action of these loads, point C is located at 2m below point A. What are the co-ordinates Y_B and Y_D in this configuration? Determine the tension in each segment of the cable.

Solution

Free-body diagram of the entire cable is shown in Fig. 4.15(a) & (b). Since the slope of the portions of cable attached at A and E is not known, the reactions at A and E must be represented by two components each.

Equilibrium equations give

$$\sum F_x = 0; \qquad R_{Ax} = R_{Ex}$$
$$\sum F_y = 0; \qquad R_{Ay} = R_{Ey} = 6$$
$$\sum M_A = 0; \qquad 47R_{Ey} = 2 \times 11 + 24 \times 3 + 36 \times 1$$

or $R_{Ey} = 2.77 \text{ kN}$

and $R_{Ay} = 3.23 \text{ kN}$

Three equations of equilibrium are not sufficient to determine the reactions R_{Ax} and R_{Ex}. Additional equation is necessary to solve this. This can be achieved by considering the equilibrium of a portion of the cable, for which co-ordinates of a point in the cable has to be known. From the problem statement, co-ordinates of point C are known.

From Fig. 4.15(c),

$$\sum M_c = 0; \ 2 \times 13 - 3.23 \times 24 + R_{Ax} \times 2 = 0$$

or $R_{Ax} = 25.76 \text{ kN}$

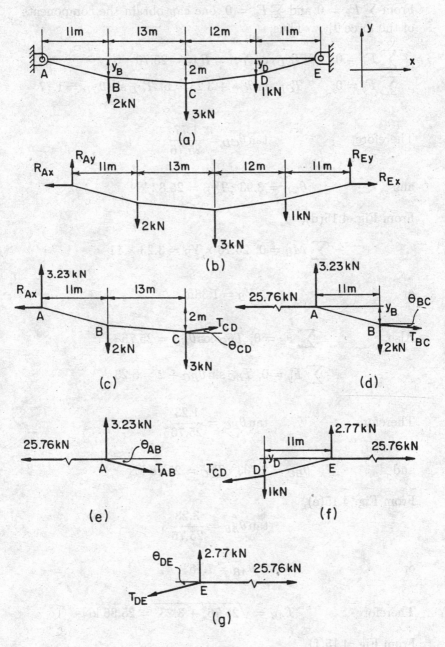

Fig. 4.15

From $\sum F_x = 0$ and $\sum F_y = 0$, one can obtain the components of the force T_{CD}. Thus,

$$\sum F_x = 0; \quad T_{CD}\cos\theta_{CD} = R_{Ax} = 25.76$$

$$\sum F_y = 0; \quad T_{CD}\sin\theta_{CD} + 3.23 = 5; \; T_{CD}\sin\theta_{CD} = 1.77$$

Therefore, $$\tan\theta_{CD} = \frac{1.77}{25.76}$$

and $$\theta_{CD} = 3.93°; T_{CD} = 25.82 \text{ kN}$$

From Fig. 4.15(d),

$$\sum M_B = 0; \; 25.76 \times Y_B = 3.23 \times 11$$

or $$Y_B = 1.38\text{m}$$

Also, $$\sum F_x = 0; \; T_{BC}\cos\theta_{BC} = 25.76$$

$$\sum F_y = 0; \; T_{BC}\sin\theta_{BC} + 2 = 3.23$$

Therefore, $$\tan\theta_{BC} = \frac{1.23}{25.76}$$

and $$\theta_{BC} = 2.733°; T_{BC} = 25.78 \text{ kN}$$

From Fig. 4.15(e),

$$\tan\theta_{AB} = \frac{3.23}{25.76}$$

or $$\theta_{AB} = 7.15°$$

Therefore, $$T_{AB} = \sqrt{25.76^2 + 3.23^2} = 25.96 \text{ kN}$$

From Fig. 4.15(f),

$$\sum M_D = 0; \; 2.77 \times 11 - 25.76Y_D = 0$$

or $$Y_D = 1.18 \text{ m}$$

Also, $$\tan \theta_{DE} = \frac{2.77}{25.76}; \ \theta_{DE} = 6.14°$$

and $$T_{DE} = \sqrt{(2.77)^2 + (25.76)^2} = 25.91 kN$$

Tension in each segment of the cable is summarised below:

$$T_{AB} = 25.96 \text{ kN}; \ \theta_{AB} = 7.15°$$

$$T_{BC} = 25.78 \text{ kN}; \ \theta_{BC} = 2.73°$$

$$T_{CD} = 25.82 \text{ kN}; \ \theta_{CD} = 3.93°$$

$$T_{DE} = 25.91 \text{ kN}; \ \theta_{DE} = 6.14°$$

From the above results it can be easily seen that $T \cos \theta$ is a constant. $T \cos \theta$ represents the horizontal component of the tension at a point. Thus, *the horizontal component of tension in a cable subjected to transverse loads is constant along the length of the cable.*

4.11 Flexible Cable Supporting Distributed Loads in the Transverse Direction

Consider a cable attached to two fixed points A and B and carrying a distributed load (Fig. 4.16). Under action of the distributed load, the cable hangs in the shape of a curve. The equilibrium condition of the cable will be satisfied if each infinitesimal element of the cable is in equilibrium. The free-body diagram of a differential element is shown in Fig. 4.16(b). The size of the differential element in the limit is shrunk to zero. By letting the element become vanishingly small, the conditions which must be satisfied at a point on the cable is determined. Now, returning back to the free-body diagram (FBD) of the differential element, let T be the tension at one end of the cable and let it make an angle θ with the x-axis. Over a length ds of the cable, the tension changes to $T+dT$ and the angle to $\theta + d\theta$. Let the distributed

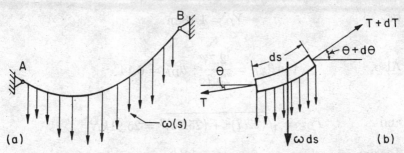

Fig. 4.16 A cable subjected to a general distributed loading.

load be expressed as a function $w(s)$ along the *length of the cable*. The resultant force due to this distributed load is $w(s)ds$ acting downwards as shown in the FBD. Applying equations of equilibrium to the element ds, one gets,

$$\sum F_x = 0; \ (T + dT)\cos(\theta + d\theta) - T\cos\theta = 0$$

$$\sum F_y = 0; \ (T + dT)\sin(\theta + d\theta) - T\sin\theta - w(s)ds = 0 \quad (4.2)$$

The above equations can be simplified, by first using the trigonometric identities for the sum of two angles. A further simplification is possible by using the approximations $\sin(d\theta) = d\theta$ and $\cos(d\theta) = 1$, since $d\theta$ is very small. The resulting equations are

$$d(T\cos\theta) = 0; \quad d(T\sin\theta) = w(s)ds \quad (4.3)$$

We are interested in expressing every quantity in terms of the length of the segment ds. From geometry, $\cos\theta = dx/ds$ and $\sin\theta = dy/ds$. Using these one gets,

$$d\left(T\frac{dx}{ds}\right) = 0 \quad (4.4)$$

$$d\left(T\frac{dy}{ds}\right) = w(s)ds \quad (4.5)$$

From Eq. (4.4) one gets,

$$T\frac{dx}{ds} = \text{constant}$$

Let this constant be denoted by T_o. Then,

$$T\frac{dx}{ds} = T_o \quad (4.6)$$

Eq. (4.5) gives

$$\frac{d}{ds}\{T\frac{dy}{ds}\} = w(s) \tag{4.7}$$

which can be written as

$$\frac{d}{ds}\{T\frac{dy}{dx}\cdot\frac{dx}{ds}\} = w(s) \tag{4.8}$$

Substituting for $[T\frac{dx}{ds}]$ as T_o,

$$\frac{d}{ds}\cdot(\frac{dy}{dx}) = \frac{w(s)}{T_o} \tag{4.9}$$

The equation of the deformed shape of the cable can be obtained by integrating the above equation.

The length of the differential element ds is

$$ds^2 = dx^2 + dy^2$$

or

$$ds = \sqrt{1 + (dy/dx)^2}dx \tag{4.10}$$

The length of the cable can be obtained by integrating Eq. (4.10), i.e.,

$$s = \int_0^L \{\sqrt{1 + (dy/dx)^2}\}dx \tag{4.11}$$

To summarise, in dealing with flexible cables, to obtain the tension in the cable one can use:

$$T\frac{dx}{ds} = T_o \tag{4.12}$$

Equation of the deformed shape of the cable can be obtained as

$$\frac{d}{ds}(\frac{dy}{dx}) = \frac{w(s)}{T_o} \tag{4.13}$$

and the length of the cable as

$$s = \int \{\sqrt{1 + (dy/dx)^2}\}dx \tag{4.14}$$

4.12   Uniformly Loaded Cables

In common engineering structures, flexible cables are often employed under two kinds of distributed loading: (i) a uniform force distributed horizontally (surface force distribution) and (ii) a uniform force distributed along the length of the cable (body force distribution). The cables of a suspension bridge are loaded approximately as mentioned in case (i). Suspension bridges are better suited for evenly spread loads than large single loads - such as railway trains with their heavy locomotives. In view of this, railway bridges are not constructed as suspension bridges. The weight of the cable which is distributed uniformly along the length of the cable is negligible in comparison to the uniform weight of the roadway, which loads the cable as mentioned in case (i). On the otherhand, cables of electrical power transmission lines are loaded, as in case (ii), due to their self weight. The above two engineering examples are not exhaustive and classification of the cables are based on whether the surface load distribution is significant or the body force distribution is significant.

In the analysis of the cables, apart from determination of the tension in the cable, it is very important to know the length of the cable for the specified application. This requires knowledge of the shape that the cable takes under the given loading. It will be shown that if the cable is loaded by a uniform force distributed horizontally (surface force distribution), the cable takes the shape of a parabola. Such cables are known as *parabolic cables*. The shape of the curve taken by a cable loaded uniformaly along its length (body force distribution) is known as a catenary or hanging-chain curve, derived from Latin—*catena* meaning a chain. Such cables are known as catenary cables. In what follows, we shall discuss these cables in detail.

4.12.1   Parabolic Cables

A cable under surface force distribution, for example the cable in a suspension bridge with uniform load distribution across the span, hangs in a parabolic arc (Fig. 4.17). The depth h (Fig. 4.17) is known as the sag of the cable and L is the *span*. Let the uniform load distribution be w_o per unit length along the horizontal. In SI units the basic unit of w_o is N/m.

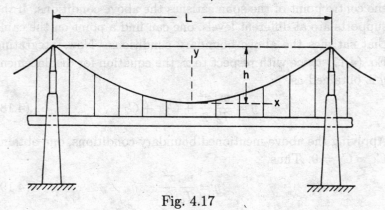

Fig. 4.17

Referring to Fig. 4.18,

$$w_o dx = w(s)ds$$

or

$$w(s) = w_o \frac{dx}{ds} \qquad (4.15)$$

Substituting this in Eq. (4.13), the equation of the deformed shape of the cable is obtained as,

$$\frac{d}{ds} \cdot \frac{dy}{dx} = \frac{d}{dx} \cdot \frac{dy}{dx} \cdot \frac{dx}{ds} \cdot = \frac{w_o}{T_o} \frac{dx}{ds} \qquad (4.16)$$

or

$$\frac{d^2 y}{dx^2} = \frac{w_o}{T_o} \qquad (4.17)$$

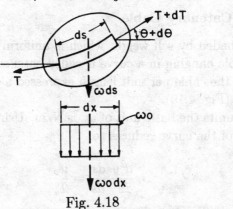

Fig. 4.18

Consider a co-ordinate system such that at $x = 0$ and $y = 0$, $dy/dx = 0$. When the supports are at the same level (Fig. 4.17), the centre point of the span satisfies the above conditions. If the supports are at different levels, one can find a point on the cable that satisfies the above boundary conditions. Now integrating Eq. (4.17) twice with respect to x, the equation for displacement y is obtained as,

$$y = \frac{w_o}{T_o}\frac{x^2}{2} + C_1 x + C_2 \qquad (4.18)$$

Applying the above mentioned boundary conditions, one obtains $C_1 = C_2 = 0$. Thus,

$$y = \frac{w_o}{T_o}\frac{x^2}{2} \qquad (4.19)$$

which is the equation of a parabola.

Equation (4.12) gives the tension in the cable. From Eq. (4.10),

$$\frac{dx}{ds} = 1/\sqrt{1 + (dy/dx)^2} \qquad (4.20)$$

Using Eqs. (4.12), (4.18) and (4.21), the tension in the cable is obtained as,

$$T(x) = \sqrt{T_o^2 + w_o^2 x^2} \qquad (4.21)$$

and the length of the cable as

$$s = \int_{x_1}^{x_2} \{\sqrt{1 + (w_o x/T_o)^2}\}dx \qquad (4.22)$$

4.12.2　Catenary Cable

A cable loaded by self weight which is uniform along the length of the cable hanging in a curve is called catenary. Let w_o be the weight of the cable per unit length expressed along the length of the cable (Fig. 4.19).

In SI units the basic unit of w_o is N/m. Using Eq. (4.13) the equation of the curve reduces to

$$\frac{d^2 y}{dx^2}\cdot\frac{dx}{ds} = \frac{w_o}{T_o} \qquad (4.23)$$

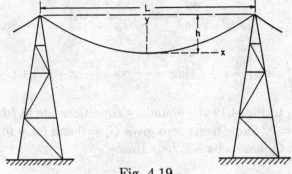

Fig. 4.19

Replacing dx/ds in terms of the slope dy/dx and making use of Eq. (4.20), gives

$$\frac{d^2y}{dx^2} = \frac{w_o}{T_o}\sqrt{1 + (dy/dx)^2} \qquad (4.24)$$

Integration of Eq. (4.24) is considerably complicated. Let dy/dx be taken as q. Then Eq. (4.24) becomes

$$\frac{dq}{dx} = \frac{w_o}{T_o}\sqrt{1 + q^2} \qquad (4.25)$$

Rewriting Eq.(4.25) gives

$$\frac{dq}{\sqrt{1 + q^2}} = \frac{w_o}{T_o}dx \qquad (4.26)$$

Integrating this equation (using the table of integrals) gives

$$\log_e\{q + \sqrt{1 + q^2}\} = \frac{w_o}{T_o}x + C_1 \qquad (4.27)$$

Let $(w_o x)/(T_o) + C_1$ be taken as A for simplicity in writing, then Eq. (4.27) becomes

$$q + \sqrt{1 + q^2} = e^A$$

$$\sqrt{1 + q^2} = -q + e^A$$

Solving for q gives

$$q = \frac{e^{2A} - 1}{2e^A} = \frac{e^A - e^{-A}}{2} = \sinh A$$

$$q = \frac{dy}{dx} = \sinh\left[\frac{w_o}{T_o}x + C_1\right] \tag{4.28}$$

$$y = \int \sinh\left[\frac{w_o}{T_o}x + C_1\right]dx = \frac{T_o}{w_o}\cosh\left[\frac{w_o}{T_o}x + C_1\right] + C_2 \tag{4.29}$$

Referring to Fig. 4.19 the boundary conditions are $dy/dx = 0$ at $x = 0$, $y = 0$. Slope being zero gives $C_1 = 0$ and for y to be zero at $x = 0$, C_2 has to be $-T_o/w_o$. Hence,

$$y = \frac{T_o}{w_o}\left[\cosh\frac{w_o}{T_o}x - 1\right] \tag{4.30}$$

This equation defines the catenary.

The length of the differential element ds is

$$ds = \sqrt{1 + (dy/dx)^2} = \sqrt{1 + \sinh^2\left(\frac{w_o}{T_o}x\right)} = \cosh\frac{w_o}{T_o}x$$

and the length of the curve can be obtained as

$$s = \int_{x_1}^{x_2} \cosh\frac{w_o}{T_o}x\,dx$$

$$s = \frac{T_o}{w_o}\sinh\frac{w_o}{T_o}x\Big|_{x=x_1}^{x=x_2} \tag{4.31}$$

Finally, the tension in the curve is given as,

$$T(x) = T_o\frac{ds}{dx}$$

or

$$T(x) = T_o\cosh\left(\frac{w_o}{T_o}x\right) \tag{4.32}$$

The tension can also be expresed in terms of a function in y by using Eq. (4.30) as

$$T(y) = T_o + w_o y \tag{4.33}$$

It is to be remembered that $T(x)$ and $T(y)$ gives the tension along the cable as a function of x and y, respectively. The expressions should not be confused with x and y components of tension.

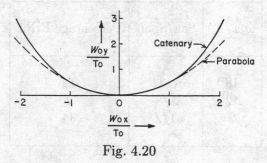

Fig. 4.20

4.12.3 Approximation of Catenary Cable as a Parabolic Cable for Small Sag to Span Ratios

The solution of catenary problems where the sag to span ratio is small may be approximated by the relationships developed for parabolic cables. For comparison, the equation of the parabola can be rewritten as

$$\frac{w_o}{T_o}y = \frac{1}{2}(\frac{w_o}{T_o}x)^2 \qquad (4.34)$$

The hyperbolic cosine of the catenary can be developed into a Taylor's power series as

$$\frac{w_o}{T_o}y = \frac{1}{2}(\frac{w_o}{T_o}x)^2 + \frac{1}{24}(\frac{w_o}{T_o}x)^4 + \frac{1}{720}(\frac{w_o}{T_o}x)^6 \qquad (4.35)$$

The first term of this expansion corresponds to the formula of the parabola. The other terms represent the difference between the two. A comparison is shown graphically in Fig. 4.20.

The difference between the two becomes significant only for large values of $w_o x/T_o$. Thus a parabola is a good approximation for a catenary for small sag to span ratio. A small sag to span ratio means a stiff cable, and the uniform distribution of weight along the cable is not much different from the same load intensity distributed uniformly along the horizontal. The percentage error by assuming a catenary cable as a parabolic one for various spans to sag ratio is given in the table below.

$\omega_o x/T_o$	Sag/Span	Percentage Error
18	1/32	0.1
12	1/8	2
2	1/2	34

4.12.4 Historical Background of the Development of Solotion for Catenary Cable

We have seen in the present section that for a stiff cable, the arc of the cable can be *approximated* by a parabola. In fact, for a long time, the cable supporting its own weight was *considered* to hang as a *parabola*. The studies of Galileo (1564-1642) in the field of mechanics cast doubt on the correctness of this view, but Galileo was unable to corroborate or refute the idea. In 1669, Jungius established, both theoretically and experimentally, that the line of suspension of the cable is not a parabola. But the mathematics of that time was not sufficiently equipped to find the true shape of the curve. Soon after Newton and Leibniz worked out the methods of infinitesimal analysis (differential calculus and integral calculus) it was possible to solve the problem of the curve of suspension of a cable. The problem was formulated in 1690 by James Bernoulli, and was thereafter solved by his brother John Bernoulli, Huyghens and Leibniz.

Example 4.6

A cable supports a uniformly distributed load of 1.5 kN/m along the horizontal and is suspended from two fixed points that are 36 m apart as shown in Fig. 4.21. Determine the maximum tension in the cable and its length.

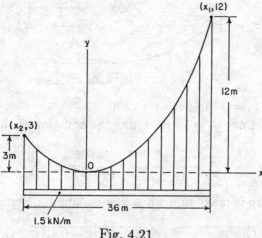

Fig. 4.21

Solution

This problem belongs to the class of parabolic cable for which one has the following equations for

$$y = \frac{w_o}{T_o}\frac{x^2}{2}$$

$$T = \sqrt{T_o^2 + w_o^2 x^2}$$

$$s = \int_{x_1}^{x_2} \sqrt{1 + (\frac{w_o}{T_o}x)^2}dx$$

curve, tension, and length, respectively. If the lowest point, as shown in Fig. 4.21, is taken as origin, then using the equation of curve one has

$$12 = \frac{1.5}{T_o}\frac{x_1^2}{2} \text{ and } 3 = \frac{1.5}{T_o}\frac{x_2^2}{2}$$

The distance $x_1 - x_2$ is given in the problem as 36 m. Using these one gets,

$$4(x_1 - 36)^2 = x_1^2$$

or

$$2(x_1 - 36) = \pm x_1$$

The solution of the above equation gives

$$x_1 = 24 \quad \text{or} \quad x_1 = 72$$

Since the entire span is only 36 m, $x_1 = 72$ is not a valid solution and hence discarded.

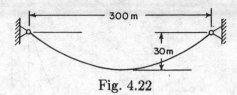

Fig. 4.22

When $x_1 = 24$ m ; $x_2 = -12$ m.
For $x_1 = 24$ m ; $y_1 = 12$ m. Using the equation of the curve one gets,

$$T_o = \frac{w_o}{y_1}\frac{x^2}{2} = \frac{1.5(24)^2}{12} = 36 \text{ kN}$$

The tension is maximum when x is maximum. Hence,

$$T_{max} = \sqrt{T_o + w_o^2 x_{max}^2} = \sqrt{(36)^2 + (1.5)^2(24)^2} = 50.91 \text{ kN}$$

The length of the cable s is

$$s = \int_{-12}^{24} \sqrt{1 + (\frac{w_o}{T_o}x)^2}dx = \int_{-12}^{24} \sqrt{1 + (\frac{1.5x}{36})^2}dx = \frac{1}{24}\int_{-12}^{24} \sqrt{(24)^2 + x^2}dx$$

From the table of integrals, the solution for the above integration is given as

$$s = \frac{1}{24}[\frac{x}{2}\sqrt{x^2 + (24)^2} + \frac{(24)^2}{2}\log_e(x + \sqrt{x^2 + (24)^2})]\Big|_{-12}^{24} = 40.03 \text{ m}$$

The results are summarised as:

$$T_o = 36 \text{ kN} \quad T_{max} = 50.91 \text{ kN} \quad s = 40.03 \text{ m}$$

Example 4.7

A flexible cable weighing 120 N/m is supported from two supports at the same level, 300 m apart as shown in Fig. 4.22. The cable has a sag of 30 m. Find the total length of the cable, tension at middlength and the maximum tension.

Solution

This problem belongs to the class of catenary cable for which

$$y = \frac{T_o}{w_o}(\cosh \frac{w_o}{T_o}x - 1)$$

$$s = \frac{T_o}{w_o}\sinh \frac{w_o}{T_o}x \Big|_{x_1}^{x_2}$$

$$T_{(y)} = T_o + w_o y$$

$$T(x) = T_o \cosh \frac{w_o}{T_o}x$$

are the equations for curve, length and tension in the cable respectively, when origin is taken as shown in Fig. 4.19. Tension T_o, at mid-length has to be determined first. This can be done by using the equation of the curve.

From Fig. 4.22 at $x = 150m, y = 30$ m.

Hence, $$30 = \frac{T_o}{120}(\cosh \frac{120 \times 150}{T_o} - 1)$$

or $$\frac{3600}{T_o} = \cosh \frac{18,000}{T_o} - 1$$

By trial and error method, the above equation can be solved for

$$T_o = 45,550 \text{ N}$$

Note: The above equation can also be solved graphically or one can approximate the catenary curve as a parabola (see Example 4.8) and get an initial value for T_o. Accurate results can then be obtained by using an iterative technique.

Maximum tension occurs at the supports. Hence,

$$T_{max} = T_o + w_o 30 = 45,550 + 120 \times 30 = 49,150 \text{ N}$$

The total length of the cable is

$$s = 2 \times \frac{T_o}{w_o}\sinh \frac{w_o}{T_o}x \Big|_0^{150} = 2 \times \frac{45,530}{120}\sinh \frac{120 \times 150}{45,550} = 307.87 \text{ m}$$

The results are summarised as:

$$T_o = 45550 \text{ N} \quad T_{max} = 49150 \text{ N} \quad s = 307.87 \text{ m}$$

Example 4.8

Approximate the curve of example Problem 4.7 as a parabola and solve for the tension at mid-length, the maximum tension and total length of the cable.

Solution

If the curve is considered as a parabola, the following equations can be written:

$$y = \frac{w_o}{T_o}\frac{x^2}{2}$$

$$T = \sqrt{T_o^2 + w_o^2 x^2}$$

$$s = \int_{x_1}^{x_2} \sqrt{1 + (\frac{w_o x}{T_o})^2}\, dx$$

The load w_o is assumed to be acting horizontally rather than along the cable. At $x = 150$ m and $y = 30$ m, $w_o = 120$ N/m, hence

$$30 = \frac{120}{T_o}\frac{(150)^2}{2}$$

or $T_o = 45,000$ N

Maximum tension occurs at the supports where $x = 150$ m. Therefore,

$$T = \sqrt{(45000)^2 + (120)^2(150)^2} = 48466.5 \text{ N}$$

The length of the cable is

$$s = 2\int_0^{150} \sqrt{1 + (\frac{120x}{45,000})^2}\, dx$$

$$= \frac{2}{375}\int_0^{150} \sqrt{(375)^2 + x^2}\, dx.$$

Using the table of integrals,

$$s = \frac{2}{375}\left[\frac{x}{2}\sqrt{x^2 + a^2} + \frac{a^2}{2}\log_e(x + \sqrt{x^2 + a^2})\right]_0^{150}$$

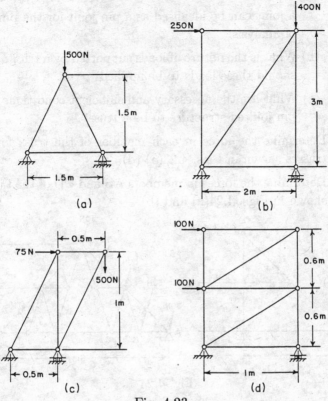

Fig. 4.23

Here $a = 375$ hence

$$s = 307.82 \text{ m}$$

The results are summarised as:

$$T_o = 45000 \text{ N} \quad T_{max} = 48466.5 \text{ N} \quad s = 307.82 \text{ m}$$

The results are quite close to the results obtained in Example 4.7. This shows that for small sag to span ratio, approximating the catenary cable to be a parabolic cable provides a good approximation.

Problems

4.1 (a) Usually the members forming the truss are bolted, riveted or welded to a gusset plate. Explain how such a joint can be idealised as a pin joint for the purpose of analysis.

(b) What is the nature of loads supported by a truss? How should these loads to be applied?

(c) What are the necessary and sufficient conditions for a pin jointed structure to be a truss?

4.2 Determine the force in each member of the truss for the trusses shown in Figs. 4.23(a)-(d).

4.3 Determine the forces in members AB and CD of the trusses shown in Figs. 4.24(a) and (b).

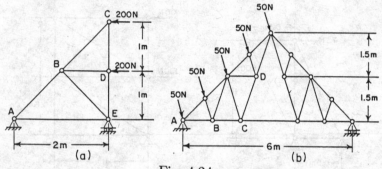

Fig. 4.24

4.4 Determine the zero-force members in the trusses shown in Figs. 4.25(a) and (b) for the given loading.

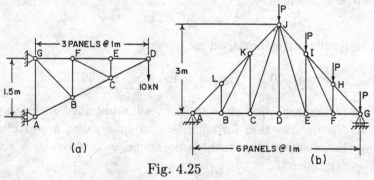

Fig. 4.25

4.5 Determine the forces in members JI and DE of the truss shown in Fig. 4.26. (Hint: Use section 1-1).

4.6 Determine the force in member GC of Fig. 4.27 using only one free-body diagram and one equation of equilibrium.

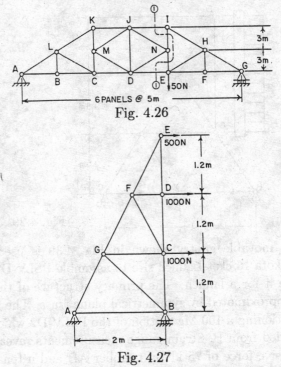

Fig. 4.26

Fig. 4.27

4.7 A small rail road bridge is constructed of steel members. A train stops on the bridge, and the loads applied to the truss on one side of the bridge is as shown in Fig. 4.28. Determine the forces in members AB, BC and CD.

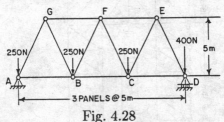

Fig. 4.28

4.8 In example problem 4.1, if the weight per unit length of the members used is (500/a) N, determine the forces in all the members.

4.9 Figure 4.29 shows a truss supporting a sign board. The board is subjected to a wind load of 10 kN. The force can be equally lumped at joints E, F and G. Determine the forces in members AG and GD.

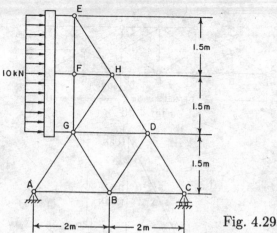

Fig. 4.29

4.10 The movable gantry shown in Fig. 4.30 is used at Sri-
harikota Rocket Launch site to assemble PSLV-D2 to prepare it for a launch. The primary structure of the gantry
is approximated by symmetrical plane truss. The gantry is
positioning a 100 Mg section of the PSLV-D2 which is suspended from B. Strain gauge measurements reveal a compressive force of 75 kN in member AB and a tensile force
of 120 kN in member GF. Calculate the forces in members
BC, CE, DE and BF.

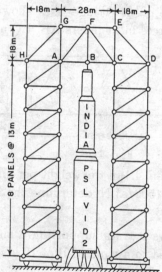

Fig. 4.30

4.11 (a) What are the various engineering applications of cables?

(b) Explain how does a cable resist lateral loads.

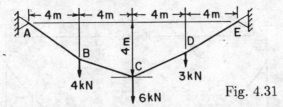

Fig. 4.31

(c) High voltage transmission lines have to be laid between Kanpur city and Panki power station which are 10 kms apart. The distance between two transmission towers is 500 m. For the application the ratio of (T_0/w_0) is to be maintained as 1500 m. The technician involved in laying the cable asks for a cable length of 10,050 m + 25 m (as wastages) from the stores for laying a single line. The person incharge of stores is not willing to release beyond 10,025 m. As a technocrat how will you convince the stores in-charge that the requirement intended by the technician is reasonable.

4.12 Figure 4.31 shows a cable supporting three concentrated loads. It has been measured that the sag at C is 4 m. Determine the reactions at the supports and the maximum tension in the cable.

4.13 A cable is suspended from two supports which are separated by 20 m horizontally and 6 m vertically as shown in Fig. 4.32. Determine the reactions at the supports and the maximum tension in the cable.

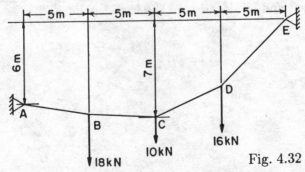

Fig. 4.32

4.14 A cable of length 110 m is supported from two points 100 m apart on the same level. If the cable supports a total load of 5000 N uniformly distributed along the horizontal

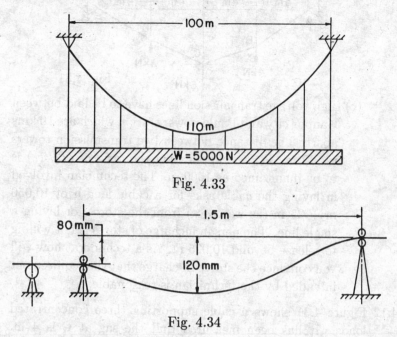

Fig. 4.33

Fig. 4.34

as shown in Fig. 4.33, what is the maximum force in the cable?

4.15 Figure 4.34 shows the feeding mechanism of paper to a printing press. The mass per unit length of paper is 300 g/m. Assume that the curve formed between points A and B as parabolic. Determine the maximum tension in the sheet and the location of the lowest point C.

5. LAWS OF FRICTION AND SIMPLE MACHINES

5.1 Introduction

Frictional forces play a very important role in our daily lives. People could not walk or drive automobiles without the beneficial effects of friction which make tractive forces possible. Belt drives, friction clutches and brakes also require frictional forces in order to function. In the above examples friction is beneficial, but in many other situations, such as in bearings, one needs to minimise friction. In such cases, appropriate lubricants are used to minimise energy loss due to friction.

In this chapter, a comprehensive definition of friction is given. Distinction is made between external and internal friction and the laws governing dry friction/Coloumb friction are then discussed. The advantage of friction in providing "self-locking" is shown for a class of simple machines. Other related concepts such as mechanical advantage, velocity ratio and mechanical efficiency are also discussed.

5.2 Definition of Friction

One of the important forces in study of mechanics is the force due to friction. The force of friction is defined as the force that develops at the surfaces of contact of two bodies, when there is a tendency of one body to slide over the other or when the body actually starts sliding. This force is exerted on the bodies, tangential to their surfaces of contact and in the direction opposite to their impending or actual relative velocity of motion. Thus this force always impedes their relative motion. Although the nature of these resisting forces is not completely understood, one can recognize them by their effects, and experimental investigations have taught us how to account for them within acceptable engineering accuracy.

Distinction can be made between *external and internal friction*. The word *friction* is almost synonymous with the phrase *friction force*, although it may signify the entire phenomena associated with the occurrence of frictional force.

5.3   External Friction

External friction is the interaction between the surfaces of two solid bodies in contact. When the surfaces are at rest but having a tendancy of relative motion with respect to each other, we speak of *static friction*; when the surfaces are in relative motion, we speak of *sliding or kinetic friction*. Of late, the term dynamic friction is used to indicate the force developed when the bodies are in relative motion. Henceforth, the term *dynamic friction* will be used in place of *kinetic friction*.

External friction phenomena involving both static and dynamic friction is termed as *dry friction* or *Coloumb friction*. In dry friction, the maximum frictional force is proportional to applied normal load. The dynamic or kinetic friction force is always less than the *maximum static friction force* and this occurs when the body is about to slip. A detailed discussion on the laws governing static and dynamic friction would follow later in this chapter.

The laws of dry friction are useful in the study of wedges, power screws, partially lubricated bearings, brakes, clutches, belt drives etc. The main objective of this chapter is to study the effects of dry friction or Coloumb friction.

When a body rolls along the surface of another body without slipping, a special kind of resistance is introduced which is called the *rolling resistance* or *rolling friction*. The frictional force due to rolling resistance is the least and this led to the development of ball and roller bearings.

5.4   Internal Friction

Internal friction can be of two types. One is *fluid friction* or *viscous friction* and the other is *solid friction*.

5.4.1   Fluid Friction

Fluid friction is developed between fluid elements when adjacent layers in a fluid (liquid or gas) are moving at different velocities. The frictional force developed is proportional to the relative velocity between the layers and viscosity of the fluid.

Fluid friction gains, importance in problems involving flow of fluids through pipes and orifices, or while dealing with bodies immersed in fluids where a relative motion exists between the body and the fluid. It is also of consequence in analysis of the motion of well lubricated surfaces in contact. For example, the study of a well lubricated bearing, where a thin film of lubricant completely separates the axle and the journal, has to be analysed using the *hydrodynamic theory* which is usually developed in a study of fluid mechanics. In the study of hydrodynamics and aerodynamics the frictional force due to fluid friction is termed as *friction drag*. The viscous friction is termed as viscous damping in vibration studies.

5.4.2   Solid Friction

Internal friction is found in all solid materials subjected to cyclic loading. Here, energy is dissipated internally within the material itself. For example, a spring oscillating in vacuum eventually comes to rest because of solid or internal friction. Experiments by several investigators indicate that for most structural metals such as steel or aluminium the energy dissipated per cycle is independent of the frequency over a wide range of frequencies and is proportional to the square of the amplitude of vibration. Thus, frictional force is proportional to the displacement. In the study of vibration problems, solid friction is termed as solid damping or material damping. In such problems, for the purpose of simplified analysis, the effect of solid friction is included as equivalent viscous friction or Coloumb friction.

5.5   The Laws of Dry Friction

Frictional force comes into play between bodies in contact when there is a tendency of one body to slide over the other, or the body actually slides (Fig. 5.1). For example, the block (Fig. 5.1(a))

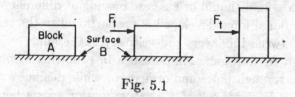

Fig. 5.1

will not experience any frictional force unless there is an external force to push it. When pushing force F_t is applied, frictional forces develop at the interface between block A and surface B. When force F_t is increased, the frictional force also increases. The effect of force F_t is balanced by the frictional force and the block remains in equilibrium and at rest. This situation does not continue indefinitely when force F_t is increased. Because there is an upper limit of the frictional force and when force F_t is increased to balance this maximum frictional force, we say that the block is about to slide or the block is in impending motion. Any further increase in force F_t will cause the block to slide. Once the block begins to slide, dynamic friction comes into play and the frictional force is less than what was when the block was in impending motion. In view of the difference between the applied force and the resisting frictional force, there is initial acceleration of the block, when the block begins to slide.

A common experience is that once the body begins to slide, the effort necessary to maintain its motion at constant velocity along a straight line is less than the original effort to cause it to move from rest. One must remember that the block still remains in equilibrium while it is moving at a constant velocity and the pushing force is equal to the frictional force. It has been found experimentally that the frictional force is more or less independent of the relative velocity between the contacting surfaces. The dependence of frictional force on the applied force and the relative velocity between contacting surfaces is shown in Fig. 5.2.

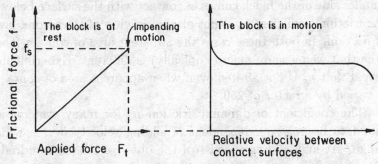

Fig. 5.2 Frictional force as a function of the applied force and the relative velocity between the contacting surfaces.

Based on experimental observations, Amonton formulated the following two empirical laws of sliding friction:

1. The magnitude of friction force (maximum static friction force) is directly proportional to the normal load between the surfaces for a given pair of materials.

2. The magnitude of friction force (maximum static friction force) is independent of the area of the contacting surfaces (apparent area) for a given normal load.

To this a third law is added which is attributed to Coloumb.

3. The magnitude of friction force (kinetic/dynamic friction force) is independent of the sliding velocity.

The above three laws are generally referred to as *Coloumb's laws of friction.*

The first law may be expressed as

$$f_s = \mu_s N \tag{5.1}$$

where μ_s is the proportionality constant, known as the *coefficient of static friction* and N is the normal load. Experiments reveal that μ_s remains constant for two mating surfaces even when the normal load is varied by a factor of nearly 10^6. A coefficient of friction applies to a given pair of mating surfaces. A coefficient of friction for a single surface is meaningless.

Referring to Fig. 5.1(c), if the block is rotated by 90° and the smaller side of the block comes in contact with the surface below, the friction coefficient μ_s is not altered in view of the second law of friction. In both these cases the apparent area of contact (the area that we measure macroscopically) is different. Experiments reveal that μ_s is not altered even when apparent area of contact is varied by a factor of 250.

The coefficient of dynamic friction μ_d for many systems is found to be nearly independent of sliding velocity over a wide range. At high sliding speeds (of the order of tens or hundreds of meters per second), for metal to metal contact, μ_d falls with increasing velocity. This phenomena is observed for unlubricated or poorly lubricated surfaces and can lead to frictional oscillations, often called as stick-slip. This phenomenon is responsible for many of the noises in our environment, including the creaking of doors, squeaking of brakes etc. When the surfaces are well lubricated, the friction coefficient increases with increase in the sliding velocity and minimizes the frictional oscillations. The coefficient of friction in this case, however, is much less in comparison with those encountered with unlubricated surfaces.

In considering the laws governing friction between two solid bodies in contact, it is necessary to distinguish between the three possible states of the surfaces ; (i) dry, (ii) greasy or partially lubricated, and (iii) film or completely lubricated. For dry or partially lubricated surfaces, the laws of Coloumb friction can be applied. If a film of lubricant is present then one has to use the laws of viscous friction (fluid friction) for the analysis.

The term impending motion is repeatedly used in the study of dry friction. One should have a clear idea of what the term impending motion actually signifies. The laws of static friction are valid only for the case of impending motion. Imagine yourself being engaged in the game of "tug of war". You loose the game when the rope begins to slide out of your hands. The laws of static friction are valid only at this crucial moment and the frictional force preventing the rope from sliding is given by Eq. (5.1). To avoid loosing the game, the normal force on the rope is increased by clasping it hard - which eventually raises the maximum limit of the frictional force. The next time you engage in this game, try applying a suitable powder on your palms to raise the coefficient of friction between your palm and the rope.

Remember that the laws of friction give only the maximum frictional force (Eq. 5.1) when the motion is impending. Equation (5.1) cannot be used to find the frictional force when the motion is not impending. In such a case, the frictional force can be found only by using the equations of equilibrium. The following example illustrates this point.

Example 5.1

A block of size $2a \times 2a$ and weight 200 N is resting on a floor. Initially a force F_1 is applied (Fig. 5.3(a)). Force F_2 is applied at a height of $1.5a$ from the floor (Fig. 5.3(b)). The force is gradually increased from zero to 50 N, 90 N and 100 N. Draw the FBD showing clearly the point of application of various forces and also state the condition of the block, if the coefficient of static friction between the block and the floor is 0.3 and the dynamic coefficient of friction is 0.25..

Solution

Let the general FBD be as shown in Fig. 5.3(c). Equilibrium equations give

$$\sum F_x = 0; F_2 = f$$

$$\sum F_y = 0; N = 300 \text{ N}$$

$$\sum M_c = 0; \ Nx + 100\frac{a}{2} - 1.5aF_2 = 0$$

or

$$x = \frac{a}{300}(50 - 1.5F_2)$$

These equations show that N is independent of F_2 and x is a function of F_2.

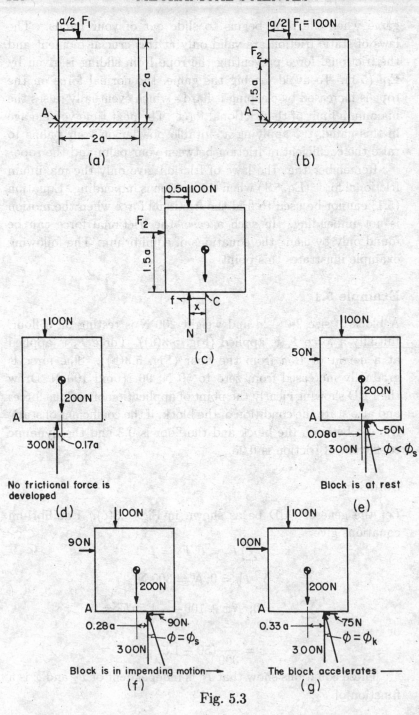

Fig. 5.3

F_2 (N)	x (m)	N (N)	f (N)	f_s (N)	Condition
0	$0.17a$	300	0	90	Block is at rest
50	$-0.08a$	300	50	90	Block is at rest
90	$-0.28a$	300	90	90	Block is in impending motion
100	$-0.30a$	300	75	90	Block accelerates

In the table above, f indicates the value of actual frictional force developed and f_s indicates the maximum static frictional force. The FBD for the above cases are shown in Figs. 5.3(d)-(g).

5.6 Mechanism of Static and Dynamic Friction

The surface of a solid body, even if finely ground, is far from smooth. On the atomic scale, all material surfaces are rough. The surface has on it microscopic projections, depressions and other irregularities. When the surfaces of two bodies come into contact, the microprojections often engage the corresponding depressions (Fig. 5.4), on a highly magnified scale. The meshing

Fig. 5.4 Magnified view of the micro-projections in contact.

of these projections impedes relative motion of the contacting bodies. Apart from this, surface adhesion at various spots on the surface, also impedes relative motion.

The engagement of the microprojections and the partial adhesion of the surfaces are promoted by the normal force at the interface and the frictional force increases when the normal force is increased.

A pushing force, less than the maximum force of static friction, leads mainly to elastic deformation of the micro projections and the contact points where the forces of intermolecular cohesion are exerted. The resulting elastic force is the force of static friction.

While static friction is mainly due to elastic deformation of the microprojections on the contacting bodies, sliding friction occurs as a result of the plastic deformation of the projections and their partial destruction.

When the pushing force is gradually raised, some microprojections are torn off. Therefore, the pulling force is concentrated on the remaining projections which can no longer withstand the increasing load. This leads to an avalanche-type destruction of microprojections and the body suddenly starts to move. The friction force decreases as motion begins because of the reduced meshing and adhesion of the surfaces. Therefore, the coefficient of sliding or dynamic friction is less than the coefficient of static friction.

The frictional forces result in a loss of energy which is dissipated in the form of heat. Friction between mating parts cause wear, due to the shearing of microprojections.

5.7   Further Comments on Static Friction Coefficient

The coefficient of static friction depends upon the quality of surface finish of the mating bodies. The friction of ground and polished surfaces is usually less than that of rough surfaces. However, this is true only within limits. Experiments show that for a very high class of finish, when contacting surfaces are carefully ground and polished (like those of standard gauge blocks), the bodies firmly adhere. The coefficient of friction is increased in such cases.

The main cause of the development of friction forces is the molecular adhesion when the surfaces are finely ground and the meshing of microprojections when the surfaces are rough.

The force of static friction is a weak function of the time of contact of the surfaces with each other. Prolonged contact and a large normal force cause plastic deformation of the contacting surfaces. This contributes to cohesion of the bodies and increases the force of static friction.

Friction coefficients are subject to considerable variation depending on exact conditions of the mating surfaces. The simplest way to determine experimentally the value of coefficient of friction is by means of an inclined plane (Fig. 5.5(a)). The plane is lined with one of the materials of the pair between which the coefficient of friction is to be determined. The surfaces in contact are made very similar to the actual conditions. The angle of inclination of the plane is slowly increased until the block starts to slide down. The angle at which, the block is in impending motion is noted. Figure 5.5(b) shows free body diagram of the

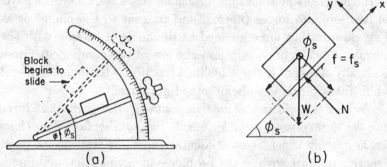

Fig. 5.5 A simple experimental setup to determine the coefficient of static friction

block under this condition. The frictional force is designated as f and it opposes the motion. Since the block is in impending motion, $f = f_s$ and is equal to the friction coefficient multiplied by the normal load as given by Eq. (5.1). Now,

$$\sum F_x = 0; \quad f_s = W \sin \phi_s$$
$$\mu_s N = W \sin \phi_s$$
$$\sum F_y = 0; \quad N = W \cos \phi_s$$

Solving one gets

$$\mu_s = \tan \phi_s$$

The angle ϕ_s is known as the *friction angle*. It is also termed as the *angle of repose* since it is the maximum value of ϕ for which the block remains at rest on the inclined plane.

5.8   Guidelines for Solving a Problem Involving Friction

An important result from the experiment mentioned in Sec. 5.7 is that μ_s for a pair of mating surfaces is determined only for the condition when the block is in impending motion. Thus, the laws of friction are valid only when the motion is impending and using Eq. (5.1) one can get the maximum static frictional force. When the motion is impending, the reaction of one body on the other at the surface where motion impends can be shown on the FBD as two separate forces (normal and tangent to the plane) or as a single resultant force inclined at the angle of friction with the normal. For algebraic solution, the use of two component forces is usually simpler. A single inclined reaction is particularly useful when the solution is to be obtained graphically.

When the motion is not impending, the frictional force can be determined only by using the equilibrium equations - a key point which most students ignore. It is always a good practice to start solving the problems involving frictional forces by representing them as unknowns with suitable symbols (f_1, f_2 etc.). Then in the next step, analyse whether the motion is impending or not. If it is impending, replace the frictional force by ($\mu \times normal\ load$) acting on the mating surfaces. Otherwise leave the unknowns as f_1, f_2 etc., and determine them using equilibrium equations. Another common mistake usually found is in representing the point of application of the normal force N. Most students wrongly represent the point of application of N passing through the centroid. As illustrated in Example 5.1, the point of application of N is determined by the moment equilibrium and hence it is always a good practice to initially show N as in Fig. 5.5(b) and determine the exact location by solving the moment equation. Referring to Figs. 5.6(a) and (b), if the point of application of N lies within the side AB then sliding

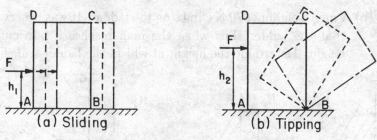

Fig. 5.6 Definition of sliding and tipping.

occurs. But if it lies outside the side AB then the block will tip before sliding. In certain class of problems, whether tipping will precede sliding or vice versa needs to be determined. Referring to Fig. 5.6(b), for the limiting case when N coincides with point B, and F is evaluated from the moment equilibrium conditions, then one can say that if

$$F > f_s$$

then sliding will precede, and if

$$F < f_s$$

then tipping will precede.

There is also another important point which one should keep in mind while solving problems involving friction. In general, unknown forces on FBD can be assumed to be acting in any direction and the algebraic result indicates whether the original assumed direction has to be reversed or not. However, this general rule does not apply here. The frictional forces are developed to oppose the motion. Hence every effort must be made to indicate the correct direction of frictional forces on the FBD. If not you will end up solving a different problem!

Example 5.2

(a) Figure 5.7(a) shows a ladder weighing 220 N resting against a wall. The coefficient of friction between the ladder and the floor is 0.3 (point A) and between the ladder and the wall is 0.2 (point B). The motion is impending at point B. Determine the frictional forces.

(b) A man weighing 670 N climbs on the ladder. It was observed that the ladder slips when the man reaches a particular height. Determine the height at which, the ladder will slip.

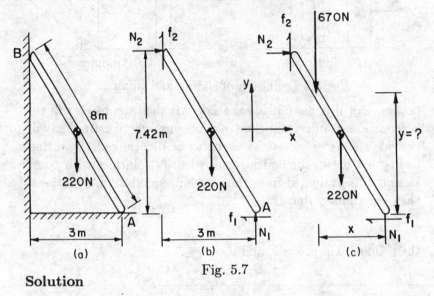

Fig. 5.7

Solution

(a) In Fig. 5.7(b), the frictional forces at A and B have the directions as shown since the tendency of the ladder is to move away from the wall due to its self-weight.

As suggested in the text, the frictional forces are labelled as f_1 and f_2. The problem states that at point B, the motion is impending. Hence,

$$f_2 = \mu_B N_2 = 0.2 N_2$$

Using $\sum F_x = 0$ gives

$$f_1 = N_2$$

$$\sum F_y = 0; \quad N_1 + f_2 = 220 \text{N}$$

$$\sum M_A = 0; \quad - \quad 7.42 N_2 - 3 f_2 + 220 \times 1.5 = 0$$
$$- \quad (7.42 + 3 \times 0.2) N_2 + 220 \times 1.5 = 0$$
$$N_2 = 41.15 \text{ N}$$

From the above equations,

$$N_1 = 211.77 \text{ N}; f_{s1} = 0.3 \times 211.77 = 63.53 \text{ N}$$

$$f_1 = 41.15 \text{N}; \ f_1 < f_{s1}$$

hence at A the motion is not impending.

(b) The ladder will slip when the motion is impending at both points A and B. It is given in the problem that when a man weighing 670 N climbs, at some stage the ladder slips. Hence, from Fig. 5.7(c), $f_1 = \mu_A N_1$ and $f_2 = \mu_B N_2$. Also,

$$\sum F_x = 0; N_2 = f_1; N_2 = 0.3N_1$$
$$\sum F_y = 0; N_1 + f_2 = 220 + 670$$
$$\frac{N_2}{0.3} + 0.2N_2 = 890$$

or $$N_2 = 251.89 \text{ N}$$

$$\sum M_A = 0; -7.42N_2 - 3f_2 + 220 \times 1.5 + x(670) = 0$$

or $$-(7.42 + 0.6)251.89 + 330 + x(670) = 0$$

and $$x = 2.52 \text{ m}; y = 6.23 \text{ m}$$

Hence, when the man reaches a height of 6.23m from the ground, the ladder begins to slip.

5.9 Analysis of Simple Machines

5.9.1 Definition of a Machine

Machines are devices which contain moving parts and are designed to transmit and modify forces and couples. The main purpose of a machine is to transform input forces (couples) into output forces (couples) and thus perform useful mechanical work. With the above definition, not only the complicated and intricate mechanisms are classified as machines but even the simple tools, system of levers and system of pulleys can be classified as machines since they perform the task of modifying the input force. For example, a simple cutting plier magnifies the applied force. The crimping tool analyzed in Chapter 3 (Example 3.3) is another example of a machine.

A system can be defined as a *machine* or a *mechanism* on the basis of its primary objective. When the objective is only to transfer and transform motion, without consideration of forces involved, the system is said to be a mechanism. On the other hand, if the primary objective is to transfer mechanical energy, then it is called a machine. Since mechanical work is always associated with movement, every machine has to transmit motion. Hence, every machine is a mechanism, but not vice versa.

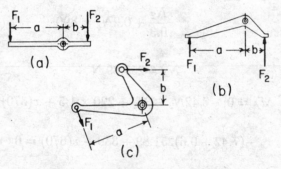

Fig. 5.8 Various forms of levers.

Levers find wide applications in engineering mechanisms. A lever is a rigid body that is hinged at one point. The hinge is called the *fulcrum*. Figure 5.8 shows the various forms of levers. The forces F_1 and F_2 are called applied forces. The perpendicular distances a and b from the fulcrum to the lines of action of the applied forces are called the *lever arms*. Appropriate combination of levers constitute a mechanism. A simple cutting plier is nothing but the combination of two levers (Fig. 5.9) and since the primary objective is to magnify the force, it is a machine.

A pulley behaves like a lever and is used for changing the direction of a force or for magnifying the effect of a force. For the purpose of analysis, the pulley wheel is assumed to turn freely on

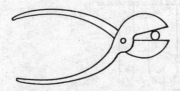

Fig. 5.9 Cutting plier: a combination of two levers.

its axle. A pulley system is a machine consisting of one or more cables wound over pulleys and configured to multiply the force applied when moving loads. Although pulleys are not truly static or motionless, they are assumed to be *quasi-static* meaning that they are essentially at *static equilibrium* at all points in the course of their motion.

5.9.2 Mechanical Advantage

Mechanical advantage is a measure of the effectiveness of a machine. It is the ratio of the output force to the input force.

In Chapter 3, after the analysis of a crimping tool, a word of caution was given that even out of curiosity do not put your finger between its jaws. Its mechanical advantage is as high as 84. Analysis of a pulley system is done here to illustrate the concept of mechanical advantage.

Example 5.3

Figure 5.10 shows a device which is called "Archimedes system of pulleys". Neglecting friction, calculate the mechanical advantage of the system.

Solution

Consider the equilibrium of the pulley supporting the load W
$\sum F_y = 0$ gives,

$$T_1 + T_2 = W$$

$$\sum M_A = 0$$

$$T_2 a = T_1 a$$

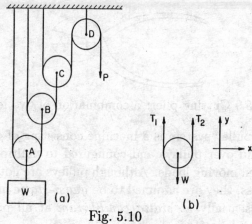

Fig. 5.10

Solving one gets

$$T_1 = T_2 = W/2$$

The above result shows that, in the absence of friction and when the pulley is in equilibrium, tension T in the rope is constant. The load supported by pulley B is now $W/2$. Following the above analysis, load supported by pulley C is $W/4$ and tension in the rope surrounding pulley C is $W/8$. Considering the equilibrium of pulley D one can easily obtain $P = W/8$. Now,

     Mechanical Advantage $= W/P = 8$

5.9.3   Velocity Ratio

We know that a machine transforms an input force into an output force and also that every machine has to transmit motion. Let input force be F_A and velocity component in its direction be V_A. Let output force be F_A and velocity component in its direction be V_B, then

$$\text{Velocity Ratio} = V_A/V_B$$

and the Mechanical advantage $= F_B/F_A$.

It is to be noted that in the case of ideal machines, the conservation of energy implies, $F_A V_A = F_B V_B$. This shows that whatever one gains in force as mechanical advantage is lost in motion.

5.9.4 Mechanical Efficiency

Every machine, has at least one point at which energy is supplied and at least one other point at which energy is delivered. The ratio of the energy delivered to energy supplied is termed as mechanical efficiency of the machine. Note the difference between the definition of mechanical advantage and mechanical efficiency. The quantities, mechanical advantage, velocity ratio and mechanical efficiency are inter related by the relation

Mechanical Efficiency = Mechanical Advantage/Velocity Ratio

In an ideal machine, energy delivered would be exactly equal to the energy supplied. In an actual machine, some of the energy supplied is absorbed in overcoming the inevitable friction at various joints and couplings so that the energy delivered is less than that supplied

5.10   Applications of Friction in Machines

Though frictional forces oppose motion and hence in general are undesirable. There are, however, certain applications wherein these forces can be used to advantage. In certain class of machines "self locking" is achieved with the help of frictional forces.

5.10.1   Wedges

A wedge is one of the oldest and simplest machine. It is used as a means of producing small adjustments in the position of a body or as a means of applying large forces. The function of a wedge is largely dependent on friction.

Example 5.4

Figure 5.11(a) shows a wedge lifting a weight of 10 kN. The friction between the contacting surface is 0.3 and the angle of the wedge is 5°. Determine the force P required to lift the load. What is the mechanical advantage, velocity ratio and mechanical efficiency ?

Solution

Figures 5.11(b) and (c) show the FBD of the wedge and the block, respectively. Note carefully the directions of the frictional forces. Since the motion is impending,

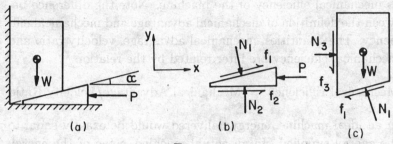

(a)   (b)   (c)

Fig. 5.11

$$f_1 = \mu_s N_1$$

$$f_2 = \mu_s N_2; \ f_3 = \mu_s N_3$$

Equilibrium of the wedge gives

$$\sum F_x = 0$$

$$-P + \mu_s N_2 + \mu_s N_1 \cos \alpha + N_1 \sin \alpha = 0$$

$$\sum F_y = 0$$

$$N_2 - N_1 \cos \alpha + \mu_s N_1 \sin \alpha = 0$$

Equilibrium of the block gives

$$\sum F_x = 0$$

$$N_3 - \mu_s N_1 \cos \alpha - N_1 \sin \alpha = 0$$

$$\sum F_y = 0$$

$$-\mu_s N_3 - W - \mu_s N_1 \sin \alpha + N_1 \cos \alpha = 0$$

We have four equations and four unknowns namely P, N_1, N_2 and N_3. Solving the above set of simultaneous equations, one can get

$$P = 8.1 \ kN$$

$$N_1 = 12 \text{ kN}; \ N_2 = 11.6 \text{ kN}; \ N_3 = 4.68 \text{ kN}$$

$$\text{Mechanical advantage} = W/P = \frac{10}{8.1} = 1.23$$

$$\text{Velocity ratio} = \frac{\text{Distance moved by } P}{\text{Distance moved by } W} = \cot \alpha = 11.43$$

$$\text{Mechanical efficiency} = \frac{1.23}{11.43} \times 100 = 10.8\%$$

The purpose of a wedge is to produce small adjustments in the position of a body. A wedge is driven under a body by applying force. What is expected is that after the wedge is driven under the body, to maintain its position no external force should further be required. This is known as self-locking. The condition of self-locking is a function of the coefficient of friction between the surfaces and the angle of the wedge.

Example 5.5

For the system shown in Fig. 5.11(a) determine whether the wedge is self-locking.

Solution

Self-locking means that, when the force P is removed, the wedge should remain in place. When the force P is removed, the block tries to push the wedge outward. If the wedge is not self-locking then impending motion will set and eventually the wedge would be pushed out. To check whether the wedge is self-locking, assume the limiting condition that the motion is impending. The frictional forces are as shown in Figs. 5.12(a and b).

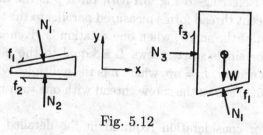

Fig. 5.12

Equilibrium of the block gives

$$\sum F_x = 0; N_3 + \mu_s N_1 \cos \alpha - N_1 \sin \alpha = 0$$

$$N_3 = N_1(\sin \alpha - \mu_s \cos \alpha) = -0.21 N_1$$

A negative value of N_3 implies that the vertical wall must apply a pulling force. This is not possible and hence the wedge is self-locking. In other words, the above equations were obtained by assuming that the motion is impending. However, the results show that the motion is not impending and hence the wedge remains in place.

5.10.2 Screw Jack

Screws represent a most clever application of the concept of wedges; each screw thread may be considered as a wedge wrapped on a circular shaft. Screw treads are made in a variety of shapes (square, trapezoidal, V-shape etc.) for numerous purposes (Fig. 5.13).

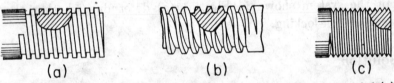

(a)   (b)   (c)

Fig. 5.13 Various thread root forms (a) square (b) trapezoidal (c) V-thread.

Of these, square threads are the simplest to analyse since they are similar to wedges (Fig. 5.14(a)). The follower in the case of the screw thread is the nut. Square-threaded screws are commonly used in jacks, presses and for transmitting power in machines. Referring to Fig. 5.14(b), pitch p is the distance between adjacent thread forms measured parallel to the thread axis. Translation of the screw, when one rotation is given is known as lead L. For single-start screws L is equal to the pitch p. For multi-start screws, $L = np$, where n is the multiplicity of threads. Figure 5.15(a) shows the screw thread with one turn unwrapped.

A major consideration required in the detailed analysis of screw threads is the evaluation of load distribution. Experimental

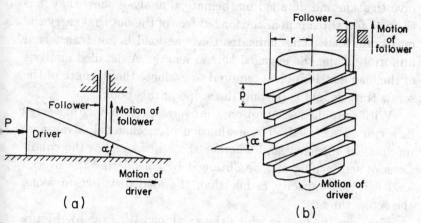

Fig. 5.14 Square thread is a wedge wound over a cylinder.

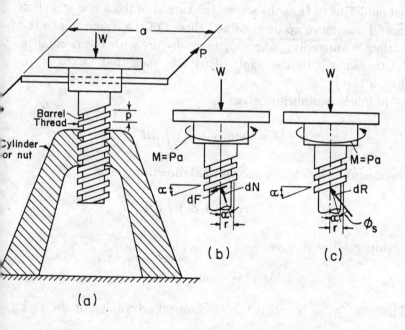

Fig. 5.15 (a) A screw jack; (b) and (c) FBD of the screw.

investigation and detailed mathematical analysis show that it is the first few threads near the loaded face of the nut that carry the maximum load. This indicates that the load is not transferred uniformly along the engaged thread length. A detailed analysis of the load distribution is required to evaluate the strength of the screw threads and is beyond the scope of this book.

With simplified assumption, the overall performance of a screw jack can be assessed. The evaluation of mechanical advantage, mechanical efficiency and whether the screw satisfies the conditions of self-locking can be obtained by considering the overall load carrying capacity rather than the load distribution along the screw threads.

Consider the case in which the axial load W is raised. Figure 5.15(b) illustrates the normal force dN and the frictional force dF that act on an infinitesimal length of the thread. If M is just sufficient to turn the screw, the thread of the screw will slide around and move up on the fixed thread of the frame. Since the motion is impending, angle ϕ_s made by dR with the normal to the thread will be the angle of friction such that $\tan \phi_s = \mu_s$ (Fig. 5.15(c)).

Moment equilibrium gives

$$M = r \sin(\alpha + \phi_s) \int dR$$

Force equilibrium along the vertical direction yields

$$W = \cos(\alpha + \phi_s) \int dR$$

Dividing M by W gives

$$M = W r \tan(\alpha + \phi_s)$$

The ratio $\frac{W}{(M/r)}$ is called the mechanical advantage of the jack. Thus,

$$\text{Mechanical Advantage} = \cot(\alpha + \phi_s)$$

If load W is lowered, the sense of the frictional force is reversed and one gets

$$M = W r \tan(\alpha - \phi_s)$$

This equation shows that if $\phi_s = \alpha$, moment M vanishes for equilibrium and the screw will support weight W without unwinding. If $\phi_s > \alpha$, a negative M is required to lower the weight. A jack that has $\phi_s \geq \alpha$ is said to be self-locking, a desirable property for a jack to have.

The work input per revolution is $2\pi M$, while the useful work of raising the weight is pW. Thus

$$\text{Mechanical Efficiency} \quad \eta = \frac{pW}{2\pi M} = \frac{\tan \alpha}{\tan(\alpha + \phi_s)}$$

For small values of α and ϕ_s, the efficiency is approximately

$$\eta \simeq \frac{\alpha}{\alpha + \phi_s}$$

For a self-locking device, $\phi_s \geq \alpha$ and the efficiency cannot exceed 50%.

Now

$$\text{Velocity Ratio} = \frac{\text{Distance moved by } (M/r)}{\text{Distance moved by } W} = \frac{2\pi r}{p} = \cot \alpha$$

Note that

$$\text{Mechanical Efficiency} = \frac{\text{Mechanical Advantage}}{\text{Velocity Ratio}}$$

from the above equations.

Problems

5.1 Define the role of *friction* in our daily lives.

5.2 (a) For what class of problems is *fluid friction* important.

  (b) Explain why a spring oscillating in vacuum eventually comes to rest?

5.3 State the laws of *dry friction*. For what class of surface conditions these laws are applicable?

5.4 What is the behaviour of poorly lubricated surfaces at high sliding speeds?

5.5 What are the main causes for the development of friction in the following cases:

   (a) When the surfaces are finely ground.

   (b) When the surfaces are rough.

5.6 What is the distinction between a machine and a mechanism? Are all mechanisms machines?

5.7 Define

   (a) Mechanical Advantage

   (b) Velocity Ratio

   (c) Mechanical Efficiency

5.8 Define *self-locking* and what is the maximum efficiency of a *self-locking* screw-jack?

5.9 A block of mass 200 kg. rests on the floor (Fig. 5.16). A force P whose direction can be varied is applied to the block. Keeping the force at 600 N , the value of angle is gradually increased and the block begins to move when θ is 30°. Determine the static coefficient of friction μ between the block and the floor.

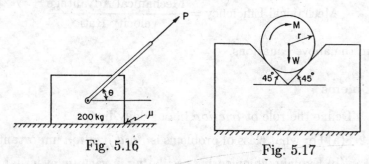

Fig. 5.16   Fig. 5.17

5.10 A circular cylinder of weight 200 N rests on a 45° V-shaped groove (Fig. 5.17). The coefficients of static friction between the cylinder and the groove is 0.25. Find the value of M for which the cylinder is on the verge of turning.

5.11 (a) Figure 5.18 shows a block of height h breadth b and weight **W** resting on a rough surface. The coefficient of friction between the surface and the block is μ. A force **P** is applied (Fig. 5.18) to move the block. It is desired that under the action of force **P** the block should slide along the surface. To satisfy this requirement what should be the relationship between b, h, W, P and μ?

(b) Let $b = 120$ mm and $h = 100$ mm. When P is gradually increased, determine how the loss of equilibrium occurs for the following cases :

  (a) Block is made of steel weighing 100 N and rests on steel surface.

  (b) Block is made of teflon weighing 5 N and rests on a steel surface. {Coefficients of friction are : $\mu = 0.61$ for steel on steel, $\mu = 0.04$ for teflon on steel.}

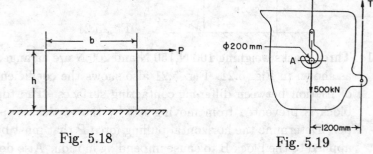

Fig. 5.18   Fig. 5.19

5.12 A hot-metal ladle with its contents of molten cast iron weighs 500 kN. The coefficient of friction between the hook and the pinion is 0.3. Determine the tension T in the cable at B (Fig. 5.19) that would be required to start tipping the ladle for pouring the molten metal.

5.13 A steel Almirah weighing 150kg needs to be shifted by 5m. The coefficient of static friction between the floor and steel is 0.25 and dynamic coefficient of friction is 0.2. Determine the least force required to slide the Almirah along the floor. Also, indicate the point of application of this force. In the new location, thin wooden planks are to be inserted under the legs(Fig. 5.20). How could you do it ? What is the point of application and the magnitude of the least force needed to achieve this.

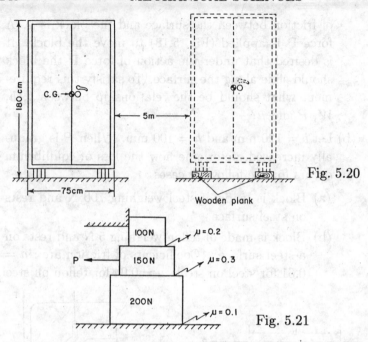

Fig. 5.20

Fig. 5.21

Wooden plank

5.14 Three blocks weighing 100 N, 150 N and 200 N are arranged. as shown in Fig. 5.21. Fig. 5.21 also shows the coefficient of friction between different contacting surfaces. The top block is prevented from moving left by the vertical stopper. Determine the horizontal pulling force P that must be applied to the block B to cause impending motion. Also determine the frictional forces developed between blocks A-B and B-C when block B is in impending motion.

5.15 Find out the Mechanical Advantage of the block and tackle arrangement shown in Fig. 5.22.

5.16 The device shown in Fig. 5.23 is called a differential pulley. The chain is suspended from two sprockets of radii r and R that are connected and turn as a unit. The chain hangs slack from the right side of the smaller sprocket. Determine the Mechanical Advantage of the system. Draw neat sketches of the FBDs of various sprockets.

5.17 In building construction it is common to build a floor or a roof on temporary supports. This permits *levelling up* before setting the permanent columns in place. Figure 5.24

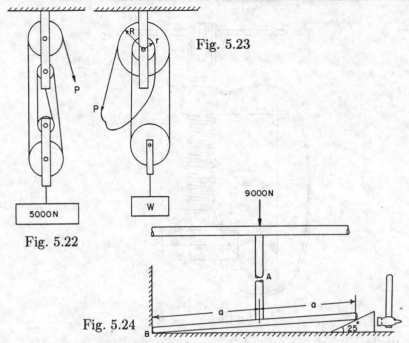

Fig. 5.23

Fig. 5.22

Fig. 5.24

shows one of the ways in which this levelling up is performed. The temporary column A supports a weight of
9000 N. Driving in the wedge at C lifts one end of the rigid
bar BC and hence lifts A.

(a) Assuming that the coefficient of friction at all the surfaces in contact is 0.25, calculate the minimum force
required to hammer the wedge to move further in.

(b) Determine the coefficient of friction required for self-
locking.

5.18 (a) The mean diameter of a square threaded jackscrew is
50 mm. The pitch of the thread is 8 mm and the coefficient of friction is 0.1. Determine the torque required
to raise a weight of 50 kN.

(b) What should be the coefficient of friction for self- locking.

(c) Determine the mechanical efficiency when the screw is
self-locking.

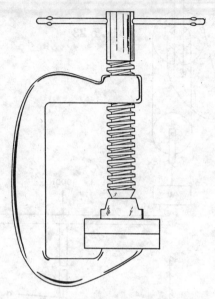

Fig. 5.25

5.19 A C-clamp is used to hold two blocks together as shown in Fig. 5.25. The clamp has a square thread with a pitch of 2.5 mm and a mean diameter of 12.5 mm. The coefficient of friction between the contacting surfaces is 0.2. If a compressive load of 200 N is required, what amount of torque needs to be applied?

5.20 The turnbuckle shown in Fig. 5.26 is used to adjust the tension of a guy wire. Each of the screws has a mean diameter of 36 mm and has a single square thread with a pitch of 8 mm. One of the screws is right-handed and the other is left-handed. By rotating the central member, it is possible to adjust the tension in the guy wire. If the coefficient of static friction is 0.25 and the tension in the guy wire is 60 kN, determine the torque required to loosen the turnbuckle.

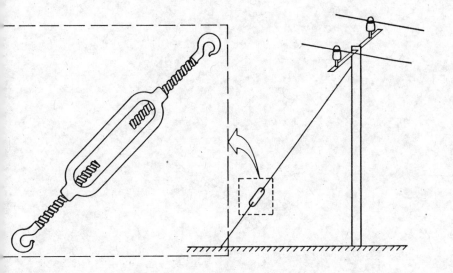

Fig. 5.26

6. DYNAMICS

6.1 Introduction

Dynamics is the mechanics of bodies in motion. More strictly speaking, it is the mechanics of bodies subjected to acceleration. The motion of such bodies is described by Newton's second law of motion, (at least in all cases of common experience and interest).

Newton's second law is about forces and momentum. It is expressed by the vector equation

$$F = \frac{d}{dt}(mv) \qquad (6.1)$$

where F represents the magnitude and direction of the net external force acting on a body of mass m, and v the magnitude and direction of its velocity. Then mv is the momentum of the body. The right hand side of Eq. (6.1) then represents the rate of change, with time, of the momentum. If the mass m does not change with time, Eq. (6.1) becomes

$$F = m\frac{dv}{dt} = ma \qquad (6.2)$$

So Newton's law relates force F with acceleration a of the body. What we need to note, specifically, is that if F is zero, a must also be zero. Thus, where we know that the body in question has a net force acting on it, we must conclude that the body is accelerating; conversely, if measurements confirm that the body is accelerating, it follows that there must be a net force acting on it, even though the source of the force may not be apparent.

153

6.2   Force

A force is something that pushes or pulls a body. That, however, is hardly a technical definition; in fact, a comprehensive technical definition is hard to come by. Fortunately, we only need to define "force" in the context of dynamics. Newton's second law, Eq. (6.2), provides the basis for that; force is the agent that causes a body to accelerate.

Where do forces come from ? They come from physical contact between bodies. When we walk, there is physical contact between us (at our feet) and the ground. The contact force from the ground is the agent that propels us forward. There are other kinds of forces that originate at a distance and act without direct physical contact. Gravity is such a force, as are those arising out of magnetic and electric fields.

We see that Newton's law (Eq. 6.2) does not distinguish between one type of force and another. Thus all we need to know in order to work out the motion of a body is the magnitude and the direction of the force; Newton's law does not ask us to specify the *nature* of F. Conversely, in those cases where the motion is known, Newton's law tells us nothing about the nature of F but does provide us with the means of working out the magnitude and direction of the net force acting on the body.

Clearly, we cannot supply F into the left-hand side of Eq. (6.2) without independent means of ascertaining F. Therefore, in order to be able to predict the motion of a body, we have to know the laws that govern the forces acting on the body. For example, in predicting the motion of a car we must know the laws of sliding and rolling friction and the law of aerodynamic drag; in predicting the motion of planets and satellites we need to know the law of gravitation. Such laws, which are independent of Newton's law, give us the net force F which goes into the left hand side of Eq. (6.2) to enable us to determine the motion of the body.

But engineering is not just a matter of predicting motion. Very often the objective is to obtain a *desired* motion. For example, we may desire to move an object along a specified path - a characteristic problem in the theory of mechanisms. We may or may not be interested in the forces acting on the body (although, the motion being known, we can use Eq. (6.2) to obtain the net force). Suppose the desired path is ensured by means of a guide, such as a slot. It may be relevant for us to know the velocities and accelerations of the body at every point on that path, whereas the associated forces on the body may not be of practical interest. This aspect of mechanics, in which the motion is studied without reference to the associated forces, is known as kinematics. Kinematics, as we shall see, is purely a matter of geometry; Newton's law does not come into it.

We shall see that the right hand side of Eq. (6.2) consists of kinematical information which can be expressed in different forms. The left hand side is the net force, which is the vector sum of all the different forces acting on the body. The forces and the kinematics of motion, together, make up the dynamics of the body. We come back to the statement with which we started this chapter: dynamics is governed by Newton's second law of motion.

6.3   Reference Frames and Coordinate Systems

The relationship between force and motion as expressed by Newton's Law (Eq. 6.2) is not an abstract notion. It is the practical basis for determining or quantifying the dynamics of a body. Thus we need to have numbers to put into Eq. (6.2), numbers that specify the value of the force or the acceleration, for example. That, in turn, means that we must make observations and take measurements. So we need an observation post, i.e. a platform on which the observer stands and from which he makes the measurements. This observation post is known as the reference frame because all measurements are relative to this frame. Any description of dynamics must begin with the unambiguous identification of the reference frame. The importance of this is recognised from the fact that the reference frame may be stationary or moving and the same motion will look different in the two cases. For example, an observer standing on the sidewalk may measure the velocity of a passing truck; another observer, travelling in a moving car, will measure a very different velocity for the same truck.

Force as well as the kinematical quantities (position, velocity, acceleration) being vectors, they can be expressed in terms of their components along specified, mutually perpendicular, directions. In fact, it is convenient to do so. The immediate consequence of expressing these vectors in terms of their components is that the single vector equation, Eq. (6.2), is broken up into one or more scalar equations, one for each of the directions along which components are taken.

The directions along which components are taken specify the coordinate system being used. The most common coordinate systems are the rectangular (cartesian) and polar coordinate systems. For motion in a plane, the coordinate system is two-dimensional, so there are only two directions. In the cartesian system, these directions are x and y. In the plane polar system, the directions are r and θ. The coordinate system has an origin O which marks the location of the observer. Note that both descriptions, cartesian as well as polar, can specify fully the position, for example, of a moving particle P. In the first instance, Fig. 6.1(a),

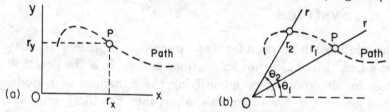

Fig. 6.1 (a) Cartesian coordinates and (b) Plane polar coordinates.

the coordinates r_x and r_y locate P uniquely; as P moves along its path, the values r_x and r_y keep changing with time whereas the x and y axes remain fixed. In plane polar coordinates, the particle is located by its radial distance r from the origin and the angle θ that this radial line makes with some arbitrary reference line. In Fig. 6.1(b), the particle P is shown in two positions along its path. These two positions are described by the polar coordinates (r_1, θ_1) and (r_2, θ_2). Thus, we see a fundamental difference between cartesian and polar coordinates. In the former, the x—y axes will remain fixed in space if the reference frame itself is fixed; in the latter, on the other hand, the r-axis always passes through the particle and hence the r-θ system is necessarily rotating. The effect of this we shall see presently.

6.4 Kinematics

The kinematical quantities that describe the motion of a particle are its position vector r, its velocity vector v and its acceleration vector a. A "particle", in dynamics, is any body whose motion consists purely of translation. The translation may be rectilinear (i.e. along a straight line) or curvilinear (i.e. along a curved path). In pure translation, all points in the body have the same velocity. Thus any body, which moves in such a way that all points within it can be said to have the same (or nearly the same) velocity, is classified as a particle. Its physical dimensions (size) are then irrelevant. What this means is that a given body may or may not be a particle, from the viewpoint of dynamics, depending on the nature of its motion. If the body has rotational motion, with or without simultaneous translation, it can no longer be looked upon as a particle and its motion has to be described by the methods of the dynamics of rigid bodies. Our interest at present is limited to such bodies as have negligible rotation and can be treated as particles.

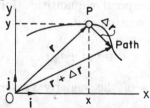

Fig. 6.2 Change of position vector with time.

In Fig. 6.2 we see a particle being observed from the fixed origin O. Its position vector (or displacement vector) is the vector r which indicates that P is at a distance r from O along the arrow of r. This is the position vector at any instant of time, t. At a slightly later instant of time, $t + \Delta t$, the particle will have moved and its position vector will have changed by Δr, giving the new position vector $r + \Delta r$. The position vector r can be described in terms of its components along x and y as

$$r = xi + yj \qquad (6.3)$$

The unit vectors i and j do not alter the sizes of r_x and r_y because a unit vector has the magnitude unity. Thus multiplication by a unit vector is like multiplication by unity. The unit vector indicates the direction of the particular component. For example, if $x = 20$ m and $y = 300$ m, then the position vector is $r = 20i + 300j$ in metres.

Velocity, by definition, is the time rate of change of the position vector. Thus,

$$v = \frac{d}{dt}(r) \qquad (6.4)$$

$$= \frac{d}{dt}(xi + yj)$$

Now, the unit vectors are constant in the cartesian coordinate system; they have the constant magnitude of unity and the constant directions (Fig. 6.2). Because they do not change with time,

$$v = \dot{x}i + \dot{y}j \qquad (6.5)$$

The dots above the symbols x and y in Eq. (6.5) represent the time derivative of the respective quantities (for example $\dot{x} = \frac{dx}{dt}$). Equation (6.5) can be written as

$$v = v_x i + v_y j \qquad (6.6)$$

where the components of the velocity vector are $v_x = \dot{x}$ and $v_y = \dot{y}$.

An important consequence of the definition of velocity is that the velocity vector is always tangent to the path. This follows mathematically from the meaning of the time derivative, but is also obvious intuitively and we shall simply accept it.

Acceleration is the time rate of change of the velocity vector. Therefore,

$$a = \frac{d}{dt}(v) \qquad (6.7a)$$

or
$$a = \frac{d}{dt}(\dot{x}i + \dot{y}j)$$

or
$$a = \ddot{x}i + \ddot{y}j \qquad (6.7b)$$
$$= a_x i + a_y j \qquad (6.7c)$$

since the unit vectors i and j are constant in time. The components of the acceleration vector are $a_x = \dot{v}_x = \ddot{x}$, $a_y = \dot{v}_y = \ddot{y}$, or simply the second derivatives of the corresponding components of the position vector with respect to time.

Equations (6.3), (6.5) and (6.7) describe the position, velocity and acceleration vectors, respectively in plane cartesian coordinates. We can easily add a z-direction, with the unit vector k, to account for a third dimension for which $z = zk$, $v_z = \dot{z}k$, $a_z = \ddot{z}k$. The simplicity of these results is due to the fact that the unit vectors do not change with time.

Example 6.1

A particle is moving on a plane so that its position varies with time as $r = 7t^2 i + 3tj$ m. Find its velocity and acceleration at the start of the motion and again after 10 s.

Solution

Comparison with Eq. (6.3) shows that, in this case, $x = 7t^2$ and $y = 3t$. Using Eq. (6.5),

$$v = 7(2t)i + 3j \quad \text{m/s}$$
$$= 14ti + 3j \quad \text{m/s}$$

Then using Eq. (6.7a),

$$a = 14i \quad \text{m/s}^2$$

At the start of motion ($t = 0$),

$$v = 3j \text{ m/s}, \quad a = 14i \text{ m/s}^2$$

At $t = 10$ s,

$$v = 140i + 3j \text{ m/s} \quad a = 14i \text{ m/s}^2$$

Thus this particle is set into motion from the origin ($r = 0$ at $t = 0$) with an initial velocity in the y−direction and a fixed acceleration in the x−direction (Fig. 6.3).

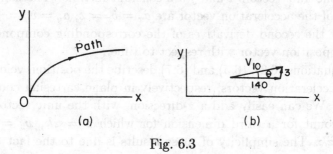

(a) (b)

Fig. 6.3

The velocity at $t = 10$ s may also be expressed as

$$v_{10} = \sqrt{140^2 + 3^2} \simeq 140 \text{ m/s}^2$$

making the angle $\theta = \tan^{-1} \frac{3}{140}$ with the x−axis.

When plane polar coordinates are used, the expressions for v and a become more complex because of the rotation of the r-θ axes and hence their unit vectors. To obtain the time rate of change of the unit vectors $\hat{r}$ and $\hat{\theta}$, consider Fig. 6.4, in which the unit vectors are seen to rotate through a small angle $\Delta\theta$ in a small time interval Δt.

The corresponding changes in the unit vectors are $\Delta\hat{r}$ and $\Delta\hat{\theta}$. Because the angle $\Delta\theta$ is very small, one can write

$$\Delta\hat{r} = |\hat{r}|\Delta\theta\hat{\theta} \quad \text{(i.e. along } \hat{\theta})$$
$$\Delta\hat{\theta} = |\hat{\theta}|\Delta\theta(-\hat{r}) \quad \text{(i.e. along } -\hat{r})$$

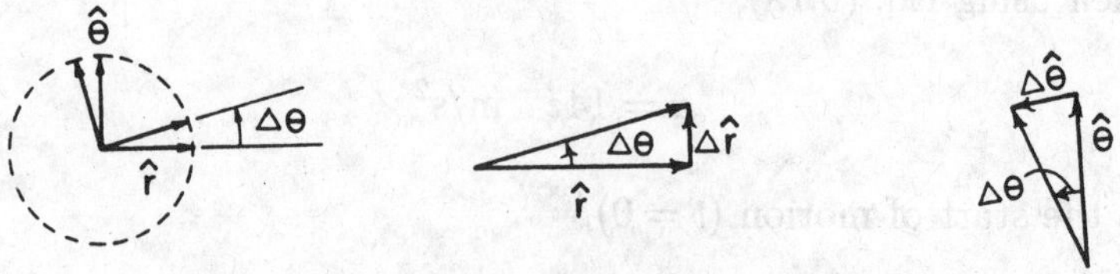

Fig. 6.4 Rotation of unit vectors r and θ.

where the modulus sign indicates the magnitude of the unit vector and so is unity. In the limit $\Delta t \to 0$,

$$\dot{\hat{r}} = \frac{d\hat{r}}{dt} = (1)\frac{d\theta}{dt}\hat{\theta} = \dot{\theta}\hat{\theta} \qquad (6.8a)$$

$$\dot{\hat{\theta}} = \frac{d\hat{\theta}}{dt} = (1)\frac{d\theta}{dt}(-\hat{r}) = -\dot{\theta}\hat{r} \qquad (6.8b)$$

Now the position vector at any instant of time (Fig. 6.5) is

$$r = r\hat{r} \qquad (6.9)$$

As mentioned in Section 6.3 with reference to Fig. 6.1, the θ - co-

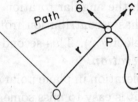

Fig. 6.5 Position vector in plane polar coordinates.

ordinate is measured from an arbitrary reference line. This allows us to choose the current value of the θ−coordinate of the position vector to be zero. Then Eq. (6.9) is the correct expression for the position vector. The velocity will then be

$$v = \frac{dr}{dt} = \left(\frac{dr}{dt}\right)\hat{r} + r\frac{d}{dt}(\hat{r}) = \dot{r}\hat{r} + r\dot{\hat{r}}$$

Finally, referring to Eq. (6.8a),

$$v = \dot{r}\hat{r} + r\dot{\theta}\hat{\theta} \qquad (6.10)$$

The velocity vector has a radial component $v_r = \dot{r}$ and a transverse component $v_\theta = r\dot{\theta}$, of which the latter arises purely out of the rotation rate $\dot{\theta}$ of the axes even when r remains constant. The acceleration is

$$a = \left(\frac{d\dot{r}}{dt}\right)\hat{r} + \dot{r}\frac{d\hat{r}}{dt} + \frac{d}{dt}(r\dot{\theta})\hat{\theta} + r\dot{\theta}\frac{d\hat{\theta}}{dt}$$

$$= \ddot{r}\hat{r} + \dot{r}\dot{\theta}\hat{\theta} + \dot{r}\dot{\theta}\hat{\theta} + r\ddot{\theta}\hat{\theta} - r\dot{\theta}^2\hat{r}$$

or

$$a = (\ddot{r} - r\dot{\theta}^2)\hat{r} + (r\ddot{\theta} + 2\dot{r}\dot{\theta})\hat{\theta} \qquad (6.11)$$

The expression for acceleration deserves special attention. Not only are there a radial component $a_r = \ddot{r} - r\dot{\theta}^2$ and a transverse component $a_\theta = r\ddot{\theta} + 2\dot{r}\dot{\theta}$, but also each component is made up of two terms or contributions. Thus, the radial component has a contribution from the rate of change of the radial velocity component and also a contribution from the angular rotation of the $r - \theta$ axes. This latter term is always in the negative r direction (i.e., towards the centre of curvature of the path) and is known as the *centripetal acceleration*. This exists even when r is a constant. The transverse acceleration a_θ has a contribution from the rate at which the angular rotation itself changes even if r is constant; it also has a contribution from the rate of change of r (i.e. the radial velocity $\dot{r}$). This second term, $2\dot{r}\dot{\theta}$, is known as the *Coriolis acceleration*.

To obtain the acceleration in plane polar coordinates, we have to be careful because it is easy to miss some of the contributions. Check carefully whether $\dot{r}$, $\ddot{r}$ and $\ddot{\theta}$ exist. Once this check has been carried out, all the terms of Eq. (6.11) can be accounted for without error.

Example 6.2

A particle is constrained to move in a slot. At a particular instant of time, its position vector is given by $r = (3t^2 + 2t)\hat{r}$ m. It is also observed that the radial line from the origin O to the particle sweeps out an angle which varies with time as $\theta = 2t^3$ radian. Determine its velocity and acceleration at time $t = 0$ and $t = 1$ s.

Solution

Given $\qquad\qquad r = 3t^2 + 2t \quad \theta = 2t^3$

Then $\qquad\qquad \dot{r} = 6t + 2 \quad \ddot{r} = 6 \quad \dot{\theta} = 6t^2 \quad \ddot{\theta} = 12t$

The velocity is given by Eq. (6.10) as

$$v = \dot{r}\hat{r} + r\dot{\theta}\hat{\theta} = (6t + 2)\hat{r} + (3t^2 + 2t)(6t^2)\hat{\theta} \qquad (a)$$

The acceleration contains all the terms of Eq. (6.11), with

$$a_r = \ddot{r} - r\dot{\theta}^2 = 6 - (3t^2 + 2t)(6t^2)^2$$
$$a_\theta = r\ddot{\theta} + 2\dot{r}\dot{\theta} = (3t^2 + 2t)12t + 2(6t + 2)(6t^2)$$

From the above results, for $t = 0$

$$\boldsymbol{r} = 0 \quad \boldsymbol{v} = 2\hat{\boldsymbol{r}} \quad a_r = 6 \quad (r\dot{\theta}^2 \text{ is zero})$$

$$a_\theta = 0 \text{ (both terms are zero)} \quad \boldsymbol{a} = 6\,\hat{\boldsymbol{r}} \text{ m/s}^2$$

For $t = 1$ s, $\boldsymbol{r} = 5\hat{\boldsymbol{r}}$ m

$$\boldsymbol{v} = 8\hat{\boldsymbol{r}} + 30\hat{\boldsymbol{\theta}} \text{ m/s}$$
$$\boldsymbol{a} = (6 - 180)\boldsymbol{r} + (60 + 96)\hat{\boldsymbol{\theta}} \text{ m/s}^2$$

Fig. 6.6. shows the acceleration components. Note that, although

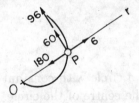

Fig. 6.6

the net $a_r = -174$ m/s^2, the radial distance between O and P does not decrease. In fact $\dot{r}$ and $\ddot{r}$ are both positive for all time. The negative contribution of 180 m/s^2 is the consequence of the particle moving along a curved slot and is large because of the large $\dot{\theta}^2$ term.

An important special application of plane polar coordinates is that of circular motion. If the particle traces out a circle of constant radius, then relative to an origin at the centre of the circle (Fig. 6.7),

$$\boldsymbol{r} = r\hat{\boldsymbol{r}}, \text{ constant}$$

$$\dot{r} = \ddot{r} = 0$$

and hence

$$\boldsymbol{v} = r\dot{\theta}\hat{\boldsymbol{\theta}}$$

$$a = -r\dot{\theta}^2\hat{r} + r\ddot{\theta}\hat{\theta}$$

If, in particular, the speed of the particle is constant, then $\dot{\theta}$ = constant and $\ddot{\theta} = 0$. In that case the acceleration is purely centripetal.

The choice of the coordinate system is a matter of convenience. There is no rule in mechanics that restricts the choice. Apart from the question of convenience, any problem may be treated by either the cartesian or the polar coordinate system.

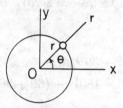

Fig. 6.7 Circular motion of a particle may be described either in cartesian or polar coordinates.

Example 6.3

A particle moves in a circle with constant speed, around an observer standing at the centre of the circle. Express the position, velocity and acceleration of this particle in terms of the x-y coordinate system illustrated in Fig. 6.7.

Solution

From Fig. 6.7,

$$r = r\cos\theta i + r\sin\theta j \qquad (a)$$

The unit vectors i and j being constant and the magnitude of the position vector, r, also being a constant, only $\sin\theta$ and $\cos\theta$ have time derivatives. Thus,

$$v = \dot{r} = r\frac{d}{dt}(\cos\theta)i + r\frac{d}{dt}(\sin\theta)j$$

or

$$v = -r\dot{\theta}\sin\theta i + r\dot{\theta}cos\theta j \qquad (b)$$

$$a = \dot{v} = -r\dot{\theta}^2 \cos\theta \boldsymbol{i} - r\dot{\theta}^2 \sin\theta \boldsymbol{j} \qquad (c)$$

That these are the correct expressions for velocity and acceleration in the x-y coordinate system is easily confirmed from Fig. 6.8.

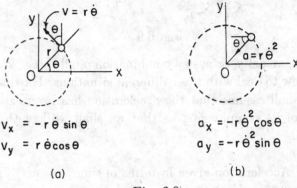

$$v_x = -r\dot{\theta}\sin\theta \qquad\qquad a_x = -r\dot{\theta}^2\cos\theta$$
$$v_y = r\dot{\theta}\cos\theta \qquad\qquad a_y = -r\dot{\theta}^2\sin\theta$$

$$\text{(a)} \qquad\qquad\qquad\qquad \text{(b)}$$

Fig. 6.8

We have transformed the kinematical quantities, r, v and a in polar coordinates to the equivalent quantities in cartesian coordinates; the above equations express the relationships for these transformations. Similar transformation relationships exist for any given pair of coordinate systems, although we shall not be concerned with them. However, note that, in the x-y coordinate system, the velocity and acceleration in circular motion exhibit sinusoidal variation with θ and hence with time, since $\theta = \dot{\theta}t$. That also means that when a particular component of velocity is the maximum, the corresponding acceleration component is zero and the acceleration component is maximum when the corresponding velocity component is zero (Fig. 6.9). This kind of motion is known as simple harmonic motion (SHM) and we shall discuss it later in this chapter.

When the position vector is known, successive differentiation gives the velocity and acceleration. This is what we have just seen. But Newton's law, Eq. (6.2), gives us the acceleration when the forces are known. To obtain velocity and position from the acceleration, we have to perform integrations, and integrating is intrinsically more difficult than differentiating.

In a kinematical problem in which the acceleration is given, the description of the acceleration may be in terms of time, or

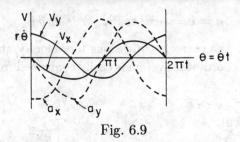

Fig. 6.9

position, or velocity, or even a combination of all three. We shall now see how to deal with these different situations. For that purpose, we shall assume that the acceleration has been expressed in terms of its components, so that we shall work with scalar quantities.

Case (i): Acceleration given in terms of time, $a = a(t)$. This is the simplest case. Here,

$$a = \frac{dv}{dt} \quad \text{or} \quad \int dv = \int a(t)dt$$

Since acceleration is a function of time, $a(t)$, the integrand is purely in terms of t and the integration can be performed to give v as a function of t as

$$v = f_1(t) + C_1$$

The constant of integration C_1 is determined from a known initial velocity, for example $v = v_0$ at $t = 0$. In that case, we end up with v as a function of time and, of course, there is also the constant v_0. The position coordinate, say x, is obtained by integrating $v(t)$. Thus,

$$v = \frac{dx}{dt} \quad \text{or} \quad \int dx = \int v(t)dt$$

Therefore,

$$x = f_2(t) + C_1 t + C_2$$

The new constant of integration C_2 is determined from a known initial position, for example $x = x_0$ at $t = 0$. Thus an initial condition is needed at each integration if the velocity and position are to be determined fully.

Case (ii): Acceleration given as a function of position rather than time, i.e. $a(x)$. To derive the velocity and position, we recast the basic definition of acceleration as given by Eq. (6.6) as

$$a = \frac{dv}{dt} = \frac{dv}{dx}\frac{dx}{dt}$$

where $dx/dt = v$

Hence

$$a = v\frac{dv}{dx} \qquad (6.12)$$

Equation (6.12) is an alternative definition of acceleration which enables us to integrate as follows:

$$\int v\,dv = \int a(x)dx \quad \text{or} \quad \frac{1}{2}v^2 = f_1(x) + C_1$$

i.e. we obtain $v(x)$. The procedure for deriving the displacement starts with Eq. (6.4). Thus,

$$v(x) = \frac{dx}{dt} \quad \text{or} \quad \int dt = \int \frac{dx}{v(x)}$$

This integration will give the time as a function of x, i.e.,

$$t = f_2(x) + C_2$$

from which x is to be derived as a function of time. The constants of integration, C_1 and C_2, will now be given by known initial conditions such as $v = v_0$ and $t = t_0$ when the particle is at $x = x_0$.

Case (iii): Acceleration given as a function of velocity, $a(v)$. The integration to be performed will depend on whether the velocity is desired as a function of time, $v(t)$ or as a function of position, $v(r)$.

If $v(t)$ is desired, one uses Eq. (6.7a) to write

$$\int \frac{dv}{a} = \int dt$$

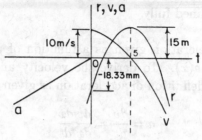

Fig. 6.10

and then obtain $f(v) = t + C_1$ from which $v = v(t)$ can be written. If $v(r)$ is desired, one uses Eq. (6.12) to write

$$\int \frac{v\,dv}{a} = \int dr$$

from which

$$f(v) = r + C_1$$

To obtain the displacement as a function of time, one needs to carry out a further integration as in the above cases. Once again, two constants of integration C_1 and C_2 have to be determined from initial or boundary conditions.

Of the three cases outlined above, case (iii) is clearly the most difficult because of the nature of the integrals.

Example 6.4

A particle moves with an acceleration which is directly proportional to the time t. It is also known that motion had begun with a velocity of 10 m/s and that, at $t = 5$ s, its position was 15 m from the origin ($r = 0$) and the velocity was zero. Obtain the equation of motion of the particle.

Solution

This problem belongs to the category of case (i) above, with $a = kt$. Thus,

$$\int dv = \int a(t)\,dt = \int kt\,dt$$

Therefore,

$$v = \frac{1}{2}kt^2 + C_1$$

The initial condition relating to the velocity is

$$t = 0; \; v = 10 \text{ m/s}$$

Therefore, $10 = C_1$ and

$$v = \frac{1}{2}kt^2 + 10$$

To find the constant of proportionality k, the condition $t = 5$ s; $v = 0$ gives

$$0 = \frac{1}{2}k(25) + 10$$

or

$$k = -10 \times \frac{2}{25} = -\frac{4}{5}$$

Therefore,

$$v = -\frac{2}{5}t^2 + 10$$

Integration of $v = dr/dt$ gives $\int dr = \int vdt$

or

$$r = -\frac{2}{5}\frac{t^3}{3} + 10t + C_2$$

For determining C_2, the initial condition is

$$t = 5 \text{ s}; \; r = 15 \text{ m}$$

which gives

$$C_2 = 15 + \frac{2}{5 \times 3}(125) - 10(5) = -18.33$$

So, finally

$$r = -0.133t^3 + 10t - 18.33$$

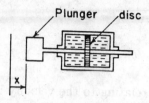

Fig. 6.11

The particle is in rectilinear motion (Fig. 6.10), i.e. it moves in a straight line, with

$$a = -\frac{4}{5}t$$
$$v = -0.4t^2 + 10$$
$$r = -0.133t^3 + 10t - 18.33$$

The acceleration, velocity and position may be plotted as shown. Since the particle moves in a straight line, the position coordinate r could have been replaced by x.

Example 6.5

The motion of the plunger is arrested by the oil dashpot which decelerates the disc attached to the plunger rod (Fig. 6.11). The deceleration is proportional to the velocity. At $t = 0$, the plunger starts moving with $v = v_0$ from $x = 0$. Determine the velocity of the plunger as functions of time and position and also its position as a function of time.

Solution

The acceleration is $a = -kv$, where k is the constant of proportionality. That is,

$$v\frac{dv}{dx} = -kv \quad \text{or} \quad \int \frac{vdv}{v} = -k \int dx$$

Therefore, $v = -kx + C_1$

Initial condition, $x = 0$; $v = v_0$ gives

$$C_1 = v_0 \quad \text{or} \quad v = v_0 - kx \qquad (a)$$

Writing this as $dx/dt = v_0 - kx$ gives

$$\int \frac{dx}{v_0 - kx} = \int dt$$

or
$$-\frac{1}{k}\ln(v_0 - kx) = t + C_2 .$$

Initial condition $t = 0$; $x = 0$ gives

$$C_2 = -\frac{1}{k}\ln v_0$$

or
$$-\frac{1}{k}\ln(v_0 - kx) = t - \frac{1}{k}\ln v_0$$

or
$$t = \frac{1}{k}\left\{\ln v_0 - \ln(v_0 - kx)\right\}$$

or
$$kt = \ln \frac{v_0}{v_0 - kx}$$

or
$$\frac{v_0}{v_0 - kx} = e^{kt}$$

which finally gives

$$v_0 - kx = v_0 e^{-kt} \quad \text{and} \quad \left[x = \frac{v_0}{k}(1 - e^{-kt})\right] \qquad (b)$$

Expression (a) gives the velocity as a function of x and expression (b) the position in terms of time. From these expressions one can obtain the time elapsed and the distance moved before the plunger comes to rest. Putting $v = 0$ in Eq. (a) gives

$$x = \frac{v_0}{k}$$

Using this in Eq. (b) gives

$$1 = 1 - e^{-kt} \quad \text{or} \quad e^{-kt} = 0$$

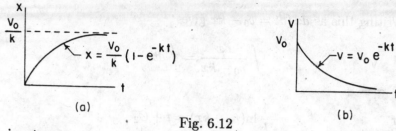

(a)

(b)

Fig. 6.12

i.e. $t \to \infty$.

The reason for $t \to \infty$ before the plunger is fully at rest is inherent in the nature of Eq. (b). This is illustrated in Fig. 6.12.

It will also be clear from the expression for the velocity in terms of t, as:

$$a = -kv \quad \text{or} \quad dv/dt = -kv$$

or

$$\int \frac{dv}{v} = -k \int dt$$

or

$$\ln v = -kt + C_1$$

Initial condition $t = 0$; $v = v_0$ gives

$$C_1 = \ln v_0$$

or

$$\ln \frac{v}{v_0} = -kt$$

or

$$v = v_0 e^{-kt} \tag{c}$$

Note that $x(t)$ could have been obtained much more easily from Eq. (c), as follows:

$$\frac{dx}{dt} = v_0 e^{-kt} \quad \text{or} \quad x = -\frac{v_0}{k} e^{-kt} + C_2$$

Initial condition $t = 0$; $x = 0$, gives

$$C_2 = \frac{v_0}{k} \quad \text{or} \quad x = \frac{v_0}{k}(1 - e^{-kt})$$

which is the same as expression (b).

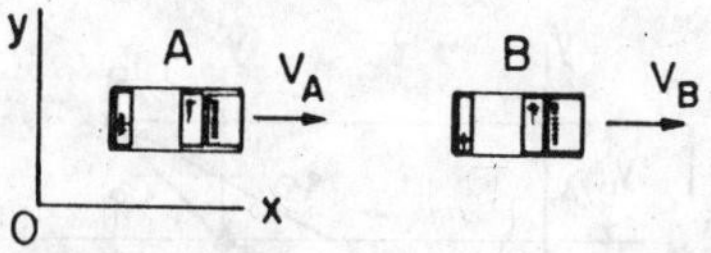

Fig. 6.13 Two cars moving along a straight road.

6.5 Relative Motion

Section 6.3 had emphasised the need to define or identify a reference frame from which all the dynamical quantities are measured (force, position, velocity, acceleration). If the reference frame itself is in motion, all measurements made from it are *relative* to the state of motion of the reference frame. Relative motion is simply the motion of one body as seen or measured from a second body which is itself in motion. To use the simplest example, consider two cars on a straight road (Fig. 6.13). Car A is travelling at a constant speed $v_A = 30$ km/h and car B has a constant speed of $v_B = 40$ km/h. A stationary observer standing at O measures the speeds v_A and v_B for these two cars and records that

$$v_A = 30i \text{ km/h}$$

$$v_B = 40i \text{ km/h}$$

He then concludes that B has a velocity 10 km/h faster than that of A and proceeds to write down the velocity of B relative to A as

$$v_{B/A} = v_B - v_A = 10i \text{ km/h}$$

The velocity of B relative to A is, of course, the velocity of B as observed by a passenger riding in car A, who sees car B going away from him at the rate of 10 km/h in the positive x-direction. Conversely, a passenger in car B sees car A moving away from him at the rate of 10 km/h in the negative x-direction. Thus,

$$v_{A/B} = -10i \text{ km/h} \quad \text{and} \quad v_{A/B} = v_A - v_B$$

This last expression is the velocity of A relative to B as inferred by the fixed observer at O, on the basis of his own measurements of v_A and v_B.

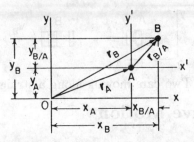

Fig. 6.14 Relative position vector Particles A and B.

The kinematical quantities measured by a fixed observer are usually referred to as "absolute" quantities; kinematical quantities measured by a moving observer are often referred to as "relative" quantities. We shall now formalise the concept of relative motion in pure translation.

Figure 6.14 shows two particles A and B. An observer standing at O measures their instantaneous absolute position vectors as r_A and r_B, where

$$r_A = x_A i + y_A j$$

$$r_B = x_B i + y_B j$$

It is known that both particles are in pure translation (rotation introduces certain special complexities) so that the reference axes x'-y' attached to A are always parallel to the fixed axes x-y. Measured from A and relative to the translating axes x'-y', the instantaneous position vector of B is $r_{B/A}$. From Fig. 6.14 we see

$$r_{B/A} = (x_B - x_A)i + (y_B - y_A)j$$

From Fig. 6.14, we can also write

$$r_B = r_A + r_{B/A} \tag{6.13}$$

or

$$r_{B/A} = r_B - r_A \tag{6.14a}$$

Equation (6.14a) is the formal statement describing the position vector of B relative to A. It is also clear that the position vector of A relative to B is the negative of $r_{B/A}$ and from Eq. (6.14a),

$$r_{A/B} = -r_{B/A} = r_A - r_B \tag{6.14b}$$

The velocities can be obtained by differentiating both sides of Eq. (6.13) as

$$v_B = \dot{r}_B = \dot{r}_A + \dot{r}_{B/A}$$

or

$$v_B = v_A + v_{B/A} \qquad (6.15)$$

A further differentiation will give

$$a_B = a_A + a_{B/A} \qquad (6.16)$$

In the case of pure translation, Eqs. (6.15) and (6.16) are straight-forward. The terms $v_{B/A}$ and $a_{B/A}$ become complicated, however, when the moving reference frame has a rotation, i.e. when x'-y' rotate relative to x-y. Kinematics in rotating reference frames is beyond the scope of our text.

Note that, while Eqs. (6.13)-(6.16) are vector equations, the corresponding scalar equations will be of the same form. Thus,

$$x_B = x_A + x_{B/A} \quad \dot{x}_B = \dot{x}_A + \dot{x}_{B/A}$$
$$y_B = y_A + y_{B/A} \quad \ddot{x}_B = \ddot{x}_A + \ddot{x}_{B/A}$$

Example 6.6

The wedge B (Fig. 6.15) starts moving to the left at time $t = 0$ with a constant acceleration $a_B = 80$ mm/s^2. At the same instant, the block A starts sliding down the wedge with a constant acceleration of 120 mm/s^2 relative to the wedge. Determine (a) the acceleration of the block A, (b) the velocity of the block A when $t = 3$ s.

Solution

(a) Choose the fixed x-y axes as shown in Fig. 6.16. Then

$$a_A = a_B + a_{A/B}$$

The vector diagram gives

$$a_A = -80i + 120\cos 30° \, i - 120\sin 30° j$$

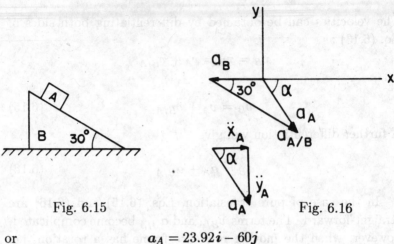

Fig. 6.15   Fig. 6.16

or $a_A = 23.92i - 60j$

and $a_A = \sqrt{(23.92)^2 + (-60)^2} = 64.59$ mm/s^2

The direction of the acceleration a_A is given by (Fig. 6.16)

$$\alpha = \tan^{-1} \frac{|\ddot{y}_A|}{|\ddot{x}_A|} = \tan^{-1} \frac{60}{23.92}$$

or $\alpha = 68.3°$

(b) Now,

$$\frac{dv_A}{dt} = a_A \quad \text{or} \quad \int_0^{v_A} dv_A = \int_0^3 a_A dt$$

This integration may be carried out in terms of the x and y components of a_A and v_A. In this case, however, the acceleration a_A is constant and so we can simply define a new set of fixed axes x'-y', where x' is along the a_A vector (Fig. 6.17).
Then,

$$v_A = a_A t|_0^3 = 64.59 \times 3 = 193.8 \text{ mm/s along x}'$$

Note that part (a) may have been done graphically by drawing the vector diagram to scale (e.g., by taking 1 cm = 20 mm/s) and then measuring off the vector a_A and the angle α.

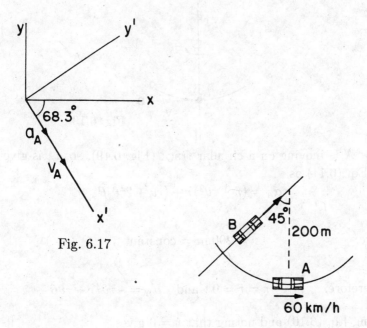

Fig. 6.17

Fig. 6.18

Example 6.7

Car A is moving along a circular road of radius 200 m and car B on a straight road (Fig. 6.18).

At the instant represented by the figure, A has a speed of 60 km/h, which is increasing. At this instant, the acceleration of A as observed from B is zero. Determine the acceleration of B and the rate $\dot{v}_A$ at which the speed of A is increasing.

Solution

It is given that

$$a_A = a_B + a_{A/B}$$

$$a_{A/B} = 0$$

Therefore

$$a_B = a_A \qquad\qquad (i)$$

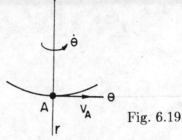

Fig. 6.19

Now A is moving on a circular track (Fig. 6.19), so a_A is given by Eq. (6.11) as

$$\boldsymbol{a}_A = (\ddot{r} - r\dot{\theta}^2)\hat{\boldsymbol{r}} + (r\ddot{\theta} + 2\dot{r}\dot{\theta})\hat{\boldsymbol{\theta}}$$

Here, $\qquad\qquad r = 200 \text{ m} = \text{constant}$

Therefore, $\qquad \dot{r} = \ddot{r} = 0 \quad$ and $\quad \boldsymbol{a}_A = -r\dot{\theta}^2\hat{\boldsymbol{r}} + r\ddot{\theta}\hat{\boldsymbol{\theta}}$

Using Eq. (6.10) and noting that $\dot{r} = 0$ gives

$$\boldsymbol{v}_A = r\dot{\theta}\hat{\boldsymbol{\theta}} \quad \text{or} \quad v_A = r\dot{\theta}$$

which gives

$$\dot{\theta} = \frac{v_A}{r} \quad \text{and} \quad \ddot{\theta} = \frac{\dot{v}_A}{r}$$

since $r = $ constant.

The expression for $\boldsymbol{a}_A$ now becomes

$$\boldsymbol{a}_A = -\frac{v_A^2}{r}\hat{\boldsymbol{r}} + \dot{v}_A\hat{\boldsymbol{\theta}}$$

Since car B is moving along a straight road, its net acceleration must be along this road.

This enables us to complete the vector diagram (Fig. 6.20) by noting, from Eq. (i), that

$$\boldsymbol{a}_B = \boldsymbol{a}_A$$

Then, $\qquad\qquad\qquad a_A \cos 45° = \frac{v_A^2}{r}$

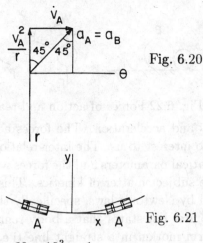

Fig. 6.20

Fig. 6.21

or $\qquad a_A = \sqrt{2}(\dfrac{60 \times 10^3}{60 \times 60})^2 \dfrac{1}{200}$ m/s^2 = 1.96 m/s^2

Finally, since $a_B = a_A$,

$\qquad a_B = 1.96$ m/s^2 (in the direction of motion of B.)

From the vector diagram (Fig. 6.20),

$$\dot{v}_A = a_A \cos 45° = \frac{a_A}{\sqrt{2}}$$

Therefore, $\qquad\qquad\qquad \dot{v}_A = 1.39$ m/s^2

Note carefully that the moving reference frame is car B, which is in pure translation on a straight road. Car A, on the other hand, rotates relative to the fixed x-y axes as it travels along the If the moving reference frame were attached to car A, the acceleration of B as observed from A would be neither zero, nor $a_{B/A}$. This is so because car A would be a reference frame in translation as well as rotation, a situation which is beyond the scope of our text. curved road (Fig. 6.21).

6.6 Kinetics of a Particle in Rectilinear Motion

Uptil now we have been looking at kinematics, which consists of the geometry of motion of a particle as described by its posi-

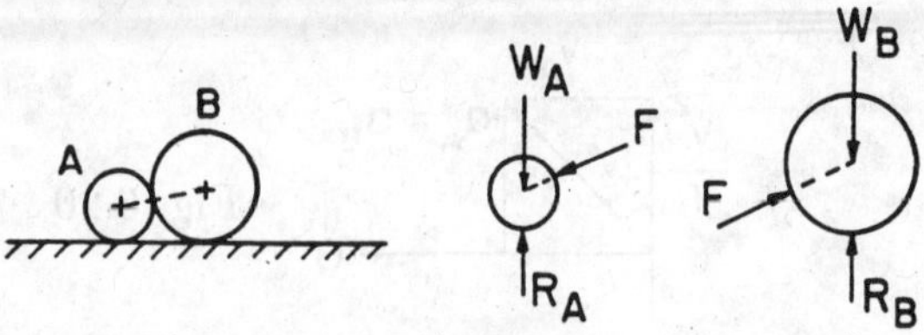

Fig. 6.22 Forces of action and reaction

tion, velocity and acceleration. The forces acting on the particle were of no interest to us. The inter-relationship between the
above kinematical parameters and the forces which cause them to
change is the subject matter of kinetics. This inter-relationship
is established by Newton's laws, specifically the second law.

Newton's first law states that a body remains in its state of
rest or uniform motion in a straight line (i.e. uniform rectilinear motion) unless a force compels a change of that state. This
is the so-called "inertia law". It postulates that all bodies have
"inertia" which can be overcome only by a force. In terms of
inertia, a body at rest is equivalent to the same body in uniform
rectilinear translation because both have zero acceleration. Let
us remember this fact because it has the profound implication
that accelerations measured relative to a body in uniform rectilinear motion are the same as those measured relative to a fixed
(stationary) body. Reference frames attached to fixed bodies and
bodies in uniform rectilinear motion are referred to as *intertial
reference frames*.

Newton's third law is the "action-reaction" law which postulates that the forces of interaction between any two bodies are
equal in magnitude and opposite in sense. If two balls on a table-
top collide with each other, the force F, which is the force of the
interaction between them, acts on each of them as shown in the
free body diagrams of Fig. 6.22. The forces of interaction are
equal and *opposite* irrespective of the size, mass, density etc. of
the two bodies in question.

The law which is of the greatest relevance to us is Newton's
second law which extends the first law and postulates the specific
manner in which a force overcomes the inertia of a body; in fact,
this law quantifies, through Eq. (6.2), the relationship between
a force and the acceleration caused by it. How it does so is
described in Sections 6.1 and 6.2 of this chapter.

We have expressed Newton's law mathematically by Eq. (6.2) as

$$F = ma$$

Either F or a must be supplied as a known quantity in order that the other may be computed. Suppose F is the known quantity and a is to be computed. The calculated a will be the acceleration in the same reference frame from which F is measured. Moreover, the reference frame must be an inertial reference frame, i.e., a reference frame which is either stationary or moves, without rotation, with constant speed along a straight path. It is only in an inertial reference frame that Newton's second law is verified to be valid. To check this out, consider a particle B (Fig. 6.23) on which a force F acts. If the absolute acceleration of B is a_B, Newton's second law applied to B gives

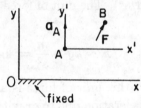

Fig. 6.23 Particle in inertial (x-y) and non-inertial (x'-y') reference frames.

$$F = m_B a_B \qquad (6.17)$$

where the acceleration is measured in the fixed x-y reference frame.

But suppose the acceleration of B is measured in the accelerating (hence *non-inertial*) reference frame x'-y' which itself is in rectilinear translation with the acceleration a_A. In that case the measured accleration of B is $a_{B/A}$ which, by Eq. (6.16), is

$$a_{B/A} = a_B - a_A$$

so that, if Newton's second law is applied in this reference frame

$$F = m a_{B/A} = m_B(a_B - a_A) \qquad (6.18)$$

Clearly, Eqs. (6.17) and (6.18) do not agree. Equation (6.18), obtained by applying Newton's second law in the non-inertial frame x'-y', is incorrect. So Newton's second law is valid only in inertial reference frames; for this reason, such frames are also called "*Newtonian reference frames*". *All inertial frames are alike* i.e., they will all give the same observed (measured) F and a for any particular body in motion.

6.7   Units

Newton's second law provides the basis for the units in which force, mass and acceleration are to be measured and expressed. The units in current use are the S.I. (International system) of units and, because Newton's law relates force to mass and acceleration, the units for these quantities are not all independent. Mass is measured in kilograms (kg) and acceleration in meters per second squared (m/s^2). The unit of force is then derived from Newton's second law as

$$F = ma \qquad F \sim kg \times \frac{m}{s^2} = kg.m/s^2$$

The name given to this unit of force is a newton (N). Thus $N = kg.m/s^2$. The acceleration due to gravity, g, has the approximate value of 9.81 m/s^2 close to the earth's surface. In the SI system, certain prefixes are used when the magnitudes are very large or very small compared to the basic units; for example, if the force is 10^6 N, it is referred to as a mega newton (MN). The most common prefixes are given in Table 1.

Table 1

Multiplication factor	Prefix	Symbol for the prefix
10^6	mega	M
10^3	kilo	k
10^{-3}	milli	m
10^{-6}	micro	μ

For example, a milligram is written as mg, a micrometer as μm, a kilo newton as kN.

6.8 Equations of Motion for Rectilinear Translation

Rectilinear motion is motion in a straight line. In such cases, the reference frame can be chosen in such a way that one of the coordinate axes is along the line of motion. For example, in Fig. 6.24

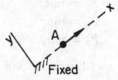

Fig. 6.24 Choice of coordinates in rectilinear motion.

the fixed x-axis has been chosen along the line of motion of particle A. Likewise, in Fig. 6.23, the moving x'-y' axes attached to the reference frame A has y' along the acceleration a_A of A; if A is in rectilinear translation, then y' represents the direction of motion of A. For another illustration, turn to Example 6.6 in which we had chosen the x-axis along the line of motion of wedge B, which was the translating reference frame from which the relative acceleration of the block A was measured. If we had so desired, we could have chosen x along the line of this relative acceleration as shown in Fig. 6.25.

From the above examples we also see that, for rectilinear motion in a fixed reference frame, Newton's second law can be written as a scalar equation. For example, the equation of motion for particle A (Fig. 6.24) is

$$F = m_A \ddot{x}$$

Where relative motion exists, however, the relative and absolute acceleration vectors may not be along the same line and the rules

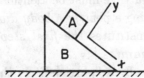

Fig. 6.25 Choice of coordinates in relative rectilinear motion.

of vector addition will still have to be invoked. We start with Eq. (6.2). This gives

$$F = ma$$

or

$$F_x i + F_y j + F_z k = m(a_x i + a_y j + a_z k)$$

Because the i, j, k unit vectors are mutually perpendicular, the above vector equation may be broken up into its scalar components as

$$F_x = ma_x$$
$$F_y = ma_y \qquad (6.19)$$
$$F_z = ma_z$$

and, in particular, if the rectilinear motion is in the direction of the x-axis, Eqs. (6.19) reduce to

$$F_x = ma_x$$
$$F_y = 0$$
$$F_z = 0$$

Equations (6.19) are the scalar equations of motion of a particle in rectilinear translation.

6.9 Application of the Equations of Rectilinear Motion

Equations (6.19) are the scalar representations of Newton's Law. The force components F_x, F_y and F_z are the x-, y- and z- components of the *net externally applied* force on the particle. External forces are applied either through direct physical contact between bodies or by agencies acting from a distance (e.g. gravity). To obtain the net external force or its components, all the forces acting on the particle must be identified and accounted for. This is done by drawing a neat *free body diagram* of the particle in question. This constitutes *the first step in the solution of any problem in dynamics*.

What do we mean by a "particle"? In the context of dynamics a particle is not necessarily a small object. A particle is any

body, large or small, having no rotation. In other words, a particle is any object to which we can correctly apply the equations of
motion for pure translation. For example, any body whose motion is fully described by Eqs. (6.19) is a "particle" in rectilinear
motion. The net force on a particle is assumed to pass through
its centre of mass and the motion of the particle is identical to
the motion of its centre of mass.

Remember that the x-y-z axes, once defined, cannot be changed.
Therefore, a_x is in the same direction as F_x, a_y in the same direction as F_y and likewise for a_z and F_z.

Sometimes the force is given and the acceleration, velocity
etc. are to be found (Sec. 6.2). In other cases the force is the
unknown to be computed from known values of the kinematical
quantities. These two situations comprise two broad classes of
problems in dynamics. The former category is more difficult to
solve because the equation of motion has to be integrated in
order to determine the velocity and position. This integration
is often difficult because the force is independently specified and
the nature of the force becomes relevant; thus the force may
be described as constant (in which case the problem becomes
simple) or as a function of time or position or velocity or even
acceleration.

Example 6.8

Determine the forces of reaction on the wedge B (Fig. 6.26) from
the ground and from the block A, and the accelerations of A
and B. The system is the one of Example 6.6. The masses are
$m_A = 5$ kg and $m_B = 50$ kg. Also, all friction forces are very
small compared to the other forces on this system.

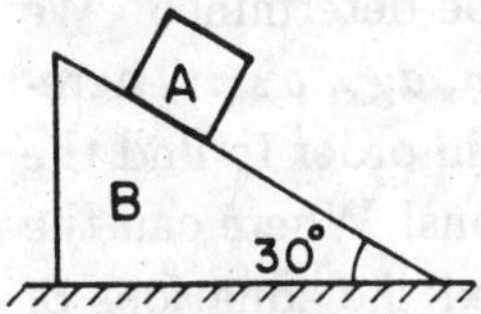

Fig. 6.26 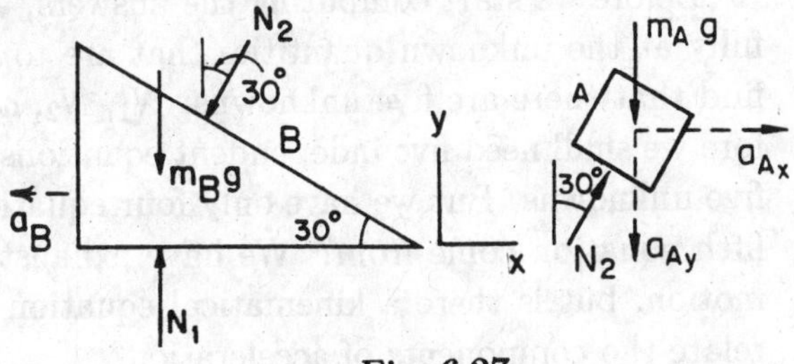   Fig. 6.27

Solution

In the free body diagrams (Fig. 6.27), N_1 is the normal reaction from the ground on B while N_2 is the normal force of interaction between A and B. Friction being negligible in comparison with other forces, we do not show any forces of reaction tangential to the surfaces. The known acceleration components of B and A are shown as broken arrows because these vectors are not forces and so are not part of the free body diagrams.

With the fixed x-y axes chosen as shown, the equations of motion are:

for B, in x-direction,

$$N_2 \sin 30° = m_B a_B \qquad \text{(i)}$$

in y-direction,

$$N_1 - m_B g - N_2 \cos 30° = 0 \qquad \text{(ii)}$$

for A, in x-direction,

$$N_2 \sin 30° = m_A a_{Ax} \qquad \text{(iii)}$$

in y-direction,

$$N_2 \cos 30° - m_A g = -m_A a_{Ay} \qquad (iv)$$

Each component of force and acceleration must be given the correct sign, positive or negative. The directions of the acceleration components, as shown, are known because of the physical constraints on the system. This is an example of "constrained motion".

Before we start computing the answers, we should look carefully at the unknown quantities that are to be determined. We find that there are five unknowns: N_1, N_2, a_B, a_{Ax}, a_{Ay}. Therefore we shall need five independent equations in order to find the five unknowns. But we have only four equations! Where can the fifth equation come from? We have exhausted the equations of motion, but is there a kinematical equation that we can use to relate the components of acceleration?

The physical constraints that determine the geometry of motion give the kinematical information. The horizontal surface compels B to move horizontally (a fact already used). Block A is constrained to slide on the inclined face of the wedge; this is the motion of A *relative* to B. From Example 6.6, we find the following relationships can be obtained (Fig. 6.28):

$$a_{Ax} = a_{A/B} \cos 30° - a_B \qquad \text{(v)}$$

$$a_{Ay} = a_{A/B} \sin 30° \qquad \text{(vi)}$$

Thus kinematics provides us with two new equations and one new unknown, $a_{A/B}$. We now have a system of six unknowns and six equations and the problem can be solved.

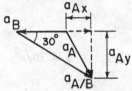

Fig. 6.28 Modified reproduction of Fig. 6.16.

Substitute from Eqs. (v) and (vi) into Eqs. (iii) and (iv) to get

$$N_2 \sin 30° = m_A(a_{A/B} \cos 30° - a_B) \qquad \text{(vii)}$$

$$N_2 \cos 30° = m_A(g - a_{A/B} \sin 30°) \qquad \text{(viii)}$$

Now Eqs. (i), (ii), (vii) and (viii) contain the four unknowns N_1, N_2, a_B, $a_{A/B}$. Combining Eqs. (i) and (vii) gives

$$N_2 \sin 30° = m_A\left(a_{A/B} \cos 30° - \frac{N_2 \sin 30°}{m_B}\right)$$

or

$$N_2 \sin 30° \left(\frac{m_B + m_A}{m_A m_B}\right) = a_{A/B} \cos 30°$$

or

$$a_{A/B} = N_2 \tan 30° \left(\frac{m_B + m_A}{m_A m_B}\right)$$

or

$$a_{A/B} = \frac{N_2}{\sqrt{3}}\left(\frac{55}{50 \times 5}\right) \qquad \text{(ix)}$$

Substitute Eq. (ix) into Eq. (viii) to get

$$\frac{\sqrt{3}}{2}N_2 = m_A\left\{g - \frac{N_2}{\sqrt{3}}(\frac{11}{50})\sin 30°\right\}$$

or $$N_2\left\{\frac{\sqrt{3}}{2} + \frac{m_A}{\sqrt{3}}(\frac{11}{50})\frac{1}{2}\right\} = m_Ag$$

Whence $$N_2 = 5 \times 9.81/\left\{\frac{\sqrt{3}}{2} + \frac{5}{\sqrt{3}} \times \frac{11}{50} \times \frac{1}{2}\right\}$$

or $$N_2 = 41.44 \text{ N}$$

Thus Eq. (ix) gives

$$a_{A/B} = 5.26 \text{ m/s}^2$$

From Eq. (ii),

$$N_1 = N_2\cos 30° + m_Bg$$
$$= 41.44 \times \frac{\sqrt{3}}{2} + 50 \times 9.81$$

or $$N_1 = 526.4 \text{ N}$$

From Eq. (i),

$$a_B = \frac{N_2\sin 30°}{m_B} = \frac{41.44 \times 1/2}{50}$$

or $$a_B = 0.41 \text{ m/s}^2$$

Finally, Eqs. (v) and (vi) give a_{Ax} and a_{Ay} as

$$a_{Ax} = 5.26 \times \frac{\sqrt{3}}{2} - 0.41 = 4.15 \text{ m/s}^2$$

$$a_{Ay} = 5.26 \times \frac{1}{2} = 2.63 \text{ m/s}^2$$

The computed answers for N_1, N_2, a_B, a_{Ax}, a_{Ay} are all positive. This means that the assumed directions for these quantities were all correct. If, for example, the answer for a_{Ax} had come out as negative, we would have concluded that the assumed direction of a_{Ax} was wrong, that a_{Ax} would be to the left and not to the right.

Finally, to complete the solution, we find the vector a_A, to specify the net magnitude and direction of the acceleration of A. Thus (Fig. 6.29),

$$a_A = \sqrt{(a_{AX})^2 + (a_{AY})^2} = 4.91 \text{ m/s}^2$$

and
$$\alpha = \tan^{-1} \frac{a_{Ay}}{a_{Ax}} = 32.36°$$

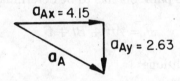

Fig. 6.29

Example 6.8 illustrates the procedure for solving any problem in dynamics. We now formalise this procedure, step-wise:

1. Draw the appropriate free body diagrams in order to account for all the forces acting on the free body (contact forces and forces acting from a distance).

2. Considering the physical constraints on the system, choose a convenient reference frame.

3. Write the component equations of motion for the system.

4. Check out the number of unknowns in the problem.

5. Write down the kinematical equations.

6. Simultaneously solve the equations of motion and the kinematical equations to obtain the desired answers.

If the problem is well-posed, the above six-step procedure will give the solution.

Example 6.9

A particle of mass m starts from rest and moves rectilinearly under the action of a time-dependent force $F(t)$. If $F(t)$ varies parabolically as shown (Fig. 6.30(a)), find the velocity of the particle at times t_0 and $2t_0$.

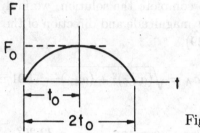

Fig. 6.30(a)

Solution

The equation of the parabola has to be determined first. Assume

$$F = At^2 + Bt + C$$

and apply the conditions

$$t = 0; \ F = 0$$

$$t = 2t_0; \ F = 0$$

$$t = t_0; \ F = F_0$$

to get

$$F = -\frac{F_0}{t_0^2}t^2 + \frac{2F_0}{t_0}t \qquad \text{(i)}$$

For rectilinear motion, this force must act along a fixed line which will also be the line along which the particle will move (Fig. 6.30(b)).

F
m   a

Fig. 6.30(b)

Thus, the equation of motion can be written simply as

$$F = ma \quad \text{or} \quad At^2 + Bt = m\frac{dv}{dt}$$

where $A = -F_0/t_0^2$; $B = 2F_0/t_0$

Thus, $$m \int_0^v dv = \int_0^t (At^2 + Bt)dt$$

or $$mv = A\frac{t^3}{3} + \frac{Bt^2}{2}$$

i.e., $$v = \frac{1}{m}\left(-\frac{F_0}{3t_0^2}t^3 + \frac{F_0}{t_0}t^2\right)$$

or $$v = \frac{ds}{dt} = \frac{F_0}{mt_0}\left(t^2 - \frac{1}{3t_0}t^3\right) \qquad \text{(ii)}$$

or $$\int_0^s ds = \frac{F_0}{mt_0}\int_0^t (t^2 - \frac{1}{3t_0}t^3)dt$$

and the displacement of the particle s is obtained as

$$s = \frac{F_0}{mt_0}\left(\frac{t^3}{3} - \frac{1}{12t_0}t^4\right) \qquad \text{(iii)}$$

The required answers are to be obtained from Eqs. (ii) and (iii). At $t = t_0$, Eq. (ii) gives

$$v_{t_0} = \frac{F_0}{mt_0}(t_0^2 - \frac{1}{3}t_0^2) = \frac{2F_0t_0}{3m}$$

From Eq. (iii),

$$s_{t_0} = \frac{F_0}{mt_0}\left(\frac{t_0^3}{3} - \frac{t_0^3}{12}\right)$$

or $$s_{t_0} = \frac{F_0t_0^2}{4m}$$

At $t = 2t_0$, Eq. (ii) gives

$$v_{2t_0} = \frac{F_0}{mt_0}(4t_0^2 - \frac{8}{3}t_0^2)$$

or
$$v_{2t_0} = \frac{4F_0 t_0}{3m}$$

From Eq. (iii),

$$s_{2t_0} = \frac{F_0}{mt_0}\left(\frac{8}{3}t_0^3 - \frac{4}{3}t_0^3\right)$$

or
$$s_{2t_0} = \frac{4F_0 t_0^2}{3m}$$

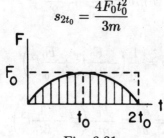

Fig. 6.31

The velocities can be determined directly by graphical integration. Thus,

$$m\int_t dv = \int_t F\,dt$$

The time-integral of the force is the area under the parabola which, in turn, is two-third of the area of the rectangle enclosing the parabola (Fig. 6.31). Thus,

$$\int_0^{2t_0} F\,dt = \frac{2}{3}F_0(2t_0) = \frac{4}{3}F_0 t_0$$

and
$$\int_0^{t_0} F\,dt = \frac{2}{3}F_0(t_0) = \frac{2}{3}F_0 t_0$$

Hence, $mv_{2t_0} = \dfrac{4}{3}F_0 t_0$ and $mv_{t_0} = \dfrac{2}{3}F_0 t_0$

Example 6.10

A steel ball of mass m is released from rest at the surface of a pool of oil. As the ball descends through the oil, it is known to experience a drag D directly proportional to the velocity. Determine the relationship between the depth y (Fig. 6.32) and the corresponding velocity.

Fig. 6.32

Fig. 6.33

Solution

The equation of motion (Fig. 6.33) is

$$mg - kv = ma \qquad \text{(i)}$$

where v is the velocity and a the acceleration of the ball.

From Eq. (i),

$$a = g - \frac{k}{m}v$$

or $\qquad v\dfrac{dv}{dy} = g - \dfrac{k}{m}v$ (using Eq. 6.12)

or

$$\int_0^v \frac{v\,dv}{g - \frac{k}{m}v} = \int_0^y dy \qquad \text{(ii)}$$

or $\qquad y = \dfrac{m^2}{k^2}\left[(g - \dfrac{k}{m}v) - g\ln(g - \dfrac{k}{m}v)\right]_0^v$

whence $\qquad y = \dfrac{m^2}{k^2}g\ln\dfrac{1}{1 - \frac{kv}{mg}} - \dfrac{mv}{k}$

Example 6.11

A block of mass m is being pulled up an incline by means of a rope and pulley system, (Fig. 6.34). The friction coefficient between the block and the incline is $\mu_s = 0.27$ (static) and $\mu_k = 0.25$ (kinetic). If $m = 30$ kg, $P = 300$ N and $\alpha = 30°$, determine the acceleration of the block.

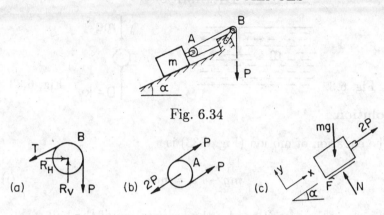

Fig. 6.34

Fig. 6.35

Solution

First, let us consider the free body diagram of the pulley B. The forces T and P (Fig. 6.35(a)) are the pulls from the rope that passes over the pulley. The reaction at the pulley bearing is represented by the components R_V and R_H. If the bearing is frictionless, there is no frictional torque at the pulley axis.

Pulleys are usually very light compared to the loads they raise, so they are taken to be massless. In that case, all the forces and moments on the pulley must be fully balanced because otherwise the pulley will have infinite acceleration ($F = ma$ means that a will be infinite when $m = 0$ and F is finite). Taking moments about the pulley axis,

$$T = P$$

The tension of a belt or rope is the same on either side of a massless and frictionless pulley. Likewise, the rope tension is P on either side of pulley A. Also, for zero net force, the reaction from the block on the centre of the pulley must be $2P$ (Fig. 6.35(b)).

Now draw the free body diagram of the block (Fig. 6.35(c)). The block has contact forces from pulley A and the ground. The former is $2P$ and the latter is represented by its normal component N and tangential component F. There is also the gravity force mg. The block moves along the incline, so its acceleration normal to the inclined surface is zero. The equations of motion are

x-direction

$$2P - F - mg\sin\alpha = ma_x \tag{i}$$

y-direction

$$N - mg\cos\alpha = 0 \tag{ii}$$

We have two equations but three unknowns F, a_x and N. The kinematics of the problem provide no additional information, there being only one acceleration, a_x. The additional information we want is the well known relationship between the friction force F and the corresponding normal reaction N.

(a) $F = \mu_s N$ at the instant when sliding is just about to begin (the *impending sliding* condition). Here, μ_s is the coefficient of static friction.

(b) $F = \mu_k N$ while sliding occurs. Here μ_k is the coefficient of kinetic friction, and $\mu_k < \mu_s$

(c) Before sliding begins and until impending sliding condition is reached, there is no known relationship between F and N. In that case, F can be found only from the equations of motion. In this problem, for example, $a_x = 0$ until the block begins to slide and F can be evaluated from Eq. (i) in terms of P, m and α.

Returning to our problem, we recognise that the block is sliding and write the third equation

$$F = \mu_k N \tag{iii}$$

Simultaneous solution of Eqs. (i), (ii), (iii) then gives

$$a_x = \frac{2P}{m} - g(\mu_k \cos\alpha + \sin\alpha)$$

Substituting the given values of P, m, μ_k and α, the above equation gives

$$a_x = 12.97 \text{ m/s}^2$$

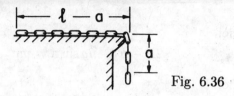

Fig. 6.36

Example 6.12

A heavy, uniform chain of length l, is released from rest with the length a hanging over the edge of a table (Fig. 6.36). The mass of the chain is ρ per metre of its length. Friction is negligible. What is the velocity of the chain when its last link leaves the table?

Solution

Since the overhanging part pulls the remainder of the chain along, the entire chain is in tension and all links move with the same speed and the same acceleration. Let us consider the situation when a length x of the chain overhangs the edge of the table.

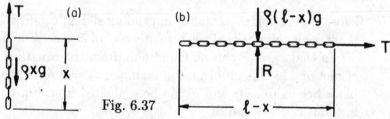

Fig. 6.37

In the free-body diagram of the overhanging length, (Fig. 6.37(a)), T is the tension between the two parts of the chain and is a function of the length x. Forces of tension between individual links are internal to the system and, by Newton's third law, self cancelling they do not appear in the free body diagram.

Since length x moves as one body, its equation of motion is

$$\rho x g - T = (\rho x)a \qquad \text{(i)}$$

where a is acceleration of the chain. Equation (i) has two unknowns, T and a. We need another equation, which is the equation of motion of the remaining part of the chain.

The length $(l - x)$ of the chain moves horizontally with acceleration a. The vertical forces, i.e., the weight and the reaction from the table, are actually distributed throughout the length $(l - x)$, but they are mutually balancing and can be shown as concentrated at the centre of mass (Fig. 6.37(b)). The equation of motion is

$$T = \rho(l - x)a \qquad \text{(ii)}$$

Substitute Eq. (ii) into Eq. (i) to get

$$\rho x g - \rho(l - x)a = \rho x a \quad \text{or} \quad a(x + l - x) = gx$$

or

$$a = \frac{g}{l}x \qquad \text{(iii)}$$

i.e.,

$$\frac{vdv}{dx} = \frac{g}{l}x$$

It is important to note that we are using the result $a = vdv/dx$ because the answer desired is the velocity corresponding to a position and the acceleration has been obtained as a function of the position using Eq. (ii). Now,

$$\int_0^v vdv = \frac{g}{l}\int_a^l xdx \qquad \text{(iv)}$$

The limits of integration are as shown because $v = 0$ when $x = a$ and the answer desired is v when $x = l$. Thus,

$$\frac{v^2}{2} = \frac{g}{l}\left[\frac{x^2}{2}\right]_a^l = \frac{g}{2l}(l^2 - a^2)$$

or

$$v = \sqrt{\frac{g}{l}(l^2 - a^2)}$$

6.10   D'Alembert's Principle – Dynamics in Non-Inertial Frames

Newton's law and hence the equations of motion are valid only in an inertial reference frame (Sec. 6.6). But sometimes, it is

easier to observe a motion and draw conclusions about it from
a non-inertial frame. In the system of Example 6.8, an observer
fixed to the ground will find it impossible to describe the motion
of the block A. An observer sitting on wedge B, which accelerates
and hence is a non-inertial reference frame, will readily describe
the motion of A along the inclined surface. His description will
be of the motion of A relative to himself (e.g., he can specify
$a_{A/B}$). The stationary observer can easily specify the motion of
B (e.g., a_B). The informations supplied by these two observers
can be easily combined to fully describe the absolute motion of
A, as was done in Example 6.6.

We now wish to formulate a rule which will permit us to
describe motions in non-inertial reference frames in pure translation. How should the equation of motion be modified so that
Newton's Law may be applied to non-inertial frames? d'Alembert's
Principle provides the answer.

Looking back at Fig. 6.23, A is the origin of the non-inertial
reference frame A-x'-y' where the observed acceleration of B is
$a_{B/A}$ and

$$a_B = a_A + a_{B/A} \qquad (6.20)$$

The acceleration of A, a_A, is measured in the inertial reference
frame 0-x-y and is known to us. The equation of motion of the
particle B is

$$F = m_B a_B \qquad (6.21)$$

where a_B is the acceleration of B in the fixed reference frame,
i.e. the absolute acceleration of B. Now multiply both sides of
Eq. (6.20) by m_B to get

$$m_B a_B = m_B a_A + m_B a_{B/A}$$

Then, fromEq. (6.21), $F = m_B a_A + m_B a_{B/A}$

or $F - m_B a_A = m_B a_{B/A}$

Note that the term $m_B a_A$ has the dimension of force, kg.m/s^2 or
N.

Define

$$F' = -m_B a_A$$

Then,

$$F + F' = m_B a_{B/A} \tag{6.22}$$

The right-hand side of Eq. (6.22) is the mass of the particle times its acceleration as measured in the non-inertial reference frame. Correspondingly, if $F + F'$ is interpreted as the force on the particle B as measured in the non-inertial frame x'-y', then Eq. (6.22) becomes the statement of Newton's second law in the non-inertial frame x'y'. To see this more clearly, define

$$F|_{x'y'} \equiv F + F'$$

$$a|_{x'y'} \equiv a_{B/A}$$

and rewrite Eq. (6.22) as

$$F|_{x'y'} = m_B a|_{x'y'} \tag{6.23}$$

which is in the familiar $F = ma$ form.

Let us review how we obtained Eq. (6.23). We first took the actual net force F on particle B, as seen from its free body diagram (Fig. 6.38). Then we took the *negative* of the product

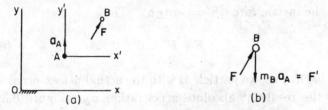

Fig. 6.38 (a) The system. (b) The equivalent force diagram in the non-inertial (x'-y') frame.

of the mass of the particle and the acceleration of the non-inertial frame x'y' and called that F'. The sum of vectors F and F' is $F|_{x'y'}$. As for $a|_{x'y'}$, it is simply the $a_{B/A}$, i.e. the accelerationof B as measured in the x'y' frame. Although F' has the units of force, it is not a real force and is sometimes referred to as a pseudo force. Engineers usually refer to it as an interia force.

A special case arises when the particle itself is treated as a non-inertial frame. Obviously, in the reference frame attached to itself, the particle has no motion at all, e.g., $a_{B/B} = 0$ In analogy with Eq. (6.22), we would expect that $F + F'$ for such a particle would be zero or, in analogy with Eq. (6.23), that $F|_{x'y'}$ would be zero.

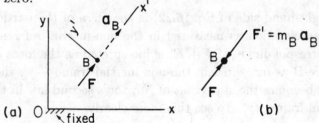

Fig. 6.39 (a) Non inertial frame attached to the particle. (b) Equivalent force diagram.

For particle B of Fig. 6.39(a), the equation of motion in the fixed (inertial) reference frame xy is

$$F = m_B a_B \qquad (6.24)$$

Transpose the right-hand side term to the left. Then,

$$F - m_B a_B = 0 \qquad (6.25)$$

Define the inertia force $F' = -m_B a_B$. Then,

$$F + F' = 0 \qquad (6.26)$$

Figure 6.39(a) shows particle B with the actual net external force F and the resultant absolute acceleration a_B. Figure 6.39(b) shows the same particle acted upon by F as well as the inertia force $F' = -m_B a_B$. Note that F' is shown in the direction opposite to that of the acceleration a_B.

Equation (6.26), and hence Eq. (6.25), is the equation of a particle in equilibrium. The equation of a dynamic system, Eq. (6.24), has been converted to that of a static system, Eqs. (6.25) and (6.26). In other words, a dynamic system has been converted to a static system by introducing the inertia force into the force diagram. Consequently, the force diagram is no longer the free body diagram; it is called, instead, the "equivalent force diagram" (Fig. 6.39(b)). Similarly, introduction of the inertia force $m_B a_A$ in Fig. 6.38 had transformed the inertial frame $x'y'$ to a static frame.

We have just illustrated d' Alembert's Principle: a dynamic particle can be converted, conceptually, to a static particle by treating the inertia force as a real force acting on the particle. Equation (6.25) represents this principle. Notice that Eqs. (6.25) and (6.24) are really the same; the direct use of d' Alembert's Principle does not simplify the solution of the problem, although the concept is intellectually stimulating and does have important applications. You are advised to use the equations of motion directly, rather than in the form of d' Alembert's Principle. Note, however, that this principle is derived from the more general concept of dynamics in non-inertial frames and that concept is often of practical value.

Example 6.13

As an illustration of the use of a non-inertial frame in conjunction with d' Alembert's Principle, in the solution of a problem in dynamics, we look again at the problem of Example 6.8.

Solution

The accelerating wedge is chosen as the non-inertial frame with the axes $x'y'$ fixed to it. The inertial frame xy is fixed to the ground (Fig. 6.40(a)).

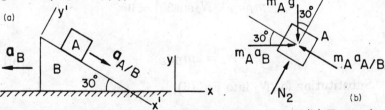

Fig. 6.40 (a) Inertial frame xy fixed to the ground. (b) Equivalent force diagram for A in $x'y'$.

The equivalent force diagram for A (Fig. 6.40(b)) shows the inertia force $m_A a_B$ which is the appropriate quantity for $\boldsymbol{F'}$ of

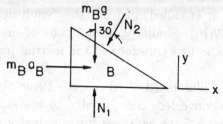

Fig. 6.41 Equivalent force diagram for B.

Eq. (6.22). In accordance with d' Alembert's Principle, the right-hand side of Eq. (6.22) has also been represented by its equivalent inertia force. Compare Figs. 6.40(a) and 6.40(b) and observe that the inertia forces have directions opposite to the accelerations from which they are derived. The block A is in equilibrium under the real and inertia forces shown and the equilibrium equations are

x'-direction : $m_A a_B \cos 30° + m_A g \sin 30° - m_A a_{A/B} = 0$ (i)

y'-direction : $N_2 + m_A a_B \sin 30° - m_A g \cos 30° = 0$ (ii)

These two equations are written by the observer sitting in the $x'y'$ reference frame and contain three unknowns $a_B, a_{A/B}$ and N_2.

The observer sitting in the xy reference frame can write the equation of motion of the wedge B after accounting for the inertia force $m_B a_B$ (Fig. 6.41).

This equation is

$$m_B a_B - N_2 \sin 30° = 0$$

or

$$N_2 = 2 m_B a_B \qquad\qquad \text{(iii)}$$

Substituting for N_2 into Eq. (ii),

$$2 m_B a_B + \frac{1}{2} m_A a_B = \frac{\sqrt{3}}{2} m_A g$$

or

$$a_B = \frac{\sqrt{3}}{2} m_A g \Big/ \left(2 m_B + \frac{1}{2} m_A\right)$$

For $m_A = 5$ kg, $m_B = 50$ kg, $g = 9.81$ m/s^2,
$a_B = 0.41$ m/s^2

Then, from Eq. (iii),

$$N_2 = 41 \text{ N}$$

and from Eq. (i),

$$a_{A/B} = a_B \cos 30° + g \sin 30° = 5.26 \text{ m/s}^2$$

Having found a_A and $a_{A/B}$, use the method of Example 6.6 to determine a_A (Fig. 6.42).

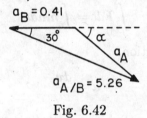

Fig. 6.42

Thus,

$$a_A = a_B + a_{A/B}$$

or

$$a_A = -0.41\boldsymbol{i} + 5.26 \cos 30° \boldsymbol{i} - 5.26 \sin 30° \boldsymbol{j}$$
$$= 4.15\boldsymbol{i} - 2.63\boldsymbol{j}$$

Therefore,

$$a_A = \sqrt{(4.15)^2 + (2.63)^2} = 4.91 \text{ m/s}^2$$
$$\alpha = \tan^{-1} \frac{2.63}{4.15} = 32.36°$$

6.11 Integrated Forms of Newton's Law

We had integrated Newton's Law with respect to time in order to determine the velocity (Eq. (ii) of Example 6.9). In Examples 6.10 (Eq. (ii)) and 6.12 (Eq. (iv)), on the other hand, we had determined the velocity by integrating Newton's Law with respect to displacements. These two approaches are formalised, respectively, as the impulse-momentum principle and the work-energy principle.

6.11.1   The Time Integral: The Impulse-Momentum Principle

We begin with Newton's law. This gives

$$\boldsymbol{F} = m\boldsymbol{a}$$

Integrate it with respect to time as

$$\int \boldsymbol{F} dt = \int m\boldsymbol{a} dt$$

and invoke the definition

$$\boldsymbol{a} = \frac{d\boldsymbol{v}}{dt}$$

When the mass does not change with time, the above equations give

$$\int \boldsymbol{F} dt = m \int \frac{d\boldsymbol{v}}{dt} dt = m \int d\boldsymbol{v}$$

If the motion is observed from time t_1 to time t_2,

$$\int_{t_1}^{t_2} \boldsymbol{F} dt = m(\boldsymbol{v}_2 - \boldsymbol{v}_1) = \boldsymbol{G}_2 - \boldsymbol{G}_1$$

where $\boldsymbol{G}_2$ and $\boldsymbol{G}_1$ are the linear momentum at time t_2 and t_1, respectively. Linear momentum is the momentum of translation, $m\boldsymbol{v}$.

The impulse-momentum equation is written as

$$\int_{t_1}^{t_2} \boldsymbol{F} dt = \boldsymbol{G}_2 - \boldsymbol{G}_1 \tag{6.27}$$

wherein the integral on the left hand side is known as the impulse of the net external force $\boldsymbol{F}$ over the time interval t_1 to t_2 . Impulse has the unit of force multiplied by time: N.s. Momentum has the units of mass times velocity: kg m/s. The impulse-momentum principle states that, over any specified interval of

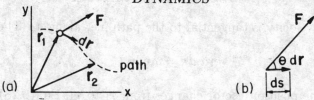

Fig. 6.43 Illustration of the definition of work.

time, the change in the momentum of the body is equal to the impulse of the net external force. Once again draw the free body diagram in order to correctly ascertain the net external force F.

The impulse-momentum equation is not a new law of dynamics; it is Newton's second law cast in a new form. When should one use it? When the data and the desired answer consist of velocities at specified instants of time, and the net force is either a constant or a function of time. Once this is recognised, there should be no difficulty in deciding when to use the impulse-momentum equation.

6.11.2 Integration over Displacement: The Work-Energy Principle

Again, we begin with Newton's law

$$F = ma$$

Suppose the point of application of the force F is displaced by a small dr (Fig. 6.43). Then

$$F.dr = ma.dr \qquad (6.28)$$

The vector dot product, gives,

$$F.dr = (F\cos\theta)ds = F_t ds \qquad (6.29)$$

which is a scalar quantity, ds being the scalar length of the displacement dr and F_t being the component of F along the displacement dr.

The right hand side of Eq. (6.28) can be treated as follows:

$$ma.dr = ma_t ds = m\frac{vdv}{ds}ds$$

since the velocity is tangential to the path, i.e. along a_t. Then,

$$ma.dr = mvdv \qquad (6.30)$$

Now, if the position vector changes from r_1 to r_2, Eq. (6.28) on integration, gives

$$\int_1^2 F_t ds = m \int_1^2 vdv$$

or

$$\int_1^2 F_t ds = \frac{1}{2}mv_2^2 - \frac{1}{2}mv_1^2 = T_2 - T_1 \qquad (6.31)$$

Equation (6.31) is the work-energy equation. The left-hand side is the work of the net external force F as the particle moves from position 1 to position 2. The unit of work is N.m, which is called a Joule. The term $\frac{1}{2}mv^2$ is the kinetic energy T of the particle, so that the right hand side of the equation is the change in the kinetic energy of the particle as it moves from position 1 to position 2. Kinetic energy is necessarily positive because it depends on the square of the velocity and has the unit of kg m^2/s^2, i.e., N.m or Joule. The change in kinetic energy, however, may be positive or negative; the kinetic energy of the particle may increase or decrease as a consequence of the work done on it by the net external force. Also note that the work energy-equation is a scalar equation, unlike the impulse-momentum equation which is a vector equation.

When the data are forces and velocity, as functions of position, and the answer desired is the velocity corresponding to a particular position, the use of the work-energy equation is indicated.

6.11.3   Power and Efficiency

Power is the rate of doing work. Thus,

$$P = \frac{dU}{dt} = F.\frac{dr}{dt}$$

wherein U represents work and $U = F.dr$

Therefore,

$$P = \boldsymbol{F}.\boldsymbol{v} \qquad (6.32)$$

Power is clearly a scalar quantity. It does not have much use in dynamics although it is defined above in terms of dynamical quantities. Power is useful in rating machines, wherein the energy or work output is the relevant measure of a machine's capacity. Power, being the rate of doing work, has the unit Joule per second (J/s), which is called a watt (W).

Efficiency is the ratio of the work output of a machine to the energy put into the machine to make it work. It is a non-dimensional quantity and can be expressed in terms of power as

$$\eta_m = \frac{P_{output}}{P_{input}}$$

where η_m stands for mechanical efficiency. Because of losses in the machine (e.g. by friction), its output is always less than the input and so η_m is always less than unity. We shall make no further reference to power or efficiency.

6.12 Conservation Theorems

A quantity is said to be conserved when it remains constant during a process. Linear momentum is conserved if the momentum vector, $\boldsymbol{G} = m\boldsymbol{v}$ does not change over the time interval t_1 to t_2. From Eq. (6.27), we see that this implies that the net impulse over this time interval is zero. The *principle of conservation of linear momentum* states that, if the net impulse on the system is zero, the linear momentum of the system does not change in the corresponding interval of time. This principle, along with the general impulse-momentum principle, is particularly useful in dealing with problems in which two or more bodies interact, such as problems involving collisions between two bodies.

The principle of conservation of energy postulates that the *net mechanical energy* of a system is conserved if the system is conservative. By net mechanical energy we mean the sum of the *kinetic* and *potential energies.* In a conservative system all forces are conservative. A mathematical test is carried out to check whether or not a given system is conservative. We shall not concern ourselves with that. Instead, we shall note simply that a system is conservative if there are no losses (dissipation) in it. In fact, the only dissipative force we are concerned about is friction. If the system is frictionless, i.e., if there is no friction force acting on the body, the system is conservative.

Potential energy is defined as the *negative* of the work of a *conservative* force. Two common examples are gravitational potential energy and the potential energy of a compressed or stretched spring.

In the case of a body falling freely under gravity, the force mg does work. If the height of fall is h, this work is mgh. The *change* in potential energy is then $-mgh$. Potential energy, whose symbol is V, is always measured as a change; for this purpose, the base value is assigned as zero. In Fig. 6.44(a), V_1 has been taken as the base value and then $V_2 = -mgh$, representing a loss of potential energy. Where has this potential energy gone? It has been converted to kinetic energy in accordance with the principle of conservation of energy, which can be written as

$$V_1 + T_1 = V_2 + T_2 \qquad (6.33)$$

or

$$-(V_2 - V_1) = T_2 - T_1 \qquad (6.34)$$

The principle of conservation of energy may be remembered in either of the two forms given above. Equation (6.33) simply says that the initial and final net energies are the same for a conservative system. Equation (6.34) is in the form of the work-energy equation; the left hand side is same as the negative of the change in potential energy, which is the same as work.

Consider the work of the force P on the spring (Fig. 6.44(b)). As the spring is compressed, the force P increases linearly with the deformation x of the spring. A spring whose P-x graph is a straight line (Fig. 6.44(c)) is called a linear spring. For such a spring,

$$P = kx \qquad (6.35)$$

where the constant of proportionality k is known as the stiffness of the spring (alternative names are *spring constant* or *spring*

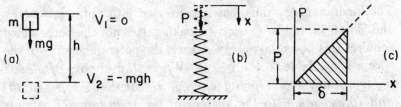

Fig. 6.44 Illustration of work and potential energy.

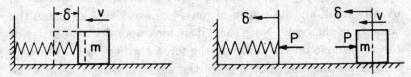

Fig. 6.45 Forces and displacements in a spring-mass sytem.

rate). The work U of this force P is computed as follows:

$$U = \int_0^\delta P dx = \int_0^\delta kx dx$$

or

$$U = \frac{1}{2}k\delta^2 = \frac{1}{2}(k\delta)\delta \tag{6.36}$$

and

$$V = \frac{1}{2}P\delta \tag{6.37}$$

Now consider the body being pushed by the spring (Fig. 6.45). For this body, the spring force P and the displacement δ are opposite in sense and so the work done $(\boldsymbol{F}.\boldsymbol{\delta})$ is negative. For the body,

$$U = -\frac{1}{2}P\delta = -\frac{1}{2}k\delta^2$$

so the change in potential energy of the body is

$$V_2 - V_1 = -U = \frac{1}{2}P\delta = \frac{1}{2}k\delta^2 \tag{6.38}$$

and, if $V_1 = 0$, then

$$V_2 = \frac{1}{2}P\delta$$

From where does the body acquire this potential energy? As the body compresses the spring, the spring force decelerates it and, as v decreases, the kinetic energy of the body decreases.

The kinetic energy thus lost is converted into potential energy and Eq. (6.33) holds. (Strictly speaking, it is not the body which acquires this potential energy; rather, the potential energy of the spring is available to the body). *Where the energy of elastic deformation, e.g. spring energy, is involved, Eqs. (6.33) and (6.34) apply to the system of the body and the spring.* The kinetic energy of the spring is neglected because its mass is usually small and its centre of mass has a velocity less than the velocity of the body. Note that, if we look upon V_2 as negative of the work done on the body by the force P, there is no difficulty in interpreting the results, since the negative of the work is the potential energy).

Example 6.14

A ball A of mass m hangs at rest, suspended by a string (Fig. 6.46). Another ball B, also of mass m and moving with velocity v normal to the string, strikes the ball A. It is observed that, after the collision, the ball B drops vertically while ball A swings up on the string. Find the maximum height h to which the ball A will rise. Both balls are perfectly elastic, which means that no energy is lost in the collision.

Solution

This is an example on collision of elastic bodies.

The first thing we note is that both balls undergo a change of momentum as a consequence of the collision. Therefore both balls are subjected to an impulse which acts for an infinitesimally small duration of time $\hat{t}$ during which the two balls are in contact (Fig. 6.47).

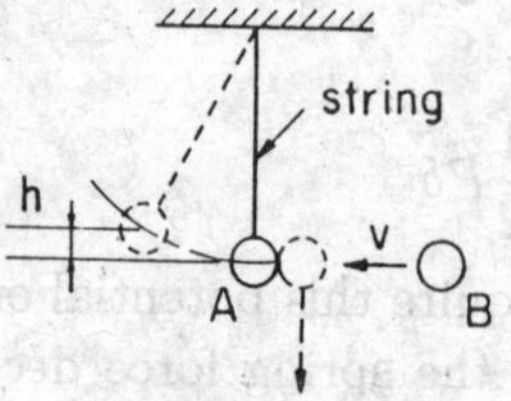

Fig. 6.46 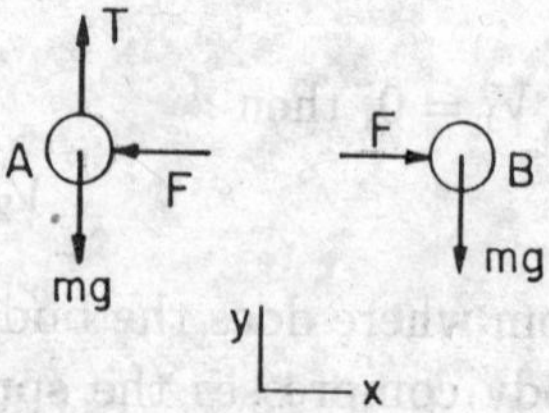   Fig. 6.47

In that small time interval $\hat{t}$, the impulse-momentum equation for A and B may be written, in vector form, as

for A:
$$\int_{\hat{t}}\left[(T - mg)\boldsymbol{j} - F\boldsymbol{i}\right]dt = \boldsymbol{G}_{A_2} - \boldsymbol{G}_{A_1} \tag{i}$$

for B :
$$\int_{\hat{t}}(F\boldsymbol{i} - mg\boldsymbol{j})dt = \boldsymbol{G}_{B_2} - \boldsymbol{G}_{B_1} \tag{ii}$$

Now mg and T are finite forces whose products with the infinitesimal $\hat{t}$ are infinitesimally small. On the other hand, the force F, which is the force of impact between the two bodies, can be very large and have a finite product with $\hat{t}$. In the integrals of Eqs. (i) and (ii), all the finite forces can be neglected in comparison with the impact force F, whereupon the equations become

$$\int_{\hat{t}}(-F\boldsymbol{i})dt = \boldsymbol{G}_{A_2} - \boldsymbol{G}_{A_1} \tag{iii}$$

$$\int_{\hat{t}}F\boldsymbol{i}dt = \boldsymbol{G}_{B_2} - \boldsymbol{G}_{B_1} \tag{iv}$$

Correspondingly, the free-body diagrams of the two balls become modified to show only the forces arising out of the collision and not the finite forces acting on the bodies (Fig. 6.48). This is customary in the solution of problems involving collisions. Note carefully that the above treatment holds only for the duration $\hat{t}$ of the collision.

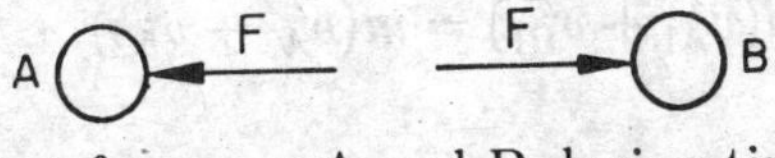

Fig. 6.48 Impact forces on A and B during time interval $\hat{t}$.

Consider Eqs. (iii) and (iv). Evidently, the change in momentum of each ball during the time interval $\hat{t}$ will be purely in the x-direction. Adding Eqs.(iii) and (iv), gives

$$\boldsymbol{G}_{A2} - \boldsymbol{G}_{A1} + \boldsymbol{G}_{B2} - \boldsymbol{G}_{B1} = 0$$

or,

$$\boldsymbol{G}_{A2} + \boldsymbol{G}_{B2} = \boldsymbol{G}_{A1} + \boldsymbol{G}_{B1} \tag{v}$$

or $$m(v_{A_2} + v_{B2}) = m(v_{A1} + v_{B1})$$

or

$$v_{A2} - v_{A1} = -(v_{B2} - v_{B1}) \qquad \text{(vi)}$$

The net momentum of the system of two balls is conserved *during* the collision.

Why this is so is apparent from Fig. (6.48) where it is clear that, if A and B are considered together as the system, the equal and opposite impact forces cancel out (i.e. they are forces *internal* to this system and not external forces) and the net impulse on this system for *the duration* $\hat{t}$ of the collision is zero. That is why the momentum of the system is conserved during $\hat{t}$.

Now let us consider the energy of the system of two balls. Since there is no loss of energy during the impact, we can use Eq. (6.33) for the duration $\hat{t}$ of the impact. This gives

$$(V_A + V_B)_1 + (T_A + T_B)_1 = (V_A + V_B)_2 + (T_A + T_B)_2 \qquad \text{(vii)}$$

In the infinitesimal time $\hat{t}$ when the bodies are actually in contact, neither ball experiences any change of gravitational potential energy. Therefore,

$$(V_A + V_B)_1 = (V_A + V_B)_2$$

Equation (vii) becomes $$(T_A + T_B)_1 = (T_A + T_B)_2$$

or $$m(v_{A1}^2 + v_{B1}^2) = m(v_{A2}^2 + v_{B2}^2)$$

or $$(v_{A2}^2 - v_{A1}^2) = -(v_{B2}^2 - v_{B1}^2)$$

or $$(v_{A2} + v_{A1})(v_{A2} - v_{A_1}) = -(v_{B2} - v_{B1})(v_{B2} + v_{B1})$$

Using the result of Eq. (vi), the above becomes

$$v_{A2} + v_{A1} = v_{B2} + v_{B1}$$

or $$v_{A2} - v_{B2} = -(v_{A1} - v_{B1})$$

or

$$\frac{v_{A2} - v_{B2}}{-(v_{A1} - v_{B1})} = 1 \qquad \text{(viii)}$$

In Eq. (viii), the numerator the relative velocity of separation of the balls at the end of the impact. The denominator is the negative of the relative velocity with which the two balls approach each other at the start of the impact. The ratio of these two relative velocities is called the *coefficient of restitution*, e. Thus,

$$e = \frac{\text{relative velocity of separation}}{-(\text{relative velocity of approach})} \qquad \text{(ix)}$$

The coefficient of restitution is a measure of the energy lost during the impact. For a perfectly elastic impact, when no energy is lost, $e = 1$, as seen from Eq. (viii). For a perfectly plastic impact, following which the bodies stick together, the relative velocity of separation is zero and $e = 0$.

Now to solve the given problem, Eq. (vi) gives

$$v_{A2} - v_{A1} = -(v_{B2} - v_{B1})$$

Since $v_{A1} = 0$ and $v_{B1} = v$, therefore

$$v_{A2} + v_{B2} = v \qquad \text{(x)}$$

Using these in Eq. (viii), one gets

$$v_{A2} - v_{B2} = v \qquad \text{(xi)}$$

Equations (x) and (xi) now give

$$v_{A2} = v$$

$$v_{B2} = 0 \qquad \text{(xii)}$$

Equation (xii) confirm that B loses all of its initial momentum at the end of the impact. The momentum of B is transferred to A, which starts to move to the left with the velocity v. The net momentum of the system of two balls remains unchanged.

At the end of the impact (i.e. for $t > \hat{t}$), B accelerates downwards under the force mg and so is seen to fall vertically

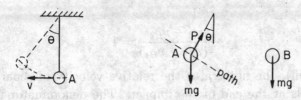

Fig. 6.49 The situtation when $t > \hat{t}$.

(Fig. 6.49). Of the two forces on A, the string tension P is always perpendicular to the displacement and so does no work ($\boldsymbol{P}.d\boldsymbol{s} = 0$). Since there is no dissipation, the total energy of A is conserved. For A, energy conservation gives

$$(T_A + V_A)_1 = (T_A + V_A)_2$$

or $$V_{A2} - V_{A1} = T_{A1} - T_{A2}$$

The ball stops after rising through a height h, i.e., $v_{A2} = 0$. Therefore, on substitution of the values of the above potential and kinetic energies,

$$mgh = \frac{1}{2}mv^2 \quad \text{or} \quad h = \frac{v^2}{2g}$$

This is the desired answer.

Example 6.15

This is the problem of the frictionless chain on a table-top which we had solved in Example 6.12 by the direct application of Newton's law. This time we shall use the principle of conservation of energy, noting that, in the absence of friction, the system is conservative.

Solution

Taking the table-top as the datum for $V = 0$, we first write the gravitational potential energies for the chain in positions 1 and 2 (Fig. 6.50). Thus,

$$V_1 = -(\rho a)g\left(\frac{a}{2}\right) = -\frac{1}{2}\rho a^2 g$$

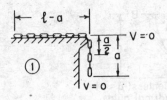

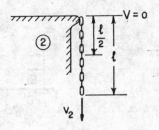

Fig. 6.50

and
$$V_2 = -(\rho l)g\left(\frac{l}{2}\right) = -\frac{1}{2}\rho l^2 g$$

Motion begins from position 1. So,

$$T_1 = 0; \quad T_2 = \frac{1}{2}(\rho l)v_2^2$$

Now, for energy conservation,

$$T_2 + V_2 = T_1 + V_1$$

Therefore,

$$\frac{1}{2}\rho l v_2^2 - \frac{1}{2}\rho l^2 g = -\frac{1}{2}\rho a^2 g$$

or
$$v_2^2 = \frac{g}{l}(l^2 - a^2) \quad \text{or} \quad v_2 = \sqrt{\frac{g}{l}(l^2 - a^2)}$$

which checks with the answer of Example 6.12.

Example 6.16

A small mass m_A starts from rest at B, slides down the frictionless guide and is embedded in the block of mass m_c which is at rest in the position D (Fig. 6.51). Thereafter m_c and m_A slide together until they come to rest at D'. Determine the distance s.

Solution

(a) The mass slides without friction upto D, so its energy is conserved from B to D.

For m_A
$$T_B + V_B = T_D + V_D \tag{i}$$

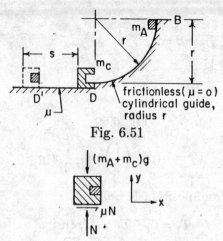

Fig. 6.51

Fig. 6.52 Free body diagram during the impact.

Choose $V_D = 0$ as the datum for V. Also, since $v_B = 0$, therefore $T_B = 0$. Then Eq. (i) gives $T_D = V_B$

or
$$\frac{1}{2}m_A(v_A)_D^2 = m_A gr$$

or
$$(v_A)_D = \sqrt{2gr} \qquad\qquad (ii)$$

(b) For the impact at D, the relative velocity of separation is zero. Therefore,
$$e = 0$$

and so $(v_c)_D = (v_A)_D$ at the end of the impact.

This result says simply that m_c and m_A move together after the impact. It supplies us with no new information. But we can take the system of m_A and m_c together and apply the impulse-momentum equation.

The impact force is internal to the system and does not appear in the free body diagram (Fig. 6.52). Over the duration $\hat{t}$ of the impact, the impulse-momentum equation for the x-direction is

$$(G_{system})_2 - (G_{system})_1 = \int_{\hat{t}}(\mu N)dt$$

Since μN is a finite force and $\hat{t}$ is infinitesimally small, the integral is negligible. Therefore,

$$(m_A + m_c)(-v_2) - \{m_c(o) + m_A(-(v_A)_D)\} = 0$$

or $$(m_A + m_c)(-v_2) = -m_A\sqrt{2gr}$$

or $$v_2 = \frac{m_A}{m_A + m_c}\sqrt{2gr}$$

Note that proper signs have been assigned to the velocities, of which v_2 is the velocity of the combined mass immediately after the impact.

(c) For the motion after the impact, friction exists and energy is not conserved. From the free-body diagram, the friction force alone does work, the other forces being perpendicular to the path. Therefore,

$$(\mu N)(-s) = (T_{system})_{D'} - (T_{system})_D$$

Again note that correct signs, with respect to the x-y axes, have been assigned to both the force and the displacement; the work of μN is thus negative. Since the system comes to rest at D', therefore $(T_{system})_{D'}$ is zero and

$$-\mu N s = -\frac{1}{2}(m_A + m_c)v_2^2$$

or $$s = \frac{m_A + m_c}{2\mu N} \cdot \frac{m_A^2}{(m_A + m_c)^2}2gr$$

From the free body diagram (Fig. 6.52),

$$N = (m_A + m_c)g$$

Therefore, $$s = \left(\frac{m_A}{m_A + m_c}\right)^2 \frac{r}{\mu}$$

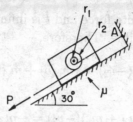

Fig. 6.53

Note carefully that the solution has been broken into three different stages of the motion: (a) the stage *before* the impact, (b) the duration of the impact and (c) the stage *after* the impact. This is the standard procedure in impact-related problems.

Note also that, for stage (b), energy is not conserved because the masses are embedded in each other and so the collision is not elastic.

Example 6.17

An assembly consisting of a block with an integral pulley rests on an inclined plane (Fig. 6.53). The two sheaves of the pulley have radii r_1 and r_2. A rope is wound around the smaller sheave and its free end is fixed at A. Another rope is wound around the larger sheave and its free end is pulled with a force P. The pulley radii are in the ratio 1:2. The block has a mass of 50 kg, $P = 500$ N and the coefficient of kinetic friction $= 0.4$. Determine the velocity of the block after it has moved 1 m along the incline.

Solution

The velocity in the initial position is given to be zero. The answer required is the velocity in a different position. Thus the use of the work-energy equation is indicated. Because of friction, energy will not be conserved. Since the motion is along the x-axis, only the x-components of the forces will do work (Fig. 6.54a and b). So N does no work.

It is now necessary to determine whether the forces P and Q drive the block up or down the incline. The friction force μN will then be opposite to the direction of motion.

The integral pulley is massless and frictionless, so the bear-

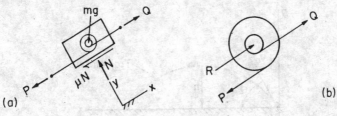

Fig. 6.54(a) and (b)

ing reaction consists of the single force R. There are also the rope tensions P and Q. About the pulley centre, moments must balance.

Therefore, $$Q(r_1) = P(r_2)$$

or $$Q = 2P$$

This also means that, for force balance, R will act in the direction opposite to the one shown. Since $Q > P$, the pulley tends to be pulled up the incline; while the block tries to hold it back and so R on the pulley is down the incline.

From the free body diagram of Fig. 6.54(a), we see that the direction of μN has been shown correctly. The work-energy equation for this system is

$$-P(\Delta x_P)+(-\mu N)\Delta x+(-mg \sin 30°)\Delta x+Q(\Delta x_Q) = \frac{1}{2}mv_2^2-\frac{1}{2}mv_1^2$$
$$\text{(i)}$$

In Eq. (i), Δx is the displacement of the block up the incline ($\Delta x = 1$ m), Δx_P and Δx_Q are the displacements of the points of application of the rope tensions P and Q respectively and v_1 and v_2 are the initial and final velocities, respectively, of the block ($v_1 = 0$). Further, from the y-direction equation of motion, $N = mg \cos 30°$.

Using above data, Eq. (i), gives

$$-P(\Delta x_P) - 415.16 + Q(\Delta x_Q) = 25v_2^2$$

Now, the point of application of Q being fixed, $\Delta x_Q = 0$. This gives

$$25v_2^2 = -P(\Delta x_P) - 415.16 \qquad \text{(ii)}$$

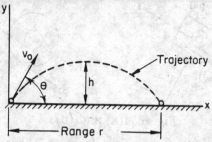

Fig. 6.55 Parameters in the motion of a projectile.

What is Δx_P? If the block moves up by 1 m, the fixed rope on the inner pulley is rolled up by 1 m. Correspondingly, since $r_2 = 2r_1$, the rope going downhill is unwrapped by 2 m even as the block moves up by 1 m. Therefore,

$$\Delta x_P = -2 \text{ m} + 1 \text{ m} = -1 \text{ m}$$

Equation (ii) now becomes

$$25v_2^2 = P - 415.16 \tag{iii}$$

Putting $P = 500$ N,

$$v_2 = 1.84 \text{ m/s} \quad \text{or} \quad v_2 = 1.84 \, i \text{ m/s}$$

6.13   The Motion of Projectiles

Projectiles undergo curvilinear motion. An artillery shell fired from a gun or a cricket ball thrown into the air are examples of projectiles. A projectile is "projected" with an initial velocity and moves under the influence of certain forces. The force of gravity is usually the dominant force and other forces such as the resistance from the air are often neglected.

As shown in Fig. 6.55, the quantities of interest in projectile motion are the range r, the maximum altitude h that is attained and the trajectory or path of motion. The initial condition consists of the velocity of projection v_o which is given in terms of its magnitude and the angle of projection θ.

If air resistance is neglected, the only force on a projectile moving through the air is the force of gravitation, mg, which is constant if the projectile remains sufficiently close to the earth's surface. The equations of motion for the x- and y-directions are

$$m\ddot{x} = 0 \qquad (6.39)$$

$$m\ddot{y} = -mg \qquad (6.40)$$

From Eq. (6.39) we see that $\ddot{x}$ is zero, so the horizontal velocity remains constant. Thus,

$$\dot{x} = v_o \cos\theta \qquad (6.41)$$

Integrating Eq. (6.40),

$$\dot{y} = -gt + C_1$$

At $t=0$, $\dot{y} = v_o \sin\theta$. Therefore,

$$C_1 = v_0 \sin\theta \quad \text{or} \quad \dot{y} = v_o \sin\theta - gt \qquad (6.42)$$

Equations (6.41) and (6.42) provide the velocity components at any time t. To obtain the position at any time t, integrate Eqs.(6.41) and (6.42), once again, to get

$$x = (v_o \cos\theta)t + D$$

At $t = 0$, $x = 0$. Therefore, $D = 0$, or

$$x = (v_o \cos\theta)t \qquad (6.43)$$

$$y = (v_o \sin\theta)t - \frac{1}{2}gt^2 + C_2$$

At $t=0$, $y=0$. Therefore $C_2 = 0$ or

$$y = -\frac{1}{2}gt^2 + (v_o \sin\theta)t \qquad (6.44)$$

The problem has been solved fully because position and velociy for all time have been determined. Equations (6.41)-(6.44) can be used to obtain other information of interest.

The trajectory is found by relating y to x. Eliminating t from Eqs. (6.43) and (6.44),

$$y = -\frac{1}{2}g\left(\frac{x}{v_o \cos\theta}\right)^2 + v_o \sin\theta\left(\frac{x}{v_o \cos\theta}\right)$$

or

$$y = -\frac{g}{2v_o^2}\sec^2\theta x^2 + x\tan\theta \qquad (6.45)$$

Thus, the trajectory is a parabola.

The maximum altitude h is found by maximising y from Eq. (6.45), i.e. by setting the slope dy/dx to zero. Thus,

$$\frac{dy}{dx} = 0 = -\frac{gx}{v_o^2}\sec^2\theta + \tan\theta$$

or

$$x = \frac{v_o^2\tan\theta}{g\sec^2\theta} \qquad (6.46)$$

Because of the symmetry of the parabola in this particular problem, the x of Eq. (6.46) is the half-range.

The range, r, will be double of this value. Substitution of x from Eq. (6.46) into Eq. (6.45) gives

$$h = -\frac{g}{2v_o^2}\sec^2\theta\left(\frac{\tan^2\theta}{g^2 v_o^4 \sec^4\theta}\right) + \frac{v_o^2\tan^2\theta}{g\sec^2\theta}$$

or

$$h = \frac{1}{2}\frac{v_o^2\tan^2\theta}{g\sec^2\theta}$$

or

$$h = \frac{(v_o\sin\theta)^2}{2g} \qquad (6.47)$$

The range may also be found by setting $y = 0$ in Eq. (6.45). Thus,

$$0 = -\frac{g}{2v_o^2}\sec^2\theta x^2 + x\tan\theta$$

The trivial solution ($x = 0$) to the above equation represents the initial position. The range r is given by the second solution, i.e.

$$\frac{g\sec^2\theta}{2v_o^2}x = \tan\theta$$

whence

$$x = \frac{2}{g}v_o^2\tan\theta/\sec^2\theta$$

i.e.

$$r = \frac{2}{g}(v_o \sin \theta)(v_o \cos \theta) \qquad (6.48)$$

Likewise, the maximum altitude may also be found by setting the vertical component of the velocity to zero. This gives, from Eq. (6.42),

$$t = \frac{v_o \sin \theta}{g} \qquad (6.49)$$

Then, from Eq. (6.44),

$$h = y|_{\dot{y}=0} = -\frac{1}{2}g\frac{(v_o \sin \theta)^2}{g^2} + \frac{(v_o \sin \theta)^2}{g}$$

or

$$h = \frac{(v_o \sin \theta)^2}{2g}$$

What is the total duration of the flight of the projectile? To find this set $x = r$ in Eq. (6.43). Thus,

$$\frac{2}{g}v_o^2 \sin \theta \cos \theta = (v_o \cos \theta)t$$

whence

$$t_{tot} = \frac{2v_o \sin \theta}{g} \qquad (6.50)$$

In this problem, because of the symmetry of the parabolic trajectory, t_{tot} may also have been found by doubling the value of t as given by Eq. (6.49).

We solved the projectile problem by determining the position and velocity for all time. From that we went on to determine the equation of the trajectory, the range, the maximum altitude and the duration of the flight. Note that $v_o \cos \theta$ and $v_o \sin \theta$ are the x and y components of the initial velocty v_o. A very important question that remains is the *maximum* range that can be attained for a given magnitude of v_o. Eq. (6.48) shows that the range can be modified by varying θ. For example, the range is zero when θ is 90°, i.e., when the projectile is fired vertically upwards. (This condition provides a check of our preceding results. Substitution of $\theta = 0$ in those results should give $r = 0$,

$h = v_o^2/2g$, $t_{tot} = 2(v_o/g)$). The maximum range will correspond
to $dr/d\theta = 0$ from Eq. (6.48). So,

$$\frac{dr}{d\theta} = 0 = \frac{2v_o^2}{g}(\cos^2\theta - \sin^2\theta)$$

or $$\cos^2\theta = \sin^2\theta$$

i.e. $$\theta = 45°$$

Thus, for any given v_o, the maximum range is obtained when the
angle of projection is 45°. Then the maximum range found from
Eq. (6.48) is

$$r_{max} = \frac{v_o^2}{g} \tag{6.51}$$

Projectiles such as artillery shells move at high speed. The wind
resistance on a high velocity projectile is quite significant and
should be accounted for. Doing so, however, introduces considerable mathematical complexities and so we shall neglect air
resistance. The effect of neglecting air resistance is that the mechanical energy is conserved. It is not customary, however, to
solve problems on projectiles by the energy conservation principle, which will give only the maximum altitude. Recall that the
energy formulation relates velocity and position, therefore its use
will not yield the trajectory or range.

Example 6.18

A bomber pilot, flying horizontally at a height $h = 1000$ m re-
leases a bomb which then goes and strikes a fuel tank. If the
airplane was flying at a constant speed of $v = 500$ km/h, determine the angle β below the horizon at which the pilot sighted
the target at the instant of release of the bomb (Fig. 6.56).

Solution

At the instant of release, the velocity of the bomb is $v = 500$
km/h, horizontal. Therefore, at that instant, the projectile is at

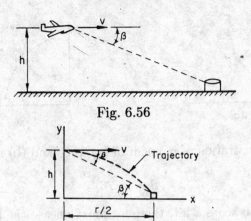

Fig. 6.56

Fig. 6.57

the vertex of its parabolic trajectory. This trajectory (Fig. 6.57) is exactly half of the one we have just analysed, so the results we have derived can be used directly. The range to the target will then be half of r.

From Eq. (6.48),

$$\frac{1}{2}r = \frac{1}{g}(v_o \sin \theta)(v_o \cos \theta)$$

Here $v_o \cos \theta$ is the constant horizontal speed $v = 500$ km/h and from Eq. (6.47),

$$v_o \sin \theta = \sqrt{2gh}$$

Therefore,

$$\frac{1}{2}r = \frac{v}{g}\sqrt{2gh} = v\sqrt{\frac{2h}{g}}$$

From Fig. 6.57,

$$\begin{aligned}
\beta &= \tan^{-1}(h/\frac{1}{2}r) \\
&= \tan^{-1}\left(\frac{h}{v}\sqrt{\frac{g}{2h}}\right) \\
&= \tan^{-1}\left(\frac{1}{v}\sqrt{\frac{gh}{2}}\right)
\end{aligned}$$

With $v = \frac{500 \times 1000}{60 \times 60}$ m/s, $g = 9.81$ m/s^2, $h = 1000$ m,

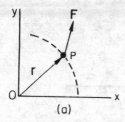
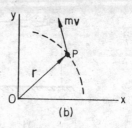

Fig. 6.58 Illustration of moment of (a) force and (b) momentum.

$$\beta = 26.76°$$

Therefore, to score a hit, the pilot must release the bomb when he sights the target at 26.76° below his horizon.

Although used, Eqs.(6.48) and (6.47), are not absolutely necessary. The problem could be solved easily from first principle by the methods used in our derivations. The reader should carry out this exercise.

6.14 Moment of Momentum and Its Conservation

Consider the particle P acted upon by a force F at a given instant of time. At that same instant, its velocity is v ; by definition, the momentum vector mv is tangent to the path.

From statics we know how to compute the moment of a force. In this case (Fig. 6.58(a)), the moment of force F about the moment centre O is given by

$$M_o = r \times F \qquad (6.52)$$

By the property of the vector cross-product, the moment vector M_o in this case is along the positive z-axis (pointing out of the paper).

We can define the moment of the momentum vector mv about the centre O. In analogy with the moment of a force, the *moment of momentum* H_o of the particle about the point O is (Fig. 6.58(b))

$$H_o = r \times mv \qquad (6.53)$$

The moment of momentum vector, in this case, also acts along the positive z-axis. Is there a relationship between M_o and H_o? To answer this question, differentiate both sides of Eq. (6.53) with respect to time. This gives

$$\frac{d}{dt}(H_o) = \frac{d}{dt}(r \times mv)$$

or

$$\dot{H}_o = \frac{d}{dt}r \times mv + r \times \frac{d}{dt}(mv)$$
$$= v \times mv + r \times F$$

where use has been made of Eqs.(6.1) and (6.4). Finally, recognising that $v \times mv$ is zero by the rules of the vector cross-product, and using Eq. (6.52),

$$\dot{H}_o = M_o \qquad (6.54)$$

Thus, the time rate of change of the vector H_o is equal to the moment M_o. This result is analogous to Eq. (6.1); if, in Eq. (6.1), we replace the force by the moment and the linear momentum by the moment of momentum, we obtain Eq. (6.54). Note carefullly, however, that the moment of the force and the moment of momentum must be expressed with respect to the same centre O. Thus Eq. (6.54) is really Newton's second law, extended to the effect of a moment. (That Eq. (6.54) is really Newton's second law becomes clear when it is applied to the dynamics of rigid bodies, in which case it describes the rotational motion of the rigid body).

In the dynamics of particles, Eq. (6.54) is used in the form of its time integral, which is analogous to the impulse-momentum equation (Eq. 6.27). Thus,

$$\int_{t_1}^{t_2} M dt = \int_{t_1}^{t_2} \frac{d}{dt}(H) dt$$

or

$$\int_{t_1}^{t_2} M dt = H_2 - H_1 \qquad (6.55)$$

The above equation states that, over any specified interval of time, the change in the moment of momentum of the particle is

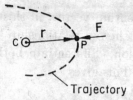

Fig. 6.59 Central force.

equal to the impulse of the moment. (The impulse of the moment is known as the angular impulse when applied to rigid bodies; we shall avoid this nomenclature for the present).

Equation (6.54) leads to the principle of *conservation of the moment of momentum*. If the moment M_o is zero, the moment of momentum H_o remains constant in time, i.e. it is conserved. From Eq. (6.55), $H_2 = H_1$. For M_o to be zero, it is not necessary for F to be zero. If r and F are always parallel, the vector cross product $r \times F$ is zero and then the moment of momentum H_o is conserved. This condition is met in motion under a central force, which constitutes an important class of problems in dynamics.

6.15   Motion Under a Central Force

The characteristic problem here is planetary or satellite motion (Fig. 6.59). The satellite is treated as a particle P and the gravitation force F acting on it is always directed towards the fixed center C. Such a force is known as a central force. Then,

$$M_c = r \times F = 0$$

Hence $$H_c = \text{constant}$$

i.e., $$r \times mv = \text{constant}$$

Thus, as r increases, v decreases and vice versa. The linear momentum mv changes along the trajectory but the moment of momentum about the centre C remains constant. Since H_c is a vector (it is a vector product), constancy implies a constant magnitude as well as a constant direction. The H_c vector is

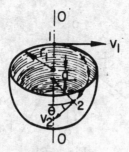

Fig. 6.60

perpendicular to the paper and always remains so, hence the path of the particle is always in the plane of the paper (the plane containing r and F). We have arrived at the very important conclusion that *motion under a central force is plane motion.*

Example 6.19

A particle of mass m moves on the inner surface of a hemispherical bowl of radius r_1 (Fig. 6.60). It starts at position 1 with a horizontal velocity v_1 tangent to the rim of the bowl. When it reaches position 2, its velocity vector v_2 makes an angle θ with the horizontal tangent to the bowl at 2. The point 2 is at a vertical height a below the rim and its radial distance from the axis O-O is r_2. Assume that there is no friction and determine the angle θ.

Solution

The first thing we notice is that the particle spirals down the side of the bowl. Therefore this is not plane motion but rather three-dimensional motion. Still, this problem has been chosen as an example to illustrate the use of the scalar components of the vector equations of motion.

From the problem, we see that the velocity v_1 at the initial position is fully defined. The velocity v_2 at a subsequent position is desired. This suggests the use of the energy formulation. Since there is no friction, mechanical energy is conserved, and

$$V_2 + T_2 = V_1 + T_1$$

or

$$T_2 = (V_1 - V_2) + T_1$$

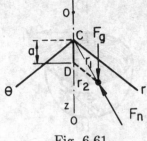

Fig. 6.61

or
$$\frac{1}{2}mv_2^2 = mga + \frac{1}{2}mv_1^2$$

or
$$v_2 = \sqrt{2ga + v_1^2} \qquad \text{(i)}$$

The energy formulation, however, can give only the magnitude of velocity. In this problem we have to determine the orientation θ of v_2. So we need one more equation. Considering the other conservation laws we recognise that linear momentum is not conserved, because the particle is acted upon by a vertical gravity force F_g and a normal reaction F_n perpendicular to the bowl (Fig. 6.61). Of these, F_n has no moment about the centre C of the hemisphere, but F_g does. So H_c is not conserved. We cannot use Eq. (6.55) either because the time interval between positions 1 and 2 has not been specified.

What options are we left with ? One is to solve for the force F_n, but that will only give us the acceleration of the particle which will still have to be integrated for the velocity. The other is to continue to search for an integrated formulation of the equation of motion that may give us the answer. The only possibility is that some *component* of the moment of momentum is conserved. Such, indeed, is the case. We see that the only non-zero moments are the moments of F_g and the vertical component of F_n and that these moments are about the θ - axis. The moment about the z-axis (axis O-O) is zero. Thus the z-component of the moment of momentum is conserved and

$$Hz_2 = Hz_1$$

or
$$r_1(mv_1) = r_2(mv_2 \cos \theta)$$

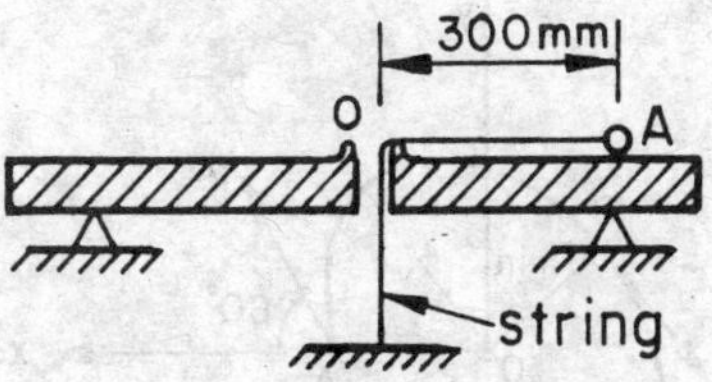

Fig. 6.62

or

$$\cos\theta = \frac{r_1 v_1}{r_2 v_2}$$

Using Eq. (i),

$$\theta = \cos^{-1}\frac{r_1 v_1}{r_2\sqrt{2ga + v_1^2}}$$

where $r_2 = \sqrt{r_1^2 - a^2}$

The above steps could be worked out by the formal vectorial approach by first adopting the r-θ-z coordinate system. Such a system is used for three-dimensional motion.

Let us review the solution. The data and the desired answer consisted of velocities, so we thought of using the integrated forms of the equation of motion. Velocity has a magnitude and a direction, so two equations were required. Straightaway we saw that energy was conserved. Careful inspection then revealed that one component of the moment of momentum was also conserved. Thus we had the two equations which yielded the answer.

Example 6.20

A small mass of 0.5 kg is attached to an elastic string which passes through a hole in a smooth table and has its other end fixed to the ground (Fig. 6.62). When the string is unstretched, the mass sits directly on the top of the hole at O. When the mass is in motion, the elastic string is stretched. The spring constant of the string is 110 N/m. When the stretch is 300 mm, it is observed that the velocity of the mass is 30 m/s at an angle of 60° with the line OA.

Determine (a) the maximum and minimum distances of the mass from O during the motion and (b) the corresponding magnitudes of the velocity.

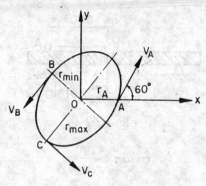

Fig. 6.63

Solution

From the problem, we note two things emerge. There is no friction, and the string pulls the mass towards the hole at O. Thus the energy of the system is conserved and, in the plane of the table, a central force acts on the mass. The other forces on the mass are its weight and an equal and opposite vertical reaction from the table.

Keeping the trajectory of the mass in mind, let us first assume that the minimum and maximum radial distances occur at two points B and C, respectively (Fig. 6.63).

At these two points, the velocity vector will be perpendicular to the corresponding radii. Because of the central force, the moment of momentum of the mass about the centre O will be conserved. When the particle is at A, the moment of momentum about O will be

$$\boldsymbol{H_{o_1}} = \boldsymbol{r_A} \times m\boldsymbol{v_A} = r_A \boldsymbol{i} \times mv_A(\cos 60° \boldsymbol{i} + \sin 60° \boldsymbol{j})$$
$$= r_A(mv_A \sin 60°)\boldsymbol{k}$$

Therefore, $$H_{o_1} = \frac{\sqrt{3}}{2} mr_A v_A$$

Then,

$$mr_{min}v_B = \frac{\sqrt{3}}{2} mr_A v_A \tag{i}$$

and

$$mr_{max}v_C = \frac{\sqrt{3}}{2} mr_A v_A \tag{ii}$$

The mechanical energy of the system consists of the kinetic energy of the mass and the potential energy of the elastic string. Therefore,

$$T_A + V_A = \frac{1}{2}mv_A^2 + \frac{1}{2}kr_A^2$$

$$T_B + V_B = \frac{1}{2}mv_B^2 + \frac{1}{2}kr_{min}^2$$

$$T_C + V_C = \frac{1}{2}mv_C^2 + \frac{1}{2}kr_{max}^2$$

For conservation of mechanical energy,

$$mv_B^2 + kr_{min}^2 = mv_A^2 + kr_A^2 \tag{iii}$$

and

$$mv_C^2 + kr_{max}^2 = mv_A^2 + kr_A^2 \tag{iv}$$

Four unknowns r_{min}, r_{max}, v_B, v_C are be obtained from Eqs. (i), (ii), (iii) and (iv).
Now,

$$mv_A^2 + kr_A^2 = (0.5)(30)^2 + (110)(0.3)^2 = 459.9 \text{ J}$$

Therefore, from Eq. (iii),

$$0.5v_B^2 + 110r_{min}^2 = 459.9 \tag{v}$$

From Eq. (i),

$$r_{min}v_B = \frac{\sqrt{3}}{2}(0.3)(30)$$

or

$$r_{min} = \frac{7.79}{v_B} \tag{vi}$$

Substituting Eq. (vi) into Eq. (v),

$$0.5v_B^2 + \frac{110 \times 7.79^2}{v_B^2} = 459.9 \tag{vii}$$

whence

$$v_B^2 = 905.05 \text{ or } 14.75$$

Note that the simultaneous solution of Eqs.(ii) and (iv) gives an equation identical to Eq. (vii), with v_B replaced by v_C. The solutions of the quadratic Eq. (vii) therefore gives v_B^2 as well as v_C^2.

Therefore,
$$v_C^2 = 14.75$$

or
$$v_C = 3.84 \text{ m/s}$$

and
$$v_B^2 = 905.05$$

or
$$v_B = 30.08 \text{ m/s}$$

Correspondingly,
$$r_{max} = \frac{7.79}{3.84} = 2.03 \text{ m}$$

$$r_{min} = \frac{7.79}{30.08} = 0.26 \text{ m}$$

Reviewing the solution, we find that velocities corresponding to position are involved, suggesting the energy formulation. Since two unknowns (r and v) are involved, a second equation is necessary. Recognising the presence of a central force we conclude that the second equation should be the equation of conservation of moment of momentum.

Can we say anything about the trajectory? For example, does the mass ever reach the centre O? If it does, the potential energy will be zero. The energy conservation equation then will be

$$mv^2 = 459.9 \qquad\qquad \text{(viii)}$$

whereas the conservation of moment of momentum requires

$$v = \frac{7.79}{r}$$

Thus, the mass will have infinite velocity at O, where $r = 0$. Furthermore, the condition of energy conservation, Eq. (viii), cannot be satisfied. The mass will never pass through O but will keep orbiting around O. To precisely determine the equation of the trajectory, we need to integrate Newton's second law and solve for the displacement vector.

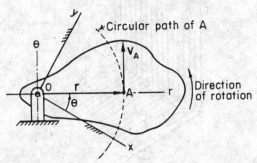

Fig. 6.64 Rigid body rotating about a fixed axis.

6.16 Fixed Axis Rotation of Rigid Bodies

Rigid bodies are distinguished from particles by the fact that they undergo rotation. The general motion of a rigid body consists of this rotation as well as a curvilinear translation of its centre of mass. We shall now derive the equation of motion of a rigid body that rotates about a fixed axis.

Figure 6.64 shows a rigid body which is hinged to a fixed support at O. The axis of the hinge is perpendicular to the plane of the paper, so the plane of rotation of the rigid body is the plane of the paper. Consider any point A in the body, at a distance r from O. Since this body is rigid, the distance between any two points in the body remains constant. Thus $\overline{OA}$ remains constant. Consequently, as the body rotates, the point A traces a circle in a plane parallel to the plane of the paper. We are interested in the kinematics of the point A, for which we refer back to (Fig. 6.7) and the expressions for velocity and acceleration in circular motion. These expressions, as obtained earlier, are

$$v_A = r\dot{\theta}\hat{\theta} \tag{6.56}$$

$$a_A = -r\dot{\theta}^2\hat{r} + r\ddot{\theta}\hat{\theta} \tag{6.57}$$

With reference to Fig. 6.7 and Fig. 6.64, we recognise that $\dot{\theta}$ is the rate of change of the angle θ between the line OA and any other line *fixed in space*, such as the x-axis in Fig. 6.64. Thus $\dot{\theta}$ is the rate of rotation of the line OA in the body with respect to any fixed set of axes x-y. This rate of rotation is known as

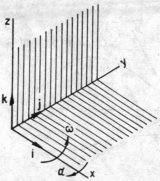

Fig. 6.65 Illustration of positive ω and negative α.

the *angular velocity* of the line OA. It can be shown that at any given instant, any line joining any pair of points in a rigid body has the same angular velocity. Because the angular velocity $\dot{\theta}$ is common to all lines in the rigid body, it is called the *angular velocity of the rigid body*. The symbol for angular velocity is ω, i.e., $\omega = \dot{\theta}$.

The quantity $\ddot{\theta}$ of Eq. (6.57) is the time rate of change of the angular velocity $\dot{\theta}$. Thus $\ddot{\theta}$ is the angular acceleration of the rigid body. Like the angular velocity, the instantaneous angular acceleration of a rigid body is also unique. The symbol for angular acceleration is α, i.e., $\alpha = \ddot{\theta}$.

We rewrite Eqs. (6.56) and (6.57) for the velocity and acceleration of any point within a rigid body whose distance *from a fixed axis of rotation* is r. Thus,

$$v = r\omega\hat{\theta} \qquad (6.58)$$

and

$$a = -r\omega^2\hat{r} + r\alpha\hat{\theta} \qquad (6.59)$$

Note carefully that whereas the x and y axes are fixed in space (i.e., fixed to the support), the r and θ axes are fixed to the body and rotate with it.

The angular velocity and angular acceleration are vectors along the axis of rotation. Their sense is given by the right-hand-rule for vector cross products. A simple way to assign the sense is to make the x-y plane the plane of rotation. If the body rotates from x towards y as shown by ω in Fig. 6.65, the angular

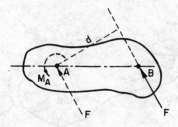

Fig. 6.66 Force and moment equivalence of a force shifted parallel to itself.

velocity vector is along the z-axis, i.e.,

$$\boldsymbol{\omega} = \omega \boldsymbol{k}$$

Likewise, a positive angular acceleration, which increases ω, will point in the positive z-direction. In Fig. 6.65, α is shown to be negative, i.e.,

$$\boldsymbol{\alpha} = -\alpha \boldsymbol{k}$$

We can rewrite Eqs. (6.58) and (6.59) in terms of vector cross-products as

$$\boldsymbol{v} = \boldsymbol{\omega} \times \boldsymbol{r} \tag{6.60}$$

$$\boldsymbol{a} = \boldsymbol{\omega} \times (\boldsymbol{\omega} \times \boldsymbol{r}) + \boldsymbol{\alpha} \times \boldsymbol{r} \tag{6.61}$$

Check and satisfy yourself that Eqs.(6.60) and (6.61) are indeed the same as Eqs.(6.58) and (6.59); for this purpose, take $\boldsymbol{\omega} = \omega \boldsymbol{k}$, $\vec{\alpha} = \alpha \boldsymbol{k}$, $\boldsymbol{r} = r \hat{\boldsymbol{r}}$.

To derive the equation of motion for the rotation of a rigid body about a fixed axis, we employ the concept of the moment of momentum, specifically Eq. (6.54). But first that, let us recall from statics that an external force acting at a point on a body can be represented as the same force and an appropriate moment at any other point in the body. For example, in Fig. 6.66, a force F acts at the point B on a body. This force can be replaced by its equivalent force system consisting of the force F and the moment M_A acting at the point A on the body where $M_A = F \times d$.

Now the body may be acted upon by many externally applied forces F_1, F_2 etc. Transferred to the axis of rotation O, the net equivalent force and moment from all of these external forces are

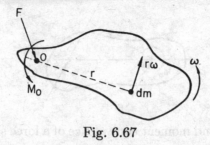

Fig. 6.67

F and M_o. Consider a particle of mass dm in the body (Fig. 6.67).

The velocity of this particle is $r\omega$, so its moment of momentum about O is

$$dH_o = (r\omega dm)r \qquad (6.62)$$

To obtain the moment of momentum H_o of the entire rigid body, we integrate dH_o over all the particles that make up the body. Thus,

$$H_o = \int dH_o = \int r^2 \omega dm$$

Now ω is unique for the body and independent of the choice of the particle dm; thus it is a constant and can be brought outside the integral sign. Then,

$$H_o = \omega \int r^2 dm \qquad (6.63)$$

The integral of Eq. (6.63) represents the moment of inertia of the body. Since r is measured from O, it is the moment of inertia about the axis of rotation, I_o. So, the moment of momentum, also called *angular momentum* in the case of rigid bodies, is

$$H_o = I_o \omega \qquad (6.64)$$

Differentiating with respect to time and noting that I_o is constant for a given body and a fixed axis of rotation,

$$\dot{H}_o = I_o \dot{\omega}$$

or

$$\dot{H}_o = I_o \alpha \qquad (6.65)$$

Now we invoke Eq. (6.54) and combine it with Eq. (6.65) to obtain

$$M_o = I_o \alpha \qquad (6.66)$$

Comparing Eq. (6.66) with Eq. (6.2), shows that they are of similar form. The angular acceleration α has replaced the linear acceleration a, the net moment M_o has replaced the net force F and, the moment of inertia has taken the place of the mass. Hence, the moment of inertia is the *rotational* inertia and Eq. (6.66) is Newton's second law for pure rotational motion about a fixed axis.

We have derived the equation of motion for rotation, Eq. (6.66), as a scalar equation purely for convenience. It is actually a vector equation,

$$\mathbf{M}_o = I_o \boldsymbol{\alpha} \qquad (6.67)$$

$\boldsymbol{\alpha}$ and $\mathbf{M}_o$ having the same sense. Likewise, Eqs. (6.64) and (6.65) should be written properly in vector form. For rotation about a fixed axis, the vector notation is superfluous.

The *angular impulse, angular momentum* equation is the time integral of Eq. (6.66). Thus,

$$\int_1^2 M_o dt = I_o \int_1^2 \alpha dt$$

or

$$\int_1^2 M_o dt = I_o \int_1^2 \frac{d\omega}{dt} dt$$

or

$$\int_1^2 M_o dt = I_o(\omega_2 - \omega_1) \qquad (6.68)$$

To derive the work-energy equation for rotational motion, we integrate the equation of motion, Eq. (6.66), with respect to the rotational displacement θ to get

$$\int_1^2 M_o d\theta = I_o \int_1^2 \alpha d\theta \qquad (6.69)$$

In analogy with Eq. (6.12),

$$\alpha = \omega \frac{d\omega}{d\theta}$$

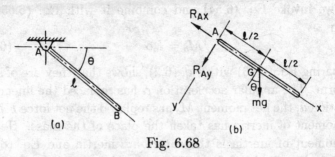

Fig. 6.68

so that Eq. (6.69) becomes

$$\int_1^2 M_o d\theta = I_o \int_1^2 \omega d\omega$$

or

$$\int_1^2 M_o d\theta = \frac{1}{2} I_o(\omega_2^2 - \omega_1^2) \qquad (6.70)$$

The integrand on the left hand side is the product of the moment and the angular displacement; this is the rotational work. The right hand side is the difference between the final and initial kinetic energies of rotation. For the special case of constant applied moment, the work energy equation, Eq. (6.70), becomes

$$M_o(\theta_2 - \theta_1) = \frac{1}{2} I_o(\omega_2^2 - \omega_1^2) \qquad (6.71)$$

Comparing Eq. (6.70) with Eq. (6.31), we see that the two are alike, with the translational quantities F, s, m, v replaced by their rotational counterparts M, θ, I and ω respectively.

Example 6.21

A uniform slender bar of mass m and length l swings in a vertical plane about the hinge at A (Fig. 6.68(a)). At a particular instant for which θ and $\dot\theta$ are known, determine the reaction at the hinge.

Solution

The free body diagram (Fig. 6.68(b)) shows the weight of the bar, mg and the x- and y-components of the reaction from the hinge. The equation of motion is

$$M_A = I_A \alpha$$

or $$(mg)\frac{\ell}{2}\cos\theta = (\frac{1}{3}m\ell^2)\alpha$$

whence

$$\ddot{\theta} = \alpha = \frac{3g\cos\theta}{2\ell} \tag{i}$$

Eq. (6.59) gives the acceleration components of the centre of mass G

$$a_{Gx} = -\frac{\ell}{2}\omega^2 = -\frac{\ell}{2}\dot{\theta}^2 \tag{ii}$$

$$a_{Gy} = \frac{\ell}{2}\alpha = \frac{3}{4}g\cos\theta \tag{iii}$$

These are the translational accelerations of the centre of mass, caused by the forces on the body

$$F_x = mg\sin\theta - R_{Ax} \tag{iv}$$

$$F_y = R_{Ay} + mg\cos\theta \tag{v}$$

Applying Newton's law to the translation of the centre of mass,

$$F_x = ma_{Gx}$$

or $$mg\sin\theta - R_{Ax} = -m\frac{\ell}{2}\dot{\theta}^2$$

or

$$R_{Ax} = mg\sin\theta + m\frac{\ell}{2}\dot{\theta}^2 \tag{vi}$$

Likewise, $$F_y = ma_{Gy}$$

or $$R_{Ay} + mg\cos\theta = m(\frac{3g\cos\theta}{4})$$

or

$$R_{Ay} = -\frac{1}{4}mg\cos\theta \tag{vii}$$

Since R_{Ay} has come out to be negative, its direction is opposite to the one shown in the free body diagram (Fig. 6.68(b)). Now,

$$R_A = \sqrt{R_{Ax}^2 + R_{Ay}^2}$$

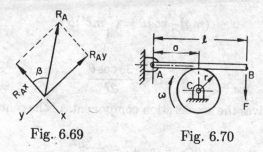

Fig. 6.69   Fig. 6.70

The direction of R_A is given by the angle β (Fig. 6.69),

$$\beta = \tan^{-1} \frac{R_{Ay}}{R_{Ax}}$$

This example illustrates how Newton's second law is applied to the curvilinear translation of the centre of mass of a rigid body. d'Alembert's Principle may also have been used. You should try that out.

Example 6.22

Figure 6.70 shows a model of a 'drum brake'. The wheel rotates about its axis at C while the bar AB is hinged at A. By applying the force F, the bar is pressed against the wheel which is then slowed down by friction. If the initial angular velocity of the wheel is ω_o and the coefficient of kinetic friction between it and the bar is μ_k, determine the time it takes to stop the wheel if its mass is m and the radius is r. The mass of the bar is negligible in comparison with the wheel.

Solution

From the free body diagram of the bar (Fig. 6.71(a)), which is in equilibrium, the normal reaction N of the wheel is $N = F\left(\dfrac{\ell}{a}\right)$.

Therefore, the friction force is $\mu_k F\left(\dfrac{\ell}{a}\right)$.

Then, from Fig. 6.71(b),

$$M_c = r\mu_k F\left(\frac{\ell}{a}\right)$$

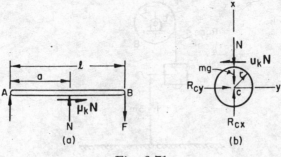

Fig. 6.71

If F is constant, M_c is also constant. Since we are dealing with velocities corresponding to time, we use the angular impulse-angular momentum equation, Eq. (6.68). Thus,

$$\int_1^2 M_c dt = I_c(\omega_2 - \omega_1)$$

Since M_c is constant, and $\omega_2 = 0$, $\omega_1 = \omega_o$, therefore

$$M_c(t_2 - t_1) = -I_c(\omega_o)$$

whence

$$t_2 - t_1 = -\frac{I_c\,\omega_o}{M_c} = -\frac{mr^2\omega_o/2}{(-r\mu_k F(\frac{\ell}{a}))} = \frac{mr\omega_o}{2\mu_k F}\frac{a}{\ell} \qquad (i)$$

Note that M_c and ω_o have opposite sense and so M_c has a negative sign.

We have determined the braking time. If we want to know the number of revolutions of the wheel during this time, we can find it by means of the work-energy equation. An alternative is to find the angular acceleration and then integrate it with respect to time.

From Eq. (6.66),

$$\alpha = \frac{M_c}{I_c} = -\frac{2\mu_k F\ell}{mra}$$

Then,

$$\omega = -\frac{2\mu_k F\ell}{mra}t + C_1$$

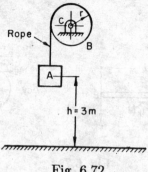

Fig. 6.72

When $t = 0$, $\omega = \omega_o$.

Therefore, $C_1 = \omega_o$

Integrating again,

$$\theta = -\frac{2\mu_k F\ell}{mra}\frac{t^2}{2} + \omega_o t + C_2$$

When $t = 0$, $\theta = 0$. Therefore $C_2 = 0$ and

$$\theta = -\frac{\mu_k F\ell}{mra}t^2 + \omega_o t$$

The time taken to arrest the wheel is the braking time $(t_2 - t_1)$, as found in Eq. (i). The corresponding angular displacement of the wheel is

$$\theta = -\frac{\mu_k F\ell}{mra}\left(\frac{mr\omega_o a}{2\mu_k F\ell}\right)^2 + \omega_o\left(\frac{mr\omega_o a}{2\mu_k F\ell}\right)$$

Therefore, $\theta = \frac{mra\omega_o^2}{2\mu_k F\ell}\left(-\frac{1}{2}+1\right) = \frac{mra}{4\mu_k F\ell}\omega_o^2$

Example 6.23

An inextensible rope wound tightly around a drum B is attached to a mass A at its free end (Fig. 6.72). The drum can rotate about a frictionless bearing at its centre C. The system is released from rest as shown. If $m_A = 4$ kg, $m_B = 14$ kg, $r = 0.3$ m, determine (a) the velocity with which A will strike the ground and (b) the corresponding time taken.

Solution

a) The system being frictionless, energy will be conserved. Choose the ground as the datum for potential energy. For the system of A and B,

$$T_1 = 0 \quad \text{(rest)}$$

$$V_1 = m_A g h + V_B$$

$$T_2 = \frac{1}{2} m_A v_A^2 + \frac{1}{2} I_B \omega_B^2$$

$$V_2 = 0 + V_B$$

$$T_2 + V_2 = T_1 + V_1$$

Therefore, $\frac{1}{2} m_A v_A^2 + \frac{1}{2} I_B \omega_B^2 + V_B = m_A g h + V_B$

or

$$m_A v_A^2 + (\frac{1}{2} m_B r^2) \omega_B^2 = 2 m_A g h \tag{i}$$

We need one more relation between v_A and ω_B. This is provided by the inextensible rope which ensures that the speed of A is the same as that of the rim of B. Thus,

$$v_A = r \omega_B \tag{ii}$$

Substituting for v_A in Eq. (i),

$$m_A r^2 \omega_B^2 + \frac{1}{2} m_B r^2 \omega_B^2 = 2 m_A g h$$

or $$\omega_B^2 = \frac{2 m_A g h}{r^2 (m_A + \frac{1}{2} m_B)}$$

Therefore, $\omega_B = \frac{1}{0.3} \sqrt{\frac{2 \times 4 \times 9.81 \times 3}{4 + 7}} = 15.42 \text{ rad/s}$

whence $v_A = 0.3 \times 15.42 = 4.63 \text{ m/s}$

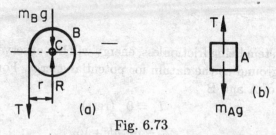

Fig. 6.73

b) Again, because of the rope being inextensible and always in
tension, the acceleration of A is the same as that of the rim
of the drum. Therefore,

$$a_A = r\alpha_B \qquad \text{(iii)}$$

The equations of motion for A and B are (Figs. 6.73(a) and (b))

$$m_A g - T = m_A a_A \qquad \text{(iv)}$$

$$(T)r = (\tfrac{1}{2} m_B r^2)\alpha_B \qquad \text{(v)}$$

Substitute for α_B from Eq. (iii) into Eq. (v) to get

$$T = \tfrac{1}{2} m_B a_A \qquad \text{(vi)}$$

Eliminating a_A from Eqs. (iv) and (vi),

$$\frac{m_A g - T}{m_A} = \frac{2T}{m_B}$$

or
$$T\left(\frac{2m_A}{m_B} + 1\right) = m_A g$$

or
$$T = \frac{m_B m_A}{2m_A + m_B} g \qquad \text{(vii)}$$

We see that the tension in the string is constant. Now use
Eq. (6.68) to get

$$\int_{t_1}^{t_2} M_c \, dt = I_c(\omega_2 - \omega_1)$$

where $\qquad M_c = (T)r = m_B m_A r / 2m_A + m_B g \text{ and } \omega_1 = 0.$

Also, $\omega_2 = \omega_B = 15.42\ \text{rad/s}$ (from part a)

Hence, $\displaystyle\int_{t_1}^{t_2} dt = \frac{(1/2)m_B r^2 (15.42)}{m_B m_A rg}(2m_A + m_B)$

or

$$t_2 - t_1 = \frac{15.42 r}{2m_A g}(2m_A + m_B) = \frac{15.42 \times 0.3}{2 \times 4 \times 9.81}(8 + 14) = 1.3\ \text{s}$$

So it takes 1.3 s for mass A to strike the ground.

6.17   Free Vibrations

A vibration is a to-and-fro motion. It may be free or forced. In forced vibration the particle or body is kept moving by a driving force, which usually varies cyclically between a maximum and a minimum value. *Free vibration* occurs in the absence of such a driving force, the body is set into motion by an initial disturbance and then continues to move back and forth under the influence of a restoring force which is part of the system itself, such as a spring on which the body is mounted. A free vibration may continue forever or may die out gradually. When it dies out, it is said to be a *damped free vibration*. The damping is caused by a dissipative force such as friction, which is also a part of the system. We shall consider only free vibration.

The simplest vibrating system is what is known as the *spring-mass system*. A mass is suspended by a spring. The system is disturbed from its equilibrium position, following which the action of the spring causes the mass to move back and forth.

Figure 6.74(a) shows the spring in its natural or undeformed length ℓ. In Fig. 6.74(b), a mass m has been suspended from the spring and the system is in equilibrium with a static deformation Δ of the spring. The corresponding free body diagram of the mass shows it in equilibrium under its weight and the spring force $k\Delta$, i.e.,

$$mg - k\Delta = 0 \tag{6.72}$$

Now the mass is given a downward displacement x from the equilibrium position (Fig. 6.74(c)). Note that we have chosen

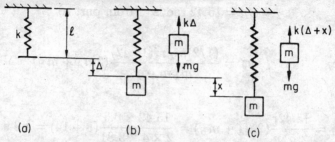

Fig. 6.74 The spring mass system, forces and displacements.

the downward displacement x as positive. The spring force is $k(\Delta + x)$ as shown in the corresponding free body diagram. This spring force is greater than the weight, as is clear from Eq. (6.72). If the system is released in the configuration of Fig. 6.74(c), the non-zero net force on the mass will cause it to accelerate. (Note that, as usual, we assume that the mass of the spring is negligible compared to m). The equation of motion of the mass when it is released (Fig. 6.74(c)) is

$$mg - k(\Delta + x) = m\ddot{x} \quad \text{or} \quad mg - k\Delta - kx = m\ddot{x}$$

Using Eq. (6.72), the above equation is reduced to

$$m\ddot{x} + kx = 0 \qquad (6.73)$$

Equation (6.73) is the equation of motion of the mass m. Observe that x is measured from the static equilibrium position of the mass (Fig. 6.74) and that the initial displacement Δ of the spring and the weight of the mass are irrelevant to the motion. Equation (6.73) is usually written as

$$\ddot{x} + \frac{k}{m}x = 0 \qquad (6.74)$$

The quantity k/m is seen to be a property of the spring-mass system and presently we shall see its importance.

Equation (6.74) is an ordinary differential equation of the second order because $\ddot{x}$ is the second derivative of x with respect to time. We can solve it by inspection by observing that

$$\ddot{x} = -\frac{k}{m}x \qquad (6.75)$$

which means that $\ddot{x}$ and x must have the same form. Recognising that the second derivatives of the sine and cosine functions are the same functions with changed signs, we write the trial solution

$$x(t) = A\cos\sqrt{\frac{k}{m}}\,t + B\sin\sqrt{\frac{k}{m}}\,t \qquad (6.76)$$

Both sine and cosine terms have been included because both are possible solutions. Also, the solution to a second order differential equation contains two arbitrary constants, here A and B. The constants A and B are found from the initial ($t = 0$) conditions on the displacement, $x(o)$ and the velocity, $\dot{x}(o)$. It is obvious that if we differentiate Eq. (6.76) twice we get Eq. (6.75), confirming that Eq. (6.76) is, indeed, the solution to the equation of motion, Eq. (6.74).

Equation (6.76) can also be written in the form

$$x(t) = x_o \sin\left(\sqrt{\frac{k}{m}}\,t + \phi\right) \qquad (6.77)$$

wherein the right hand side represents the sum of the sine and cosine terms of Eq. (6.76). The two constants of integration are now x_o and ϕ, where ϕ is the phase angle of the sine wave and x_o its amplitude. The relationships between x_o and ϕ of Eq. (6.77), and A and B of Eq. (6.76) are seen as

$$A = x_0 \sin\phi \quad \text{and} \quad B = x_0 \cos\phi$$

sothat $\qquad x_o = \sqrt{A^2 + B^2} \quad \text{and} \quad \phi = \tan^{-1}\frac{A}{B}$

The displacement $x(t)$ is plotted as a function of time (Fig. 6.75), which also shows how a rotating radial line (or rotating vector) of length x_o gives $x(t)$. If the rotating vector $\dot{x}_o$ has an angular velocity ω_n, it is seen that the sinusoidal function $x(t)$ represents the solution given by Eq. (6.77) when

$$\omega_n = \sqrt{\frac{k}{m}} \qquad (6.78)$$

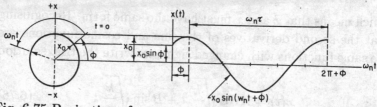

Fig. 6.75 Projection of a rotating vector to generate a sinusoidal function.

Then, since the vector x_o sweeps out an angle 2π per cycle, the time period of the vibration (or oscillation) is

$$\tau = \frac{2\pi}{\omega_n} \qquad (6.79)$$

It follows that the frequency of the oscillation is $f = 1/\tau$, i.e.

$$f = \frac{\omega_n}{2\pi} = \frac{1}{2\pi}\sqrt{\frac{k}{m}} \qquad (6.80)$$

The quantities ω_n and f have the same dimension, T^{-1}. So ω_n is called the *circular frequency*, i.e., the frequency of the circular motion of the tip of the rotating vector x_o. The two frequencies ω_n and f represent the same physical quantity, the only difference being that whereas ω_n is in radians per second, f is in cycles per second or Hertz. Note that the parameters of the system, mass and spring constant, determine the frequency. For this reason, it is known as the *natural frequency* of the system. External inputs such as $x(o)$ and $\dot{x}(o)$ have no effect on the natural frequency.

We have analysed the motion of a system that can be modeled as a mass supported on a spring. Such a system, if disturbed from its position of equilibrium, will oscillate with a natural frequency determined only by the mass m and the spring stiffness k. We have also determined the amplitude and phase of the oscillation subject to initial conditions on x and $\dot{x}$. This is *simple harmonic motion* as described in Example 6.3.

If the system is conservative, i.e., there is no dissipative agent like friction, the free spring-mass system will continue to oscillate forever, satisfying the principle of conservation of energy. That, of course, is not realistic. In practice the oscillation is damped out with time. Damping is caused either by internal friction

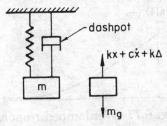

Fig. 6.76 Spring-mass-dashpot system.

in the spring or by a damper deliberately incorporated into the system. Such a damper is called a dashpot and the system is then referred to as a *spring-mass-damper* or *spring-mass-dashpot* system. The dashpot is placed in parallel with the spring as shown in Fig. 6.76.

Most dashpots consist of a piston and cylinder containing a viscous fluid Example 6.5. The force from a dashpot is usually proportional to the velocity of the piston, which in this case is the same as the velocity of the mass, $\dot{x}$. With reference to the free body diagram shown in Fig. 6.76, the equation of motion is

$$mg - kx - c\dot{x} - k\Delta = m\ddot{x}$$

which, in view of Eq. (6.72), becomes

$$m\ddot{x} + c\dot{x} + kx = 0 \tag{6.81}$$

Equation (6.81) is the equation of motion for a damped free oscillator. It differs from Eq. (6.73) by the presence of the damping force term $c\dot{x}$ in which the constant c is the *damping coefficient*.

Equation Equation (6.81) shows that x and its first two derivatives, $\dot{x}$ and $\ddot{x}$, must have the same form. A function which satisfies this requirement is the exponential function. Allowing for two arbitrary constants of integration,

$$x(t) = C_1 e^{s_1 t} + C_2 e^{s_2 t} \tag{6.82}$$

(Note that the solution to Eq. (6.74) could also have been written in this form). In Eq. (6.82), s_1 and s_2 are constants that depend on the system properties k, c and m as follows:

$$s_1 = -\frac{c}{2m} + \frac{1}{2}\sqrt{\left(\frac{c}{m}\right)^2 - 4\omega_n^2} \tag{6.83}$$

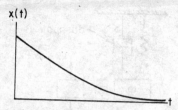

Fig. 6.77 Overdamped response.

$$s_2 = -\frac{c}{2m} - \frac{1}{2}\sqrt{\left(\frac{c}{m}\right)^2 - 4\omega_n^2} \qquad (6.84)$$

The time-dependent behaviour of the displacement, $x(t)$, depends on s_1 and s_2. Damping does not change the natural frequency ω_n of the system, which is still given by Eq. (6.78). There are three kinds of motion that arise out of the above solution.

Case (i)

If $(c/m)^2 > 4\omega_n^2$, both s_1 and s_2 are real but negative. The displacement $x(t)$ decays exponentially with time and approaches $x(o)$ without any oscillation (Fig. 6.77) and the system is said to be *overdamped*. This is the kind of motion we had seen in Example 6.5, in which the initial conditions were $x(o) = 0$, $\dot{x}(o) = v_o$.

Case (ii)

If $(c/m)^2 = 4\omega_n^2$, both s_1 and s_2 are still real and negative. The physical characteristic of the response $x(t)$ is similiar to that of case (i). This system is said to be *critically damped* because it is the threshold between the overdamped and underdamped cases. With critical damping,

$$c = 2m\omega_n$$

which is written as

$$c_c = 2m\omega_n \qquad (6.85)$$

in which c_c is called the *critical damping coefficient*.

Case (iii)

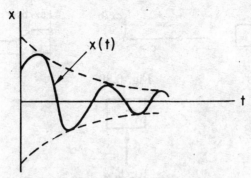

Fig. 6.78 Underdamped response.

If $(c/m)^2 < 4\omega_n^2$, both s_1 and s_2 are no longer real. We shall not pursue the mathematics of this case, but simply note that the response $x(t)$ is sinusoidal with an exponentially decaying amplitude (Fig. 6.78). The frequency of the oscillations is not ω_n but something else. The system is said to be *underdamped*.

Example 6.24

A weighing platform of mass m is connected to a spring of stiffness k through a system of four links (Fig. 6.79). When a load is placed on the platform, the deflection of the spring indicates the weight. Assuming that the links and the spring have negligible mass, find the natural frequency and time period for small oscillations of this system.

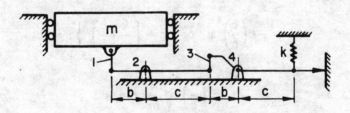

Fig. 6.79

Solution

Because of the leverage provided by the links, the vertical displacements and the forces are magnified (Fig. 6.80). A displacement x of the platform is magnified to $(c/b)^2 x$ at the spring.

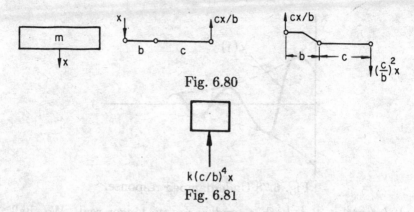

Fig. 6.80

Fig. 6.81

Likewise, a force F_s at the spring is magnified to $(c/b)^2 F_s$ at the platform. This provides the basis for defining the force on the mass for any displacement x of the platform.

$$\text{Displacement of the platform} = x$$

$$\text{Displacement of the spring} = (c/b)^2 x$$

Therefore, spring force

$$F_s = k(c/b)^2 x$$

and the force on the platform is $k(c/b)^4 x$ (Fig. 6.81).

Observe that the weight of the platform does not come into this because x is measured from the static equilibrium position (Eq. 6.72).

Now,

$$m\ddot{x} = -k\left(\frac{c}{b}\right)^4 x \quad \text{or} \quad \ddot{x} + \left\{\frac{k}{m}\left(\frac{c}{b}\right)^4\right\}x = 0$$

i.e.

$$\omega_n = \left(\frac{c}{b}\right)^2 \sqrt{\frac{k}{m}} \ \text{rad/s}$$

or

$$f = \frac{1}{2\pi}\left(\frac{c}{b}\right)^2 \sqrt{\frac{k}{m}} \ \text{hertz}$$

and

$$\tau = 2\pi\left(\frac{b}{c}\right)^2 \sqrt{\frac{m}{k}} \ \text{s}$$

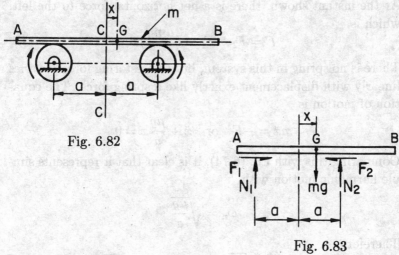

Fig. 6.82

Fig. 6.83

Example 6.25

A flat board of mass m is placed horizontally on two rollers (Fig. 6.82) that rotate in opposite directions with the same angular speed. Show that if the centre of mass G of the board is displaced by x from the mid-plane C-C and then released, the board will undergo simple harmonic motion. Determine the natural frequency and the period of this motion. Take the coefficient of kinetic friction as μ.

Solution

The free-body diagram (Fig. 6.83) shows the forces on the board. Since the board remains horizontal, the moment on it must be zero. Therefore,

$$N_1(2a) = mg(a - x) \quad \text{or} \quad N_1 = mg\frac{a - x}{2a}$$

Then, for vertical equilibrium,

$$N_2 = mg - N_1 \quad \text{or} \quad N_2 = mg\frac{a + x}{2a}$$

The friction forces are

$$F_1 = \mu mg\frac{a - x}{2a} \quad \text{and} \quad F_2 = \mu mg\frac{a + x}{2a}$$

At the instant shown, there is a net horizontal force to the left, which is

$$F = F_2 - F_1 = (\frac{\mu mg}{a})x$$

There is no spring in this system, but the restoring force F varies linearly with displacement exactly like a spring force. The equation of motion is

$$m\ddot{x} = -F \quad \text{or} \quad \ddot{x} + \frac{\mu g}{a}x = 0$$

Comparing this with Eq. (6.74), it is clear that it represents simple harmonic motion with

$$\omega_n = \sqrt{\frac{\mu g}{a}}$$

Therefore,

$$f = \frac{1}{2\pi}\sqrt{\frac{\mu g}{a}} \quad \text{and} \quad \tau = 2\pi\sqrt{\frac{a}{\mu g}}$$

As the board shifts, N_1 and N_2 and hence F_1 and F_2 change. The net force F changes direction and reverses the motion of the board.

Problems

6.1 The position co-ordinate of a particle in rectilinear motion is $x = 2t^3 - 24t + 6$. Compute (a) the time to reach a velocity of 72 m/s, (b) the acceleration when the velocity is 30 m/s and (c) the displacement over the time interval $t=1$ s to $t=4$ s.

6.2 A projectile is fired vertically with a velocity of 200 m/s. What is the maximum height it will reach? Also find the time it will take, from the instant of firing, to return to the ground. Take the acceleration due to gravity as a constant 9.81 m/s².

6.3 Water leaks from a leaky tap at the constant rate of n drops every second. Determine the distance between two consecutive drops as a function of the time t, where t is measured from the instant the trailing drop starts to fall. Neglect air resistance.

6.4 The displacement of a particle in rectilinear motion varies with time as $x = x_0(2e^{-kt} - e^{-2kt})$, where x_0 is the initial displacement and k is a constant. Sketch the displacement-time and velocity-time characteristics of the motion. Determine the maximum velocity and the corresponding time.

6.5 A particle starts from rest and moves with the acceleration given by $a = k \sin(t/T)$, where T is a constant. Determine (a) the maximum velocity, (b) the position and velocity at $t = 2T$.

6.6 A jet aircraft starts its take-off run with full engine thrust, which corresponds to a constant acceleration of $0.7g$, which remains constant until take-off. Take-off occurs at a ground speed of 200 km/h. Determine the distance covered and the time taken from rest to take-off.

6.7 A certain vehicle whose mass is m can generate an acceleration of $a = P/(mv)$, where P is the constant power output of the propulsion unit. Neglecting frictional resistance, determine the distance travelled and the time taken while the speed increases from v_1 to v_2.

6.8 A body in rectilinear motion is retarded in such a way that the velocity and displacement are related as $v^2 = k/s$, where k is a constant. At $t = 0$, it is known that $s = 225$ mm and $v = 50$ mm/s. Find its speed at $t = 3$ s.

6.9 A particle moves with the acceleration $a = -k/x$, where k is a constant and x is the displacement. It is known that $v = 4$ m/s when $x = 200$ mm and that $v = 2$ m/s when $x = 600$ mm. Determine the position of the particle when its velocity is zero and also the velocity when $x = 800$ mm.

6.10 The displacement-time variation for small oscillations of a simple pendulum is given by $s = s_0 \cos nt$, where s_0 is the amplitude of the small oscillation and $n = (g/l)^{1/2}$, l being the length of the pendulum. Determine the maximum values of the velocity and the normal and tangential accelerations.

6.11 The car A is moving on its circular road at a constant speed of 60 km/h while car B approaches it with the velocity 81

km/h and a deceleration of 3 m/s^2 (Fig. 6.84). Determine
the velocity of car A as observed from car B.

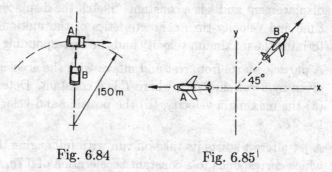

Fig. 6.84   Fig. 6.85

6.12 Aeroplanes A and B (Fig. 6.85) are flying at the same altitude, with constant speeds. The speed of A is 450 km/h,
while that of B is 750 km/h. Determine the change in the
position of B relative to A after a 3-minute interval.

6.13 The car A (Fig. 6.86) is passing the road intersection with
a constant speed of 30 km/h. At that instant another car B
starts from rest and drives towards the intersection with a
constant acceleration of 1.5 m/s^2. Determine the position,
velocity, and acceleration of B relative to A five seconds
later.

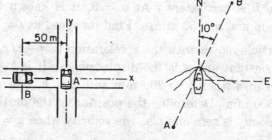

Fig. 6.86   Fig. 6.87

6.14 A boat is heading north at its maximum speed of 5 knots
(Fig. 6.87). A current drags it towards the east so that,
after 3 hours, the boat reaches the point B. If the actual
course of the boat is the line AB, whose length is 12 nautical
miles (1 nautical mile = 1.852 km), determine the velocity
of the current.

6.15 A small motorboat maintains a constant speed of 2.5 knots relative to a current. When the boat points due east, an observer on the shore sees it moving due south; when it points due northeast, the same observer sees it moving due west. What is the velocity of the current?

6.16 Determine the minimum velocity with which a shell must be fired from a gun in order that it might reach a target located 10 km away on the same horizontal level.

6.17 An anti-aircraft gun is fired when an aircraft is directly overhead and flying horizontally at a constant speed of 1000 km/h at an altitude of 5 km (Fig. 6.88). If the shell leaves the gun with a velocity of 500 m/s, what must be the firing angle in order that a direct hit will result?

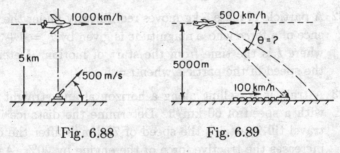

Fig. 6.88   Fig. 6.89

6.18 A bomber pilot intends to hit a train moving along the same direction (Fig. 6.89). The plane is at an altitude of 5000 m and has a speed of 500 km/h, while the speed of the train is 100 km/h. Determine the line of sight below the horizon at the instant that the pilot should release the bomb in order that he may score a hit.

6.19 The figure shows a warship firing a missile at a fixed target, with the firing angle α relative to the line of sight (Fig. 6.90). The speed of the ship is 25 knots (1 knot = 1.852 km/h) and the projectile is fired at 75 m/s relative to the ship and at a vertical angle of 30° relative to the surface of the ocean. Determine the angle for a direct hit. (Hint: Try with the ship as the reference frame.)

6.20 For problem 6.17, determine the velocity and acceleration of the shell relative to the plane at the instant of the impact.

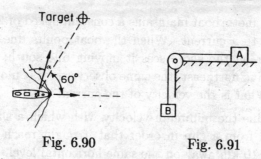

Fig. 6.90   Fig. 6.91

6.21 Figure 6.91 shows a system of two blocks tied together by an inextensible string passing over a frictionless pulley. If the masses A and B are 6 kg and 8 kg respectively, and the coefficient of friction between the block A and the table is 0.28, determine the accelerations of the blocks and the tension in the string.

6.22 A particle of mass 5 kg moves rectilinearly under the influence of a force whose magnitude is given by $F = 4 + 3t^2$ N, where t is the time from the start of motion. Determine the speed of the particle when $t = 5$ s.

6.23 A train is travelling along a horizontal and straight track with a speed of 60 km/h. Determine the distance it will travel till it attains the speed of 75 km/h after the driver increases the tractive force of the engine by 20%. Assume that all resistances amount to 1/150 of the weight of the train.

6.24 The force on a particle of mass m is given by $F = F_0 \cos pt$. Obtain the expressions for the displacement and velocity as functions of time if the initial values of both are zero.

6.25 A particle moving along the x-axis is subjected to a force $F = F_0 - kt$ where $F_0 = 30$ N and $k = 5$ N/s. If it had started from rest at the origin, at what time t will it return to the origin?

6.26 Two masses of 50 kg and 40 kg are suspended from pulleys (Fig. 6.92). Assume that the pulleys are light and frictionless and determine the acceleration of the 40 kg mass.

6.27 A man of mass m stands on a cart of mass M which is moving to the right with a speed v_0 (Fig. 6.93). He then runs

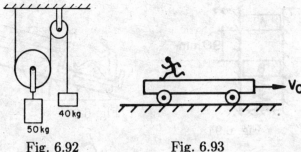

Fig. 6.92   Fig. 6.93

to the left and jumps off the cart with a speed u relative to the cart. Assuming that the cart rolls without friction, determine the increment in the speed of the cart. (Note that the velocity changes as the man runs on the cart).

6.28 Repeat Problem 27, this time with 2 men on the cart. Determine the speed increment of the cart if (a) both men run and jump off together and (b) if the men run and jump off one at a time. Take $M = 450$ kg, $m = 70$ kg, $v_0 = 1$ m/s, $u = 0.7$ m/s.

6.29 A weight of mass m rests on a massless spring and produces a static deflection δ in the spring. If the same mass drops on the spring from a height h, what will be the maximum deflection of the spring?

6.30 A particle of mass m moves along the x-axis under a force given by $F = kx$, where k is a constant. Determine the velocity as a function of the displacement x if the initial displacement is zero and the initial velocity is u.

6.31 A block of mass 2 kg is given an initial velocity of 3 m/s down an inclined plane making 30° with the horizontal. If the coefficient of friction is 0.3, find the velocity of the mass after it has moved 10 m down the plane.

6.32 Return to the system of Problem 21. If the system is released from rest, what will be the velocity of the block B after it has moved down by 0.75 m? How much time will have elapsed?

6.33 A ball is dropped from a height of 3 m and rebounds off the ground to a height of 2.5 m. What is the coefficient of

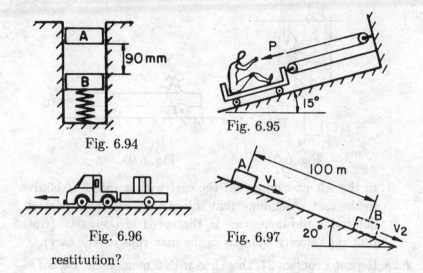

Fig. 6.94

Fig. 6.95

Fig. 6.96

Fig. 6.97

restitution?

6.34 The block A of mass 12 kg falls from rest on to the identical block B through a height of 90 mm (Fig. 6.94). If the stiffness of the spring is 3600 N/m, and if the two blocks stick together after the impact, what is the maximum displacement of the spring from the static equilibrium position as shown in the figure? Neglect friction.

6.35 A man sits in the cart and pulls the end of the rope with a force P of 300 N (Fig. 6.95). If the mass of the cart and the man together is 120 kg, find the acceleration with which the cart moves up the incline. Neglect friction and the masses of the pulleys and rope.

6.36 A box rests on the bed of a truck which moves with a speed of 60 km/h (Fig. 6.96). If the coefficient of friction between the box and the truck-bed is 0.4, determine the minimum stopping distance for which the box will not slip when the truck slows with a constant deceleration.

6.37 The block on the inclined plane (Fig. 6.97) has the velocities $v_1 = 10$ m/s and $v_2 = 5$ m/s at the positions A and B, as shown. Determine the coefficient of friction between the block and the plane.

6.38 The ball of mass m is suspended by two strings inside the frame shown (Fig. 6.98). Determine the acceleration of the frame for which the tension in the string A will be twice that in the string B.

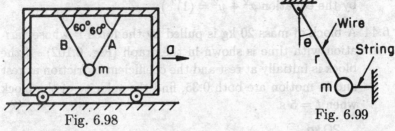

Fig. 6.98

Fig. 6.99

6.39 A wire and a string are used to suspend a small mass (Fig. 6.99). Determine the ratio of the tension in the wire before and after the string is cut. (Hint: What is the velocity of the mass at the instant the string is cut?)

6.40 Two identical blocks A and B can slide without friction in their vertical and horizontal guides (Fig. 6.100). They are connected by the light link of length 1.0 m, as shown. They are released from rest with the link at 45° to the horizontal. Determine the velocity of the mass A as it passes the intersection.

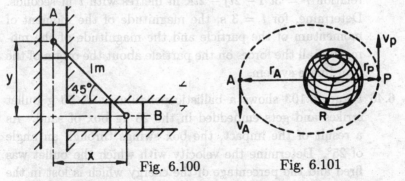

Fig. 6.100

Fig. 6.101

6.41 Return to Problem 6.35. If the cart starts moving from rest, determine its velocity after it has moved 3 m up the incline. Assume that the wheels of the cart are stuck and a friction coefficient of 0.25 exists between the locked wheels and the inclined surface.

6.42 A satellite in elliptical orbit around the earth has a velocity v_P at the perigee P (Fig. 6.101). Find its velocity v_A at the apogee A in terms of the distances r_A and r_P.

6.43 Compute the maximum velocity of the slider B of Problem 40. (Hint: the displacements of the two sliders are related by the equation $x^2 + y^2 = (1)^2$).

6.44 A block of mass 20 kg is pulled by the force P whose variation with time is shown in the graph (Fig. 6.102). If the block is initially at rest and the coefficient of friction at rest and in motion are both 0.35, find the velocity of the block when $t = 5$ s.

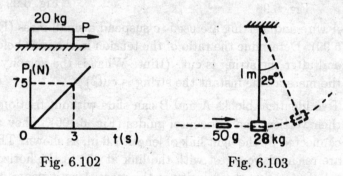

Fig. 6.102   Fig. 6.103

6.45 A particle of mass 3 kg has the position vector given by the relation $r = 3t^2\mathbf{i} - 2t\mathbf{j} - 2t\mathbf{k}$ in metres with t in seconds. Determine, for $t = 3$ s, the magnitude of the moment of momentum of the particle and the magnitude of the moment of all the forces on the particle about the origin of the coordinate system.

6.46 Figure 6.103 shows a ballistic pendulum. A 50 g bullet strikes and gets embedded in the 28 kg box of sand. As a result of the impact, the box swings through an angle of 25°. Determine the velocity with which the bullet was fired and the percentage of the energy which is lost in the impact.

6.47 A small mass m slides without friction on a smooth tabletop (Fig. 6.104). It is connected to a larger mass M through a light, inextensible, string passing through a hole as shown. At a certain instant, when the mass m is a distance a from the hole, its velocity vector v_1 is perpendicular to the string. Determine the distance x for which the velocity vector will again be perpendicular to the string.

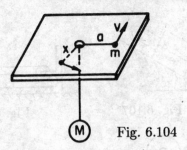

Fig. 6.104

6.48 Determine the velocity vector v with which the steel ball rebounds from the plate shown in Fig. 6.105 if the speed v_0 of approach is 30 m/s and the coefficient of restitution is 0.75.

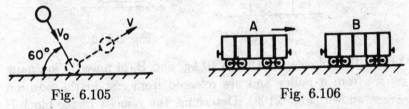

Fig. 6.105   Fig. 6.106

6.49 A railway wagon A of mass m_A (Fig. 6.106) rolls to the right and hits wagon B, of mass m_B, after which the two move together. If wagon B was initially at rest, determine the fractional loss of energy for the system.

6.50 A car is braked suddenly, whereupon it skids to a stop after 3 m. If the speed before the application of the brakes was 70 km/h, determine the time taken to stop and the coefficient of friction between the tyres and the road.

6.51 The blocks A and B have masses of 20 kg and 15 kg respectively (Fig. 6.107). The coefficients of static and kinetic friction are 0.2 and 0.15 at all the surfaces. For the applied force of 200 N, determine the acceleration of the block A and the tension in the cord.

6.52 The bracket A of mass 15 kg (Fig. 6.108) rests on a frictionless surface while the coefficients of static and kinetic friction between the block B of mass 10 kg and the bracket are 0.35 and 0.3. After the force P of 25 N is applied as shown, and B has moved from rest by 400 mm relative to A, determine the velocity of B relative to A.

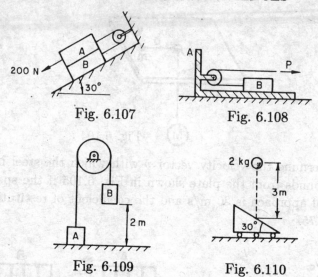

Fig. 6.107   Fig. 6.108

Fig. 6.109   Fig. 6.110

6.53 Two blocks A, of mass 10 kg, and B, of mass 15 kg, hang from a pulley and are released from rest in the position shown (Fig. 6.109). Determine the velocity of the block B as it hits the ground and also the tension in the cord on either side of the pulley during the motion. It is known that frictional loss in the axle of the pulley during the motion is 12 J.

6.54 A 2 kg ball drops from rest through a height of 3 m and strikes the inclined surface of the 5 kg wedge resting on frictionless rollers (Fig. 6.110). If the coeffiecient of restitution is 0.8, determine the velocities of the ball and the wedge immediately after the impact.

6.55 Derive the equation of motion for free undamped vibration of the spring-mass system by using the principle of conservation of energy.

6.56 Figure 6.111 shows a mass suspended by a system of two springs in series and parallel arrangements. Determine the equation of motion in each case, noting that the mass is guided to move in a straight line. If the two springs are replaced by a single spring, determine, for both cases, the stiffness of the single spring so that the motion of the mass will have the same natural frequency.

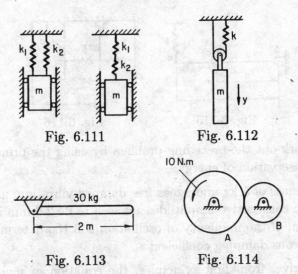

Fig. 6.111   Fig. 6.112

Fig. 6.113   Fig. 6.114

6.57 A uniform cylinder of mass m is suspended by means of a massless pulley and spring (Fig. 6.112). The cylinder is given a small displacement y_0 from its equilibrium position. Derive the equation of motion, neglecting friction in the pulley, and determine the period.

6.58 A uniform bar of mass 30 kg is supported by a hinge and released from rest in the horizontal position (Fig. 6.113). If the hinge is rusty and exerts a constant resistive torque of 5 Nm on the bar, determine the reaction on the bar at the hinge as motion begins.

6.59 Two gears in mesh are shown in Fig. 6.114. Gear A has a mass of 20 kg and gear B of 5 kg. Their radii of gyration are, respectively, 140 mm and 70 mm. If a torque of 10 Nm acts on the shaft of gear A, find the angular acceleration of the gear B.

6.60 A mass m is attached to a light (massless) rod AB which is hinged at A (Fig. 6.115). The spring of stiffness k supports the end B so that the rod is horizontal when the system is in static equilibrium. Derive the equation of motion of this system if it is given a small disturbance and then released. (Note that the rod, being massless, has no net moment or force on it; then directly apply Newton's second law to the mass m).

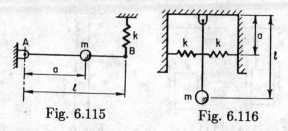

Fig. 6.115   Fig. 6.116

6.61 Work out the preceding problem by using the principle of conservation of energy.

6.62 A mass of 5 kg undergoes free damped vibration, in which two consecutive amplitudes are seen to be 2.0 mm and 1.8 mm. If the frequency of oscillation is 3 Hz, determine the viscous damping coefficient.

6.63 Derive, from first principles, the equation of motion for small oscillations of the pendulum (Fig. 6.116). Consider the hinge to be frictionless and the angular displacement θ as the displacement coordinate. In the equilibrium position, with $\theta = 0°$, each spring has a compressive force F_0. Neglect the mass of the rod and determine the period of the oscillation.

6.64 A 1 kg mass is supported by a spring and a dashpot whose damping constant is 4 Ns/m. The mass is displaced from its equilibrium position by 15 mm and released, whereupon it is seen to vibrate with a frequency of 4 Hz. Determine the amplitude of the fourth cycle of oscillation.

6.65 A mass of 4 kg is suspended from a spring of stiffness $k = 800$ N/mm. The mass is displaced 40 mm downward from its equilibrium position and released. Determine (a) the time required for the mass to move up by 60 mm, (b) the corresponding velocity and acceleration of the mass.

7. STRESS AND STRAIN

7.1 Introduction

In the design of any structure or its components, it is necessary to know the internal and/or external forces acting on the members. In the earlier chapters, various kinds of forces that generally act on a structure has been described. The size of the members and the material behaviour also play an important role. To evaluate the stresses and strains at any point in the body, we require an understanding of nature of forces that act on a member.

7.1.1 External Loads (Forces)

The loads that act external to the structure are defined as external loads. These can be in the form of forces or moments and can be considered as concentrated or distributed. Strictly speaking, there exists no concentrated forces in nature. All real forces act over a finite area. For example, the pressure of a wheel on a rail is actually transmitted through a small area.

7.1.2 Static Loads

Static loads are those which change their magnitude or point of application (direction) at a very slow rate.

7.1.3 Dynamic Loads

Dynamic loads are those whose magnitude varies with time at a fast rate. The action of such loads set oscillations. Due to changes in the speed of vibrating masses inertial forces are developed. The magnitude of these inertia forces may be many times the static loads.

7.1.4 Surface Forces

Surface forces are caused by the direct contact of one body with the surface of the other. These forces are distributed over the area of contact between the bodies.

7.1.5 Body Forces

A body force occurs when one body exerts a force on another body without direct physical contact between the bodies. Examples of this kind of forces are due to gravitation or electromagnetic fields.

In Chapter 2, we described the free body diagram, which is drawn to describe the forces acting on an element. These can be external forces, support forces and body forces.

7.2 Stress

Figure 7.1(a) represents a body in equilibrium under the action of external forces. If the body is sectioned under the action of external forces (Fig. 7.1(b)), the two loadings shown represent the resultant effect of the distributed force acting at a specified cross-sectional area of the body.

The resultant internal force F_R and internal moment M_R acting over the sectioned area is obtained by considering the equilibrium of forces and moments. In general, internal forces are not uniformly distributed over the section. If the sectioned area is subdivided into small areas (Fig. 7.1(c)), then on each of these areas the force distribution will become uniform as the area gets smaller and smaller. The force ΔF_R is resolved into two components ΔF_n normal to the cross-sectional area and ΔF_s, parallel to the cross-sectional area.

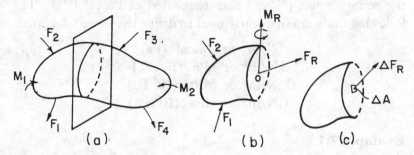

Fig. 7.1 (a) 3D body subjected to external loads; (b) A sectional view of the body under external forces; and G (c) Forces acting on an elemental area.

The average intensity of the normal force acting over the small area is ΔF_n divided by ΔA, and is defined as *average normal stress*. Thus,

$$\sigma_{av} = \frac{\Delta F_n}{\Delta A} \tag{7.1}$$

In the limit, as $\Delta A \to 0$, the *normal stress* acting at a point is defined as

$$\sigma = \lim_{\Delta A \to 0} \frac{\Delta F_n}{\Delta A} = \frac{dF_n}{dA} \tag{7.2}$$

On the other hand, the intensity of force parallel to the small area is ΔF_s divided by ΔA, and is defined as *average shear stress*. Thus,

$$\tau_{av} = \frac{\Delta F_s}{\Delta A} \tag{7.3}$$

In the limit, as $\Delta A \to 0$, the *shear stress* acting at a point is defined as

$$\tau = \lim_{\Delta A \to 0} \frac{\Delta F_s}{\Delta A} = \frac{dF_s}{dA} \tag{7.4}$$

Note that the normal and shear stresses, in general, may vary from point to point across the cross-section.

There are two types of normal stresses: *tensile* and *compressive stresses*. Tensile stress is produced by a force directed outward from an area and is considered positive. Compressive stress is produced by a force directed towards an area and is considered negative. The SI system of units for stress are Newtons per square meter (N/m^2) and designated as Pascal (Pa). The following units are frequently used to define the stress at a point:

$$N/m^2 \text{ or Pascal (Pa)}$$
$$kN/m^2 \text{ or kPa } (10^3 \text{ Pa})$$
$$MN/m^2 \text{ or MPa } (10^6 \text{ Pa})$$
$$GN/m^2 \text{ or GPa } (10^9 \text{ Pa})$$

Example 7.1

A steel bar 10mm by 20mm in cross-section is subjected to axial loads (Fig. 7.2(a)). Obtain the internal normal stresses in segments AB, BC and CD.

Solution

Figure 7.2(a) shows that the axial forces balance each other, hence the bar is in equilibrium.

To find the internal forces at a section of the bar, a plane parallel to the cross-section is passed through the section, separating the bar into two parts. Internal force at the section is then determined by considering the equilibrium of either part of the bar. Figures 7.2(b), (c) and (d) show the free body diagrams with internal forces taken as tensile. From the free body diagrams we get,

$$F_1 = -20 \text{ kN}$$
$$F_2 = 8 \text{ kN}$$
$$F_3 = 32 \text{ kN}$$

A positive value of the forces F_1 or F_2 or F_3 indicate that the section is indeed in tension, while a negative value indicates that the section is in compression. The axial force at any section between A and B is constant and is equal to 20 kN in compression. The axial force between B and C is 8 kN and is in tension and at any section between C and D is 32 kN and in tension. With the knowledge of forces acting in each section, it is now possible to determine the normal stresses acting at each section in the bar.

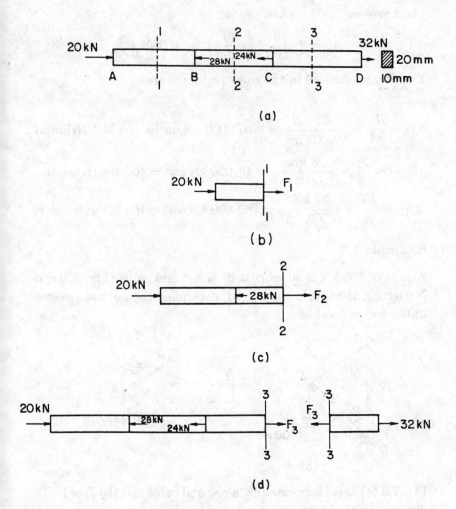

Fig. 7.2 (a) Rectangular bar under axial loads; (b) Free body diagram of the bar at section 1.1; (c) Free body diagram of the bar at section 2.2 and (d) Free body diagram of the bar at section 3.3.

The cross-sectional area of the bar is

$$A = \left(\frac{10}{1000}\right)\left(\frac{20}{1000}\right) = 0.0002 \text{ m}^2$$

The normal stresses in the three segments are

$$\sigma_{AB} = \frac{F_1}{A} = \frac{-20 \text{ kN}}{0.0002 \text{ m}^2} = -100,000 \text{ kN/m}^2 = 100 \text{ MPa (comp.)}$$

$$\sigma_{BC} = \frac{F_2}{A} = \frac{8 \text{ kN}}{0.0002 \text{ m}^2} = 40,000 \text{ kN/m}^2 = 40 \text{ MPa (tensile)}$$

$$\sigma_{CD} = \frac{F_3}{A} = \frac{32 \text{ kN}}{0.0002 \text{ m}^2} = 160,000 \text{ kN/m}^2 = 160 \text{ MPa (tensile)}$$

Example 7.2

A load of 500 kN is supported by a rod and cable (Fig. 7.3(a)). Neglecting the weight of the rod, determine the normal stresses in the rod and cables.

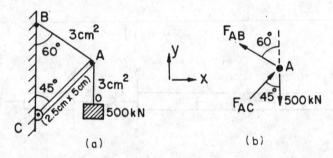

Fig. 7.3 (a) Load supported by a rod and cable and (b) Free body diagram of the system at A.

Solution

Rod AC is a two force member, hence the forces exerted on the rod AC must be along the axis of the rod. The tension in the cable AO is 500 kN.

To determine the axial forces in the cable AB and rod BC, we consider the equilibrium of the joint A and draw the free body diagram (Fig. 7.3(b)).

Writing up the equilibrium of forces along X and Y direction gives

$$\sum F_X = 0 = F_{AC} \cos 45° - F_{AB} \cos 30°$$

$$\sum F_Y = 0 = F_{AC} \sin 45° + F_{AB} \cos 60° - 500 \text{ kN}$$

From which

$$F_{AB} = 366.03 \text{ kN}, \quad F_{AC} = 448.02 \text{ kN}$$

Therefore, the normal stresses in the rod and in the cables are

$$\sigma_{AB} = \frac{366.03}{3 \times 10^{-4}} = 122.01 \times 10^4 \text{ N/m}^2 \text{ (tensile)}$$

$$\sigma_{AC} = \frac{366.03}{12.5 \times 10^{-4}} = 29.28 \times 10^4 \text{ N/m}^2 \text{ (compressive)}$$

$$\sigma_{AO} = \frac{500}{3 \times 10^{-4}} = 166 \times 10^4 \text{ N/m}^2 \text{ (tensile)}$$

Example 7.3

Figure 7.4 shows a 50 mm diameter shaft. The torque is transmitted to the shaft by means of a shear key that has a cross-section of 15 mm × 15 mm and a length of 25 mm. Determine the shear stresses developed in the key.

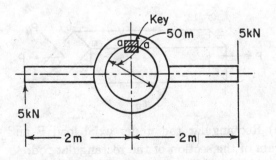

Fig. 7.4 Shaft with a key.

Solution

The magnitude of the torque T, transmitted to the shaft is

$$T = 5 \text{ kN} (4) = 20 \text{ kNm}$$

The shearing stress τ, developed at the key cross-section has to balance this torque T. The resisting torque is equal to

$$\tau \left(\frac{15}{1000} \right) \left(\frac{25}{1000} \right) \frac{50}{2} \frac{1}{1000} = 9.375 \times 10^{-6} \tau$$

Therefore,

$$9.375 \times 10^{-6} \tau = 20 \times 10^3$$
$$\tau = 2.133 \text{ GPa}$$

7.2.1 Stresses on an Inclined Plane

Figure 7.5(a) shows a rectangular rod subjected to an axial force P. In it abcd is a plane inclined at an angle α and A is the cross-sectional area of the rod.

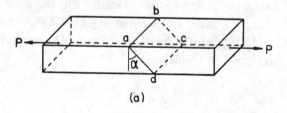

(a)

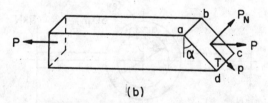

(b)

Fig. 7.5 (a) Rectangular rod under axial load P and (b) Free body diagram of the section of the rectangular rod.

Figure 7.5(b) shows the free body diagram and the forces acting on the inclined plane. The resultant force P acting on the section abcd, can be resolved into two components, one acting normal to the cross-section and the other along the cross–section.

Component normal to the inclined section $= P_N = P \cos \alpha$

Component along the inclined section $= P_T = P \sin \alpha$

The area of the inclined cross-section is $A/\cos \alpha$. Therefore, the normal stress σ is

$$\sigma = \frac{P \cos \alpha}{\frac{A}{\cos \alpha}} = \frac{P}{A} \cos^2 \alpha = \frac{P}{2A}(1 + \cos 2\alpha)$$

and the shear stress acting on the plane is

$$\tau = \frac{P \sin \alpha}{\frac{A}{\cos \alpha}} = \frac{P}{A} \cos \alpha \sin \alpha = \frac{P}{2A} \sin 2\alpha \qquad (7.5)$$

The maximum value of normal stess σ is obtained when $\alpha = 0°$ or $180°$. Similarly, the maximum value of τ is obtained when $\alpha = 45°$ or $135°$.

7.3 Strain

7.3.1 Normal Strain

A load carrying member deforms under the influence of the applied forces. The amount of deformation depends on the magnitude of the stress induced in the member and also on the mechanical properties of the material of the structural member.

There are three types of deformation. An increase in length of the member is called *elongation*. A decrease or shortening in length is called *contraction*. An angular distrotion is called *shear deformation*.

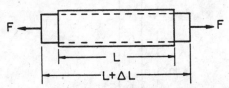

Fig. 7.6 Rectangular rod subjected to axial load.

Consider a rod of length L, subjected to an axial tensile force F. After application of the load, the rod undergoes some deformation and the length will become $L + \Delta L$ (Fig. 7.6). The axial elongation, in the direction of the applied load is equal to ΔL. Strain is defined as the change in length per unit length of the rod. For the given situation the average normal strain, $\epsilon_a v$ is defined as

$$\epsilon_a v = \frac{L + \Delta L - L}{L} = \frac{\Delta L}{L} \tag{7.6}$$

Strain is a dimensionless quantity, but it is customary to refer axial strains as mm/mm or m/m.

In many cases the deformation may be non-uniform along the length of the member. A good example of this is a rod hanging under its own weight. In such a case, the actual strain at a point may be different from the average strain.

Example 7.4

A rod is subjected to an axial force F (Fig. 7.7). The lengths between the points A, B, C, and D after the application of the force are also shown in the figure. The distance between the points before the application of the force is 10cm. Determine the axial strains between points $A - B$, $B - C$ and $C - D$. Also obtain the average axial strain in the rod.

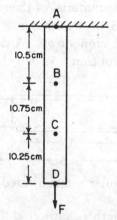

Fig. 7.7 Rod under axial force.

Solution

The axial strains between the points $A - B$, $B - C$, and $C - D$ are

$$\epsilon_{AB} = \frac{10.5 - 10}{10} = 0.05 \text{ cm/cm}$$

$$\epsilon_{BC} = \frac{10.75 - 10}{10} = 0.075 \text{ cm/cm}$$

$$\epsilon_{AB} = \frac{10.25 - 10}{10} = 0.025 \text{ cm/cm}$$

However, average axial strain of the rod is,

$$\epsilon_{av} = \frac{31.5 - 30}{30} = 0.05 \text{ cm/cm}$$

Since

$$\text{Initial length} = 30 \text{ cm}$$
$$\text{Final length} = (10.5 + 10.75 + 10.25) = 31.5 \text{ cm}$$

7.3.2 Shear Strain

Let us now consider the deformation produced under the action of a shear force. In this case, there is a change in the geometry of the body (Fig. 7.8). The solid and broken lines represent the initial and deformed shapes of the body.

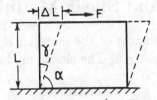

Fig. 7.8 Deformed configuration of a rectangular section under shear.

Shearing strain is defined as,

$$\tan \gamma = \frac{\Delta L}{L}$$

Since the deformation is very small, $\tan \gamma = \gamma$ and

$$\gamma_{av} = \frac{\Delta L}{L} \qquad (7.7)$$

The angle γ is measured in radians. Only very small values of shearing strain are observed in practical problems and thus the dimensions of the element are virtually unchanged. As in the case of normal strains, the shearing strain may be non-uniform. The shearing strain γ is also defined as,

$$\gamma = \frac{\pi}{2} - \alpha \qquad (7.8)$$

7.4 Poisson's Ratio

The tensile force causes an elongation in the direction of the load (Fig. 7.6). At the same time, the width of the bar gets shortened. Thus, there is a simultaneous elongation along longitudinal direction and contraction in the lateral direction. The ratio of the lateral strain to the longitudinal strain is termed as *Poisson's ratio*, and is a property of the material of the rod. It is defined by a symbol ν. For most of the metallic material, ν is approximately 0.33. For rubber like material, it is of the order 0.5.

7.5 Elastic and Shear Moduli

A measure of the stiffness of the material is represented by the ratio of stress to the strain. The elastic modulus E, is defined as

$$E = \frac{\text{normal stress}}{\text{normal strain}} = \frac{\sigma}{\epsilon} \qquad (7.9)$$

In a similar manner, the shearing stress to shearing strain is termed as shear modulus G.

$$G = \frac{\text{shear stress}}{\text{shear strain}} = \frac{\tau}{\gamma} \qquad (7.10)$$

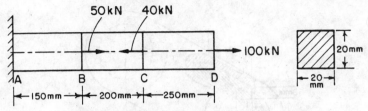

Fig. 7.9 Bar under axial loads.

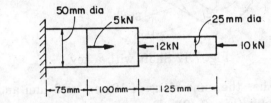

Fig. 7.10 Stepped circular rod under axial loads.

Problems

7.1 A bar carries a series of axial loads (Fig. 7.9). Compute the stress in each segment of the bar.

7.2 For the stepped circular rod (Fig. 7.10), obtain the stresses in each segment of the bar.

7.3 Determine the deformation of the bar at each segment and total deformation between segments A and D. The bar is of rectangular cross-section of 10 mm × 25 mm and is subjected to the axial loads (Fig. 7.11). The modulus of elasticity E of the bar material is 210 GPa.

7.4 A circular steel bar 0.5 m long is clamped at one end and subjected to an axial tensile load of 20 kN. Determine the required diameter of the bar if the allowable tensile stress is 138 MPa and the total deformation is limited to 0.15 mm. The modulus of elasticity of the steel is 210 GPa.

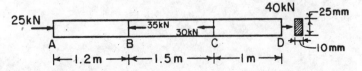

Fig. 7.11 Rectangular bar under axial load.

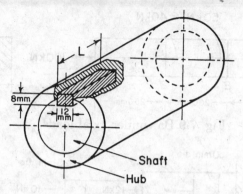

Fig. 7.12 Shaft with a key.

7.5 A key has the dimensions of 12 mm × 8 mm and length of 25 mm (Fig. 7.12). Determine the shear stress in the key when a torque 100 Nm is transmitted from the 40 mm diameter shaft to the hub.

8. GENERALIZED HOOKE'S LAW

8.1 Introduction

In Chapter 7, we discussed uniaxial stresses, both normal and shear, and the associated strains acting at a point in the cross-section. It was also pointed out that in an inclined plane under the action of axial load, both normal and shear stresses act at a point. In general, in a body the state of stress or strain is triaxial and can be easily described with reference to three mutually perpendicular planes. Depending on the nature of loading and the relative dimensions of the body, the state of stress and strain can either be represented in 3D, 2D or 1D.

All structural material possess to a certain extent the property of elasticity, i.e., if the external forces acting on the body causing the deformation are removed, the structure will return back to its original state. In this chapter, we assume that: (a) all materials are linearly elastic in their behavior up to a certain stress level; (b) the material is continuous and does not have any inclusions or voids and (c) at macro level the elastic body is homogenous so that the smallest elements cut from the body possess the same physical properties as the body. To simplify the discussions it will also be assumed that the body is *isotropic*, i.e., the elastic properties are same in all directions. Man-made materials like fiber reinforced composites, wood etc., are not isotropic. Their properties depend on the orientation of fibers in the fiber reinforced composites or orientations of the grains in the case of wood. Thus, the properties are direction dependent. Such a material is termed as *anisotropic* and in specific cases *orthotropic*, i.e., its properties change in two orthogonal directions. We shall talk about this towards the end of this chapter.

8.2   Three-dimensional Stress

Figure 8.1 shows an element of an elastic body described with respect to three mutually perpendicular axes X,Y and Z. The planes perpendicular to X,Y and Z axes are referred to as X-plane (plane ABCD and OFEG), Y-plane (planes CBFE and OGDA) and Z-plane (planes GDCE and OABF), respectively. The shear stress acting on a plane are resolved into two perpendicular components (Fig. 8.1). The stresses are acting on the positive face of the element and are shown as positive. Subscripts to σ and τ indicate the direction of the plane on which these stresses are acting.

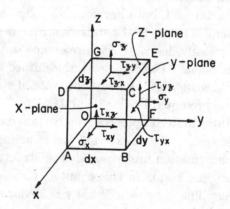

Fig. 8.1 Stresses acting on 3-D elastic body.

For example, the subscript x to σ indicates that the stress is acting on a plane normal to X-axis. The normal stress is taken as positive when it produces tensile stresses and negative when it produces compressive stresses. Two subscripts are used to indicate the shearing stress τ, the first indicating the direction of the normal to the plane under consideration and the second indicating the direction of the component of stress. For example,

τ_{xy} indicates that the shear stress is acting on the X-plane and along Y direction. Here again, the shear stresses are taken as positive along positive X,Y and Z axes. These stresses will be positive on the negative face along the negative X,Y and Z directions. Thus, at a given point in the body there are nine stress components. These are three normal stresses σ_x, σ_y and σ_z and six shear stresses τ_{xy}, τ_{yx}, τ_{xz}, τ_{zx}, τ_{yz} and τ_{zy}.

Before going further, let us consider the kind of forces that may act on bodies. Forces distributed over the surface of the body, such as pressure of one body over the other or hydrostatic pressure are called *surface forces*. Forces distributed over the volume of the body, such as gravitational forces, magnetic forces or in the case of body in motion, inertia forces are called *body forces*. The surface forces per unit area are resolved into three components parallel to X,Y and Z axes. Similarly, the body forces per unit volume are resolved into three components.

It can be easily be shown that all the six components of shear stresses are not independent, by taking moment of all the surface and body about X, Y and Z axes. The moment produced by the body forces such as the weight of the element reduces as the dimension of the element is reduced, as it is proportional to the cube of the linear dimensions. The moment produced by the surface forces is proportional to the square of the linear dimensions. One can therefore, neglect the moment produces by the body forces (gravity forces) as compared to the surface forces. Hence,

$$\tau_{xy} = \tau_{yx}, \ \tau_{xz} = \tau_{zx}, \ \tau_{yz} = \tau_{zy} \qquad (8.1)$$

Thus, at a given point in a body there are six unknown stress components; these will be called the state of stress at a point in a three dimensional body.

8.3 Three-dimensional Strain

While discussing the deformation of an elastic body we assume that: (a) enough constraints have been introduced to prevent the body from rigid body motion, so that no displacements of particles of the body are possible without its deformation and (b) only small deformations are possible. The small displacements of particles of a deformed body is resolved into three components u,v and w along the three coordinate axes X,Y and Z, respectively.

Consider a small element *dxdydz* of an elastic body (Fig. 8.1). If the body undergoes deformation and u, v and w are the components of the displacement of the point O, the displacement in the X-direction of the neighboring point A on the X-axis is $u + \frac{\partial u}{\partial x}dx$. As we move from point O to point A, the displacement increases by an amount $\frac{\partial u}{\partial x}dx$. The increase in length of the element OA is therefore,

$$u + \frac{\partial u}{\partial x}dx - u = \frac{\partial u}{\partial x}dx \qquad (8.2)$$

This increase has taken place over a length dx. Hence, ϵ_x the normal strain along X-direction is $\frac{\partial u}{\partial x}$. Similarly, one can define normal strains along Y and Z directions as $\frac{\partial v}{\partial y}$ and $\frac{\partial w}{\partial z}$, respectively. The partial derivatives are used since u, v and w are in general functions of x,y and z.

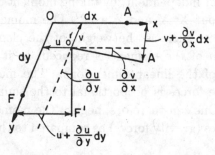

Fig. 8.2 Deformed configuration of a rectangular element.

As stated in Eq. (7.8), the shearing strain is defined as the change in the angle from 90°. In order to determine the shearing strains, consider the change in the angle between the lines OA and OF of Fig. 8.2. The displacement of point A in Y-direction and displacement of point F in X-direction are $\frac{\partial v}{\partial x}dx$ and $\frac{\partial u}{\partial y}dy$, respectively. Due to these displacements, the new direction of O'A' of the element OA is inclined to the initial direction by a small angle $\frac{\partial v}{\partial x}$. In a similar manner, the inclination of O'F' to OF is $\frac{\partial u}{\partial y}$. Thus the original angle OAF which was 90° is reduced by an amount equal to $\frac{\partial u}{\partial y} + \frac{\partial v}{\partial x}$. This is the *shearing strain* between the planes X and Z. Following similar arguments, the shearing strains between the planes Y and Z and the planes X and Z are obtained. These are

$$\frac{\partial w}{\partial y} + \frac{\partial v}{\partial z} \quad \text{and} \quad \frac{\partial u}{\partial z} + \frac{\partial w}{\partial x}$$

The six strain components are therefore,

$$\epsilon_x = \frac{\partial u}{\partial x}, \quad \epsilon_y = \frac{\partial v}{\partial y}, \quad \epsilon_z = \frac{\partial w}{\partial z} \qquad (8.3a)$$

$$\gamma_{xy} = \frac{\partial u}{\partial y} + \frac{\partial v}{\partial x}, \quad \gamma_{xz} = \frac{\partial u}{\partial z} + \frac{\partial w}{\partial x}, \quad \gamma_{yz} = \frac{\partial w}{\partial y} + \frac{\partial v}{\partial z} \qquad (8.3b)$$

8.4   Generalized Hooke's law

The relations between the components of stress and the components of strain have been established experimentally and are known as Hooke's law. Experiments have shown that in the case of isotropic material, the normal stresses do not produce any distortion of the element, only the sides of the element elongate and contract. Consider a cube of isotropic material (Fig. 8.3).

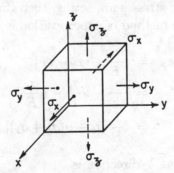

Fig. 8.3 A cubic element subjected to normal stresses.

Since the deformation is assumed to be small, principle of superposition holds good, i.e., we can apply one stress at a time and find the total strain by adding up the individual strains. Let us assume $\sigma_x \neq 0$ and $\sigma_y = 0$, $\sigma_z = 0$, then it is observed that the extension along X-direction is accompanied by a lateral contraction. From Section 7.4, the Poisson's ratio is

$$\nu = -\frac{\text{lateral strain}}{\text{longitudinal strain}} \qquad (8.4)$$

The axial strain in the X-direction due to σ_x alone is σ_x/E. However, this axial strain will be reduced due to the action of stress σ_y in the Y-direction and due to σ_z in the Z-direction, when applied. Due to σ_x alone, the lateral strain developed along Y and Z directions are equal and is given by

$$\epsilon_y = \epsilon_z = -\nu\epsilon_x \qquad (8.5)$$

Similarly, when $\sigma_y \neq 0$ and $\sigma_x = 0$, $\sigma_z = 0$, the axial strain along Y-direction is σ_y/E and the lateral strains along X and Z directions are

$$\epsilon_x = \epsilon_y = -\nu\epsilon_y \qquad (8.6)$$

When $\sigma_z \neq 0$ and $\sigma_x = 0$, $\sigma_y = 0$, the axial strain along Z-direction is σ_z/E and the lateral strains along X and Y directions are

$$\epsilon_x = \epsilon_y = -\nu\epsilon_z \qquad (8.7)$$

When all the three stresses are acting, then the net strain in the X-direction by the method of superposition is,

$$\begin{aligned}
\epsilon_x &= \frac{\sigma_x}{E} - \nu\epsilon_y - \nu\epsilon_z \\
&= \frac{\sigma_x}{E} - \nu\frac{\sigma_y}{E} - \nu\frac{\sigma_z}{E} \\
&= \frac{1}{E}\left[\sigma_x - \nu(\sigma_y + \sigma_z)\right]
\end{aligned} \qquad (8.8)$$

The net strain along Y-direction is

$$\begin{aligned}
\epsilon_y &= \frac{\sigma_y}{E} - \nu\epsilon_z - \nu\epsilon_x \\
&= \frac{\sigma_y}{E} - \nu\frac{\sigma_z}{E} - \nu\frac{\sigma_x}{E} \\
&= \frac{1}{E}\left[\sigma_y - \nu(\sigma_z + \sigma_x)\right]
\end{aligned} \qquad (8.9)$$

Similarly, the net strain along Z-direction is

$$\epsilon_z = \frac{\sigma_z}{E} - \nu\epsilon_x - \nu\epsilon_y$$
$$= \frac{\sigma_z}{E} - \nu\frac{\sigma_x}{E} - \nu\frac{\sigma_y}{E}$$
$$= \frac{1}{E}\left[\sigma_z - \nu(\sigma_x + \sigma_x)\right] \qquad (8.10)$$

Equations (8.8) to (8.10) are strain-stress relationship for a 3D isotropic elastic material. In many analysis, we may need stress-strain relations. The above equations can be rewritten as

$$\sigma_x = \frac{E}{1-\nu^2}\left[\epsilon_x + \nu(\epsilon_y + \epsilon_z)\right]$$
$$\sigma_y = \frac{E}{1-\nu^2}\left[\epsilon_y + \nu(\epsilon_x + \epsilon_z)\right] \qquad (8.11)$$
$$\sigma_z = \frac{E}{1-\nu^2}\left[\epsilon_z + \nu(\epsilon_x + \epsilon_y)\right]$$

Equations (8.8) to (8.11) show that the relation between normal strains and normal stresses are completely defined by two material constants E and ν. The same constants can be used to define the relations between shearing strains and stresses. In order to understand this, consider the deformation of a rectangular parallelopiped in which $\sigma_y = -\sigma_z$ and $\sigma_x = 0$ as shown in Fig. 8.4. Cutting out an element ABCD parallel to X-axis and at 45° to

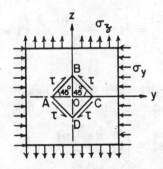

Fig. 8.4 Rectangular element subjected to biaxial state of stress.

the Y and Z axes, and summing up all the forces along and normal to the inclined plane BC, we get the normal stress to be zero and the shearing stress τ, which is along CB as

$$\tau = \frac{1}{2}(\sigma_z - \sigma_y) \tag{8.12}$$

Substituting the value of σ_y from Eq. (8.12) gives

$$\tau = \sigma_z \tag{8.13}$$

This is called *pure shear* condition. In this situation, the lengths of the element AB and BC donot change. However, the angle between them is not preserved. The change in the angle which corresponds to the magnitude of the shearing strain γ can be found from the triangle OBC. Now

$$\frac{OC}{OB} = \tan\left(\frac{\pi}{4} - \frac{\gamma}{2}\right) = \frac{1 + \epsilon_y}{1 + \epsilon_z} \tag{8.14}$$

Substituting for ϵ_y and ϵ_z from Eqs. (8.9) and (8.10) respectively with $\sigma_x = 0$,

$$\tan\left(\frac{\pi}{4} - \frac{\gamma}{2}\right) = \frac{\tan\frac{\pi}{4} - \tan\frac{\gamma}{2}}{1 + \tan\frac{\pi}{4}\tan\frac{\gamma}{2}} = \frac{1 - \frac{\gamma}{2}}{1 + \frac{\gamma}{2}} \tag{8.15}$$

since $\tan\gamma = \gamma$ for small values of γ. Hence,

$$\frac{1 - \frac{\gamma}{2}}{1 + \frac{\gamma}{2}} = \frac{1 + \frac{(1+\nu)\sigma_z}{E}}{1 - \frac{(1+\nu)\sigma_z}{E}}$$

$$\left(1 - \frac{\gamma}{2}\right)\left\{1 + \frac{\gamma}{2}\right\}^{-1} = \left[1 + \frac{(1+\nu)}{E}\sigma_z\right]\left\{1 - \frac{(1+\nu)}{E}\sigma_z\right\}^{-1} \tag{8.16}$$

Expanding the quantities within the flower bracket by Taylor series expansion and retaining only the first order terms gives

$$\left(1 - \frac{\gamma}{2}\right)\left(1 - \frac{\gamma}{2}\right) = \left[1 + \frac{(1+\nu)}{E}\sigma_z\right]\left[1 + \frac{(1+\nu)}{E}\sigma_z\right] \tag{8.17}$$

Equation (8.17), when further simplified results in

$$\gamma = \frac{2(1+\nu)}{E}\sigma_z \tag{8.18}$$

Substituting for σ_z from Eq. (8.13) gives

$$\gamma = \frac{2(1+\nu)}{E}\tau \tag{8.19}$$

Thus, the relation between shearing strain and shearing stress can be expressed in terms of ν and E. In terms of notation,

$$G = \frac{E}{2(1+\nu)} \tag{8.20}$$

and Eq. (8.19) becomes

$$\gamma = \frac{\tau}{G} \tag{8.21}$$

G is called the *shear modulus or modulus of rigidity*. Since the shearing strains are independent of normal strains (for small deformations), the shearing stresses give rise to only shearing strains and the relations are given by

$$\gamma_{xy} = \frac{\tau_{xy}}{G}, \quad \gamma_{yz} = \frac{\tau_{yz}}{G}, \quad \gamma_{xz} = \frac{\tau_{xz}}{G} \tag{8.22}$$

Thus for a 3D elastic, isotropic body, the strain-stress relations or constitutive relations are given by Eqs. (8.8), (8.9), (8.10) and (8.12). For the sake of clarity, they are reproduced below:

$$\left.\begin{aligned}
\epsilon_x &= \frac{1}{E}\left[\sigma_x - \nu(\sigma_y + \sigma_z)\right] \\[4pt]
\epsilon_y &= \frac{1}{E}\left[\sigma_y - \nu(\sigma_x + \sigma_z)\right] \\[4pt]
\epsilon_z &= \frac{1}{E}\left[\sigma_z - \nu(\sigma_x + \sigma_y)\right] \\[4pt]
\gamma_{xy} &= \frac{\tau_{xy}}{G} \\[4pt]
\gamma_{yz} &= \frac{\tau_{yz}}{G} \\[4pt]
\gamma_{xz} &= \frac{\tau_{xz}}{G}
\end{aligned}\right\} \tag{8.23}$$

In an alternate form, they are given as

$$\sigma_x = \frac{E}{1-\nu^2}\left[\epsilon_x + \nu(\epsilon_y + \epsilon_z)\right]$$

$$\sigma_y = \frac{E}{1-\nu^2}\left[\epsilon_y + \nu(\epsilon_x + \epsilon_x)\right]$$

$$\sigma_z = \frac{E}{1-\nu^2}\left[\epsilon_z + \nu(\epsilon_x + \epsilon_y)\right]$$

$$\tau_{xy} = G\gamma_{xy} = \frac{E}{2(1+\nu)}\,\gamma_{xy}$$

$$\tau_{yz} = G\gamma_{yz} = \frac{E}{2(1+\nu)}\,\gamma_{yz}$$

$$\tau_{xz} = G\gamma_{xz} = \frac{E}{2(1+\nu)}\,\gamma_{xz}$$

$$(8.24)$$

8.5   Plane Stress and Plane Strain

In many practical situations, the 3D problem can be reduced to 2D plane stress or 2D plane strain problems depending on the geometry of the structure and the loading. This reduces the complexity of the problems and also the computation time required for analyzing the problems.

8.5.1   Plane stress

If a thin plate is subjected to a system of uniform forces along the periphery (Fig. 8.5), the stress components σ_z, τ_{yz}, τ_{xz} are zero on both faces of the plate. One can safely assume that they are zero everywhere inside the plate. The state of stress at any point inside the plate is specified by σ_x, σ_y and τ_{xy} and these components are function of x and y only. For such a case, the stress-strain relation Eqs. (8.23 and 8.24) become

$$\epsilon_x = \frac{1}{E}\left[\sigma_x - \nu\sigma_y\right]$$

$$\epsilon_y = \frac{1}{E}\left[\sigma_y - \nu\sigma_x\right]$$

$$\epsilon_z = -\frac{\nu}{E}\left[\sigma_x + \sigma_y\right]$$

$$\gamma_{xy} = \frac{\tau_{xy}}{G}$$

$$(8.25)$$

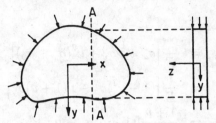

Fig. 8.5 Thin plate under in-plane loading.

and

$$\left.\begin{array}{l} \sigma_x = \frac{E}{1-\nu^2}\left[\epsilon_x + \nu(\epsilon_y + \epsilon_z)\right] \\[2mm] \sigma_y = \frac{E}{1-\nu^2}\left[\epsilon_y + \nu(\epsilon_x + \epsilon_z)\right] \\[2mm] \sigma_z = 0 \\[2mm] \tau_{xy} = G\gamma_{xy} = \frac{E}{2(1+\nu)}\,\gamma_{xy} \end{array}\right\} \qquad (8.26)$$

Thus, the problem is two-dimensional in terms of stresses and three-dimensional in terms of strains, since ϵ_z exists (Eq. 8.25).

8.5.2 Plane strain

Consider an elastic body in X, Y and Z coordinate frame. Let us assume that the dimension of the body in the direction is very large as compared to the other two dimensions, e.g., a long uniform cylinder whose axis coincides with Z-axis. It is loaded by forces which are perpendicular to the axis and do not vary along the length of the cylinder. It can, therefore, be safely assumed that all cross-sections are in the same state of stress. Since there is no axial displacement at the ends, and, by symmetry at the mid-section, it may be assumed that the same is valid at every cross-section. Some of the practical examples are; a cylindrical tube under internal pressure, a tunnel etc.

For such a case,

$$\epsilon_z = 0, \quad \gamma_{xz} = 0, \quad \gamma_{yz} = 0 \qquad (8.27)$$

The stress-strain relation Eqs. (8.23) and (8.24) get modified for this case, since $\epsilon_z = 0$ and Eq. (8.23) gives

$$\sigma_z = \nu(\sigma_x + \sigma_y) \qquad (8.28)$$

Therefore,

$$\epsilon_x = \frac{1}{E}\left[\sigma_x - \nu\sigma_y + \nu^2(\sigma_x + \sigma_y)\right]$$

$$= \frac{1}{E}\left[(1-\nu^2)\sigma_x - \nu(1+\nu)\sigma_y\right]$$

$$= \frac{1+\nu}{E}\left[(1-\nu)\sigma_x - \nu\sigma_y\right] \qquad (8.29)$$

$$\epsilon_y = \frac{1}{E}\left[(1-\nu^2)\sigma_y - \nu(1+\nu)\sigma_x\right]$$

$$= \frac{1+\nu}{E}\left[(1-\nu)\sigma_y - \nu\sigma_x\right] \qquad (8.30)$$

$$\left.\begin{array}{l} \sigma_x = \frac{E}{1-\nu^2}\left[\epsilon_x + \nu\epsilon_y\right] \\[2mm] \sigma_y = \frac{E}{1-\nu^2}\left[\epsilon_y + \nu\epsilon_x\right] \\[2mm] \sigma_z = \frac{E\nu}{1-\nu^2}\left[\epsilon_x + \epsilon_y\right] \end{array}\right\} \qquad (8.31)$$

Although the problem is 2D in terms of strains, it is 3D in terms of stresses.

Thus, it is seen that the constitutive equations (stress-strain equations) for plane stress and plane strain idealizations of 3D problems are different.

8.6 Lames' Coefficient and Relation with Engineering Constants

Adding the three components of the normal strains given by Eq. (8.25), and using the notations,

$$e = \epsilon_x + \epsilon_y + \epsilon_z$$

$$\Theta = \sigma_x + \sigma_y + \sigma_z \qquad (8.32)$$

we get following relation between the volume expansion e and sum of normal stress Θ as

$$e = \frac{1-2\nu}{E}\Theta \qquad (8.33)$$

Using Eqs. (8.32) and (8.25) for normal strains, the normal stresses in terms of normal strains are

$$\left.\begin{array}{l} \sigma_x = \frac{\nu E}{(1+\nu)(1-2\nu)}e + \frac{E}{1+\nu}\epsilon_x \\[2mm] \sigma_y = \frac{\nu E}{(1+\nu)(1-2\nu)}e + \frac{E}{1+\nu}\epsilon_y \\[2mm] \sigma_z = \frac{\nu E}{(1+\nu)(1-2\nu)}e + \frac{E}{1+\nu}\epsilon_z \end{array}\right\} \quad (8.34)$$

Using the notation,

$$\lambda = \frac{\nu E}{(1+\nu)(1-2\nu)} \quad (8.35)$$

and Eq. (8.20), Eq. (8.34) become

$$\left.\begin{array}{l} \sigma_x = \lambda e + 2G\epsilon_x \\ \sigma_y = \lambda e + 2G\epsilon_y \\ \sigma_z = \lambda e + 2G\epsilon_z \end{array}\right\} \quad (8.36)$$

Thus the stress-strain relations for isotropic elastic material can be written in terms of two constants λ and G. λ is called *Lame's coefficient* and is related to the engineering constants E, ν.

8.7 Thermal Strains

Most materials expand when temperature increases and contract when temperature decreases. When no restrictions are placed on dimensional changes, the dimensions of the structural part changes but no stresses will be developed. However, in cases where the deformations are restrained, stresses will generate which could be significant.

Different materials expand at different rates when subjected to temperature changes. The property of the material that indicates a change due to temperature is its coefficient of thermal expansion. The symbol α_T denotes this coefficient. For isotropic material α_T is independent of direction. It is a measure of the change in length of a material per unit length for a one degree change in temperature. The units of α_T are m/(m°C) or °C^{-1}.

Let the deformation due to the change in temperature ΔT over the length L be δ_{Th}. The ratio δ_{Th}/L is called *thermal strain* (ϵ_{Th}). The thermal strain is proportional to the change in temperature ΔT. Thus,

$$\epsilon_{Th} \propto \Delta T \qquad (8.37)$$

Defining the constant of proportionality as α_T,

$$\epsilon_{Th} = \alpha_T \Delta T \qquad (8.38)$$

If we heat a rectangular plate, its length, breath and thickness will expand but the shape remains unchanged. The thermal strains can be evaluated as

$$\left.\begin{aligned} \epsilon_x^{Th} &= \alpha_T \Delta T \\[6pt] \epsilon_y^{Th} &= \alpha_T \Delta T \\[6pt] \epsilon_z^{Th} &= \alpha_T \Delta T \end{aligned}\right\} \qquad (8.39)$$

$$\left.\begin{aligned} \gamma_{xy}^{Th} &= 0 \\[6pt] \gamma_{xy}^{Th} &= 0 \\[6pt] \gamma_{xy}^{Th} &= 0 \end{aligned}\right\} \qquad (8.40)$$

The components of strain due to externally applied load and change of temperature can now be written as

$$\left.\begin{aligned} \epsilon_x &= \tfrac{1}{E}\left[\sigma_x - \nu(\sigma_y + \sigma_z)\right] + \alpha_T \Delta T \\[6pt] \epsilon_y &= \tfrac{1}{E}\left[\sigma_y - \nu(\sigma_x + \sigma_z)\right] + \alpha_T \Delta T \\[6pt] \epsilon_z &= \tfrac{1}{E}\left[\sigma_z - \nu(\sigma_x + \sigma_y)\right] + \alpha_T \Delta T \\[6pt] \gamma_{xy} &= \tfrac{\tau_{xy}}{G} \\[6pt] \gamma_{yz} &= \tfrac{\tau_{yz}}{G} \\[6pt] \gamma_{xz} &= \tfrac{\tau_{xz}}{G} \end{aligned}\right\} \qquad (8.41)$$

The constitutive relations for 2D plane stress and plain strain problems can be written on the lines discussed in Section 8.5.

Example 8.1

A thin steel plate is subjected to a biaxial state of stress (Fig. 8.6). The axial strain due to the stresses are, $\epsilon_x = 0.6 \times 10^{-4}$, $\epsilon_y = -0.3 \times 10^{-4}$. Assuming that $E_s = 210$ GPa and $\nu = 0.3$, determine σ_x and σ_y.

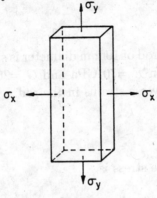

Fig. 8.6

Solution

This is basically a two dimensional plane stress problem with $\sigma_z = 0$. From Eq. (8.25),

$$\epsilon_x = \frac{1}{E}\left[\sigma_x - \nu\sigma_y\right]$$
$$\epsilon_y = \frac{1}{E}\left[\sigma_y - \nu\sigma_x\right]$$

which gives

$$\sigma_x = \frac{E}{1 - \nu^2}\left[\epsilon_x + \nu\epsilon_y\right]$$
$$\sigma_y = \frac{E}{1 - \nu^2}\left[\epsilon_y + \nu\epsilon_x\right]$$

Substituting the values of ϵ_x and ϵ_y,

$$\begin{aligned}
\sigma_x &= \frac{210 \times 10^9}{1 - 0.3^2}\left[0.6 \times 10^{-4} + 0.3(-0.3 \times 10^{-4})\right] \quad \text{Pa} \\
&= \frac{21}{0.91}[0.6 - 0.09] \quad \text{MPa} \\
&= 11.77 \quad \text{MPa}
\end{aligned}$$

$$\sigma_y = \frac{210 \times 10^9}{1 - 0.3^2}\left[-0.3 \times 10^{-4} + 0.3(0.6 \times 10^{-4})\right] \quad \text{Pa}$$

$$= \frac{21}{0.91}[-0.12] \quad \text{MPa}$$

$$= -2.77 \quad \text{MPa}$$

Example 8.2

An aluminum alloy rod of 10 mm diameter is subjected to an axial force of 6 kN. Given $E = 70$ GPa and $G = 26.3$ GPa, determine the axial and transverse strains in the rod.

Solution

Applied axial tensile stress is

$$\sigma_x = \frac{P}{A} = \frac{6 \text{ kN}}{\frac{\pi}{4}(0.01^2)} = 76400 \text{ kN/m}^2 = 76.4 \text{ MPa}$$

Since $\sigma_y = 0$, from Eq. (8.25)

$$\epsilon_x = \frac{\sigma_x}{E} = \frac{76.4 \text{ MPa}}{70 \text{ GPa}} = 0.00109 \quad \text{m/m}$$

From Eq. (8.20),

$$\nu = \frac{E}{2G} - 1 = \frac{70}{2 \times 26.3} - 1 = 0.33$$

Once again, from the second equation of (8.25),

$$\epsilon_y = -\nu\epsilon_x = -0.33 \times 0.00109 = -3.6 \times 10^{-4} \text{ m/m}$$

Example 8.3

A uniform cross-section bar is rigidly fixed at the two ends and maintained at temperature T_o. If the temperature is uniformly raised to T, calculate the stresses generated in the bar.

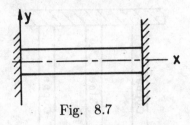

Fig. 8.7

Solution

Consider a uniform bar (Fig. 8.7). Since the movement along X-direction is constrained, $\epsilon_x = 0$. Further, the clamped ends can apply only σ_x and not σ_y and σ_z. From the first equation of (8.41),

$$\frac{\sigma_x}{E} + \alpha_T(T - T_o) = 0$$

or

$$\sigma_x = -E\alpha_T(T - T_o)$$

The stress generated along x is therefore compressive. If A is the cross-sectional area of the bar, the load

$$P = -AE\alpha_T(T - T_o)$$

From the second and third of Eq. (8.41),

$$\epsilon_y = -\nu\frac{\sigma_x}{E} + \alpha_T(T - T_o) = \alpha_T(1 + \nu)(T - T_o)$$

and

$$\epsilon_z = \alpha_T(1 + \nu)(T - T_o)$$

Further,

$$\gamma_{xy} = 0 \quad \gamma_{xy} = 0 \quad \gamma_{xy} = 0$$

Example 8.4

Three metal rods each of length 1 m and cross-sectional area 250 mm^2 are connected to the rigid end plates (Fig. 8.8). The outer rods are made of copper and the central rod is made of steel. The rods are parallel to each other, as shown, with no stresses at room temperature. Obtain the stresses in the rods if the assembly is subjected to a compressive load of 5 kN and the temperature of the assembly is raised by 25°C. Given $E_s = 210$ GPa, $E_c = 120$ GPa, $\alpha_{sT} = 12.0 \times 10^{-6}/°C$, $\alpha_{cT} = 18.5 \times 10^{-6}/°C$.

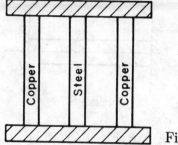

Fig. 8.8

Solution

Let the final distance between the plates be L meters. Only normal stress σ_x is developed in the rods. Therefore, the total strain ϵ is obtained as

$$\epsilon = \frac{\sigma}{E} + \alpha_T(T - T_o)$$

For copper rods,

$$\epsilon = \frac{L-1}{1} = \frac{\sigma}{120 \times 10^9} + 18.5 \times 10^{-6} \times 25$$

Hence,

$$\sigma = \left[(L-1) - 462.5 \times 10^{-6}\right] 120 \times 10^9$$

The load in the copper rod (two nos.) is

$$P_c = \left[(L-1) - 462.5 \times 10^{-6}\right] 120 \times 10^9 \times 2 \times 250 \times 10^{-6}$$

Similarly for the steel rod,

$$P_s = \left[(L-1) - 300 \times 10^{-6}\right] 210 \times 10^9 \times 250 \times 10^{-6}$$

From the equilibrium of forces,

$$F_s + F_c = -5000$$

(Negative sign, because load applied is compressive). Hence,

$$-5000 = \left\{ \left[(L-1) - 462.5 \times 10^{-6}\right] \times 60 \right.$$
$$\left. + \left[(L-1) - 300 \times 10^{-6}\right] \times 52.5 \right\} \times 10^{-6}$$
$$-5000 = (L-1) - 112.5 \times 10^{-6} - 43500$$
$$(L-1) = 343.75 \times 10^{-6} \text{ m}$$

Now,

$$F_c = (343.75 \times 10^{-6} - 462.5 \times 10^{-6}) \times 60 \times 10^6$$
$$= -7.14 \text{ kN}$$
$$F_s = (343.75 \times 10^{-6} - 300 \times 10^{-6}) \times 52.5 \times 10^6$$
$$= 2.297 \text{ kN}$$

Normal stress in copper rods is

$$\sigma_c = \frac{F_c}{2A_c} = \frac{-7.14 \times 10^3}{2 \times 250 \times 10^{-6}} = -14.28 \text{ MPa (Compressive)}$$

Normal stress in the steel rod is

$$\sigma_s = \frac{F_s}{2A_s} = \frac{2.297 \times 10^3}{2 \times 250 \times 10^{-6}} = 9.188 \text{ MPa (Tensile)}$$

Problems

8.1 A push rod in the valve mechanism of an automotive engine has a length of 200 mm. Compute the strain when temperature change from $-20°C$ to $140°C$. Given $\alpha_T = 11.2 \times 10^{-6}/°C$.

8.2 An aluminum alloy rod in a machine is held at its ends while being cooled from $100°C$. At what temperature will the tensile stress in the rod equal half of the yield strength of aluminum alloy if it is originally at zero stress. Given $E = 73$ GPa, $\alpha_T = 23 \times 10^{-6}/°$ C and $\sigma_{yield} = 414$ MPa.

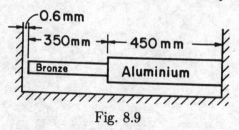

Fig. 8.9

8.3 Determine the forces developed in the bars (Fig. 8.9) after a temperature rise of $120°$ C. The material properties of bronze are $A_b = 1500$ mm^2, $E_b = 105$ GPa, $\alpha_{Tb} = 18 \times 10^{-6}/°$ C and that of aluminum are, $A_{al} = 1800$ mm^2, $E_{al} = 70$ GPa, $\alpha_{Tal} = 25 \times 10^{-6}/°$ C.

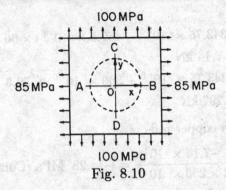

Fig. 8.10

8.4 A circle of diameter $d = 250$ mm is scribed on an unstressed square steel plate of thickness 20 mm (Fig. 8.10). The forces acting on the plate result in a stress $\sigma_x = 85$ MPa, $\sigma_y = 100$ MPa distributed uniformly across the edges of the plate. The elastic modulus $E_s = 210$ GPa and $\nu = 0.33$. Obtain the changes in the lengths of diameter AB and CD, thickness of the plate and the volume of the plate.

9.

TRANSFORMATION OF STRESSES AND STRAINS

9.1 Introduction

Chapter 8, observed that the most generalized state of stress at a given point in a body may be expressed in terms of the six components of stress viz., three normal σ_x, σ_y and σ_z; and three shear τ_{xy}, τ_{yz} and τ_{zx} stresses with respect to the Cartesian coordinate system x, y and z (Fig. 9.1(a)). If the coordinate system is rotated, the same state of stress will be represented by a different set of components as shown in Fig. 9.1(b). If the body moments are zero, then $\tau_{xy} = \tau_{yx}$, $\tau_{xz} = \tau_{zx}$ and $\tau_{yz} = \tau_{zy}$. In this section, we shall confine ourselves to the problem of plane stress in which the stresses are confined to the x-y plane, z-axis being normal to the x-y plane. Even though the problem is two-dimensional as far as the stresses are concerned, but with respect to strains the problem is three-dimensional as all the three normal stresses, ϵ_x, ϵ_y and ϵ_z exist.

9.2 Transformation of Plane Stress

When an element is subjected to normal and in plane shear stresses only (Fig. 9.2(a)), it is said to be in a state of plane stress.

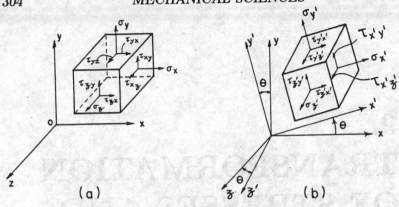

Fig. 9.1 Stresses on an 3D rectangular element.

Applied uniaxial stress, biaxial stress and pure shear stress are special cases of plane stress state.

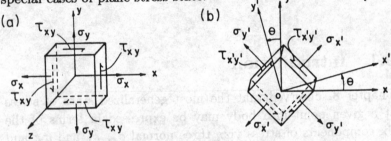

Fig. 9.2 Normal and shear stresses on a 2D element.

If now the coordinate axis is rotated (Fig. 9.2(b)), our interest is to relate the stresses $\sigma_{x'}$, $\sigma_{y'}$ and $\tau_{x'y'}$ in the element with respect to the new coordinate system to the stresses σ_x, σ_y and τ_{xy} referred to the old coordinate system. Fig. 9.2(b) shows the stresses acting on the element with respect to the x'-y' coordinates. In order to determine the normal stress $\sigma_{x'}$ and shear stress $\tau_{x'y'}$ exerted on the face normal to x'-axis, consider an element with faces normal to x, y and x' axes. If ΔA is the area of the inclined surface (Fig. 9.3), then the area of vertical and horizontal faces are $\Delta A \cos \theta$ and $\Delta A \sin \theta$, respectively. The equilibrium of forces along x' and y' directions gives

$$\Sigma F_{x'} = 0$$

or $\sigma_{x'}\Delta A - \sigma_x(\Delta A \cos \theta)\cos \theta - \tau_{xy}(\Delta A \cos \theta)\sin \theta$

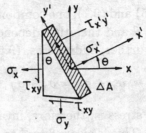

Fig. 9.3 Stresses on an inclined plane.

$$- \sigma_y(\Delta A \sin \theta) \sin \theta - \tau_{xy}(\Delta A \sin \theta) \cos \theta = 0 \qquad (9.1)$$

$$\Sigma F_{y'} = 0$$

or $\quad \tau_{x'y'}\Delta A + \sigma_x(\Delta A \cos \theta) \sin \theta - \tau_{xy}(\Delta A \cos \theta) \cos \theta$

$$- \sigma_y(\Delta A \sin \theta) \cos \theta - \tau_{xy}(\Delta A \sin \theta) \sin \theta = 0 \qquad (9.2)$$

Since $\Delta A \neq 0$, simplification gives

$$\sigma_{x'} = \sigma_x \cos^2\theta + \sigma_y \sin^2\theta + 2\,\tau_{xy} \sin \theta \cos \theta \qquad (9.3)$$

$$\tau_{x'}y' = (\sigma_y - \sigma_x) \sin \theta \cos \theta + \tau_{xy}(\cos^2\theta - \sin^2\theta) \qquad (9.4)$$

Making use of the trigonometric relations

$$\sin 2\theta = 2 \sin \theta \cos \theta; \quad \cos 2\theta = \cos^2\theta - \sin^2\theta$$

$$\cos^2\theta = \frac{1 + \cos 2\theta}{2}; \quad \sin^2\theta = \frac{1 - \cos 2\theta}{2}$$

Eqs. (9.3) and (9.4) can be rewritten as

$$\sigma_{x'} = \frac{\sigma_x + \sigma_y}{2} + \frac{\sigma_x - \sigma_y}{2} \cos 2\theta + \tau_{xy} \sin 2\theta \qquad (9.5)$$

$$\tau_{x'y'} = \frac{\sigma_x + \sigma_y}{2} \sin 2\theta + \tau_{xy} \cos 2\theta \qquad (9.6)$$

The relationship for $\sigma_{y'}$ can be obtained from Eq. (9.5) by replacing θ by $(\theta + 90°)$. Therefore

$$\sigma_{y'} = \frac{\sigma_x + \sigma_y}{2} - \frac{\sigma_x - \sigma_y}{2} \cos 2\theta - \tau_{xy} \sin 2\theta \qquad (9.7)$$

Thus, from Eqs. (9.5) and (9.7), one component of stress is invariant with respect to θ and the other depends on twice the angle between the axis. Further, adding Eqs. (9.5) and (9.7) gives

$$\sigma_{x'} + \sigma_{y'} = \sigma_x + \sigma_y \tag{9.8}$$

The sum of normal stress is, therefore, invariant with respect to rotation of the axis. This is also referred to as the First invariant of stress components.

9.2.1 Principal Stresses and Maximum Shear Stresses

From Eq. (9.6), it is possible to define planes oriented at an angle θ_P to the x-axis at which shear stresses are zero. Such planes are called *principal planes*. The value of θ_P of the parameter θ is obtained by setting $\tau_{x'y'} = 0$ in Eq. (9.6). Thus,

$$\tan 2\theta_P = \frac{2\tau_{xy}}{\sigma_x - \sigma_y} \tag{9.9}$$

This equation defines two values of $2\theta_P$ which are 180° apart, i.e., two values of θ_P separated by 90°. The stresses acting on these planes are called the *principal stresses* and the corresponding values will be σ_{max} and σ_{min}, respectively.

If we choose σ along x-axis and τ along y-axis and choose a point p with coordinates $(\sigma_{x'}, \tau_{x'y'})$ for any value of the parameter θ, all the points thus obtained will lie on a circle. To establish this, Eq. (9.5) is rewritten as

$$\sigma_{x'} - \frac{\sigma_x + \sigma_y}{2} = \frac{\sigma_x - \sigma_y}{2} \cos 2\theta + \tau_{xy} \sin 2\theta \tag{9.10}$$

Squaring both sides of Eqs. (9.6) and (9.10) and adding gives

$$\left(\sigma_{x'} - \frac{\sigma_x + \sigma_y}{2}\right)^2 + \tau_{x'y'}^2 = \left(\frac{\sigma_x - \sigma_y}{2}\right)^2 + \tau_{xy}^2 \tag{9.11}$$

Defining

$$\sigma_{av} = \frac{\sigma_x + \sigma_y}{2}; \quad R = \left[\frac{\sigma_x - \sigma_y}{2}^2 + \tau_{xy}^2\right]^{1/2} \tag{9.12}$$

Eq. (9.11) can be rewritten as

$$(\sigma_{x'} - \sigma_{av})^2 + \tau_{x'y'}^2 = R^2 \tag{9.13}$$

which represents a circle of radius R and center C at a distance of σ_{av} along σ-axis (Fig. 9.4).

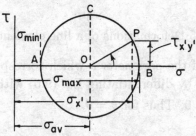

Fig. 9.4 Mohr's circle for 2D state of stress.

Points A and B, where the circle intersects the σ-axis, are of special interest. Point A corresponds to the minimum value of normal stress σ_x and point B corresponds to the maximum value of normal stress σ_x. Further, at these points the shear stress is zero. Thus, at any point among all the planes, there can be only two planes on which normal stresses exist and shear stress on these planes are zero. Such planes are called *principal planes* and the stresses acting on them, *principal stresses*. There is yet another point C on the circle, where the shear stresses are maximum, but normal stress σ_{av} exists at this point. The plane on which such a state of stress exists is called plane of maximum shear stress. Figure 9.4 shows that

$$\sigma_{max} = \sigma_{av} + R \quad \text{and} \quad \sigma_{min} = R - \sigma_{av} \tag{9.14}$$

Substituting for R and σ_{av} from Eq. (9.11), Eq. (9.13) can be written as

$$\sigma_{max,min} = \frac{\sigma_x + \sigma_y}{2} \pm \sqrt{\left(\frac{\sigma_x + \sigma_y}{2}\right)^2 + 4\tau_{xy}^2} \tag{9.15}$$

The maximum value of shear stress τ_{max} is given by the radius R of the circle. Therefore,

$$\tau_{max} = \sqrt{\left(\frac{\sigma_x + \sigma_y}{2}\right)^2 + 4\tau_{xy}^2} \tag{9.16}$$

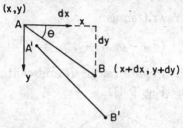

Fig. 9.5 Deformations of a line element.

The values of θ_s, of the parameter θ which corresponds to these points, is obtained by differentiating Eq. (9.6) with respect to θ and setting $\frac{d\tau_{x'y'}}{d\theta} = 0$. Thus for $\theta = \theta_s$

$$\tan 2\theta_s = -\frac{\sigma_x - \sigma_y}{2\tau_{xy}} \qquad (9.17)$$

Equation (9.17) gives two values of θ_s which are 90° apart. Either of these values can be used to determine the plane corresponding to the maximum shearing stress.

9.3   Transformation of Plane Strain

Now consider the transformation of plane strain due to rotation of coordinate axes confined to the plane strain state, i.e., the deformations are to take place only within parallel planes and are same on each of these planes. If z-axis is perpendicular to the planes, on which the deformation is taking place, then we have $\epsilon_z = \gamma_{xz} = \gamma_{yz} = 0$ and the only non-zero components of strain are ϵ_x, ϵ_y and γ_{xy}.

When the strain components at any point are available then it is possible to determine normal and shear strains along any direction at the point. Consider a line element AB between points (x, y) and $(x+dx, y+dy)$ as shown in Fig. 9.5. This line element is translated and rotated to line element $A'B'$ when deformation takes place. Let the displacement of A be (u,v) along x and y directions and that of B are (from chain rule)

$$u + \frac{\partial u}{\partial x}dx + \frac{\partial u}{\partial y}dy, \quad v + \frac{\partial v}{\partial y}dy + \frac{\partial v}{\partial x}dx$$

Taking $AB = ds$,

$$\frac{dx}{ds} = \cos\theta, \quad \frac{dy}{ds} = \sin\theta \qquad (9.18)$$

Now move $A'B'$ parallel to itself, such that A' coincides with A (Fig. 9.6).

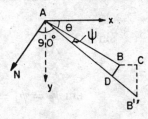

Fig. 9.6 Total deformation of a line element.

The new position of AB is AB''. BC and CB'' represents the component of displacement of B relative to A. Thus,

$$BC = \frac{\partial u}{\partial x}dx + \frac{\partial u}{\partial y}dy$$

$$CB'' = \frac{\partial v}{\partial y}dy + \frac{\partial v}{\partial x}dx \qquad (9.19)$$

The components of these displacements, normal to AB'' and along AB'' is obtained by taking components of BC and CB'' Thus,

$$BD = CB''\cos\theta - BC\sin\theta$$
$$DB'' = CB''\sin\theta + BC\cos\theta \qquad (9.20)$$

Substituting for BC and CB'' from Eq. (9.19),

$$BD = \left(\frac{\partial v}{\partial y}dy + \frac{\partial v}{\partial x}dx\right)\cos\theta - \left(\frac{\partial u}{\partial x}dx + \frac{\partial u}{\partial y}dy\right)\sin\theta \qquad (9.21)$$

$$DB'' = \left(\frac{\partial v}{\partial y}dy + \frac{\partial v}{\partial x}dx\right)\sin\theta + \left(\frac{\partial u}{\partial x}dx + \frac{\partial u}{\partial y}dy\right)\cos\theta$$

Since $AB = BD$, for small deformation, the strain $\epsilon_\theta \left(= \frac{DB''}{AB}\right)$ is

$$\epsilon_\theta = \left(\frac{\partial v}{\partial y}\frac{dy}{ds} + \frac{\partial v}{\partial x}\frac{dx}{ds}\right)\sin\theta + \left(\frac{\partial u}{\partial x}\frac{dx}{ds} + \frac{\partial u}{\partial y}\frac{dy}{ds}\right)\cos\theta \qquad (9.22)$$

Substituting for $\frac{dx}{ds}$ and $\frac{dy}{ds}$ from Eq. (9.18) in Eq. (9.22),

$$
\begin{aligned}
\epsilon_\theta &= \frac{\partial u}{\partial x} \cos^2\theta + \left(\frac{\partial u}{\partial y} + \frac{\partial v}{\partial x} \right) \sin\theta \cos\theta + \frac{\partial v}{\partial y} \sin^2\theta \\
&= \epsilon_x \cos^2\theta + \gamma_{xy} \sin\theta \cos\theta + \epsilon_y \sin^2\theta \quad (9.23)
\end{aligned}
$$

Equation (9.23) gives the normal strain along any direction θ. From Fig. 9.6,

$$
\sin\psi_\theta = \frac{BD}{AB}
$$

For small angles,
$$
\sin\psi_\theta = \psi_\theta = \frac{BD}{AB}
$$

Thus, angle ψ through which AB is rotated is

$$
\psi_\theta = \left(\frac{\partial v}{\partial x} \frac{dx}{ds} + \frac{\partial v}{\partial y} \frac{dy}{ds} \right) \cos\theta + \left(\frac{\partial u}{\partial x} \frac{dx}{ds} + \frac{\partial u}{\partial y} \frac{dy}{ds} \right) \sin\theta
$$

Substituting for $\frac{dx}{ds}$ and $\frac{dy}{ds}$ from Eq. (9.18),

$$
\psi_\theta = \frac{\partial v}{\partial x} \cos^2\theta - \frac{\partial u}{\partial y} \sin^2\theta + \left(\frac{\partial v}{\partial y} - \frac{\partial u}{\partial x} \right) \sin\theta \cos\theta \quad (9.24)
$$

The line element AN makes an angle of $90°$ with AB and its rotation $\psi_{\theta+\frac{\pi}{2}}$ is obtained by substituting $\theta+\frac{\pi}{2}$ for θ in Eq. (9.24). Therefore,

$$
\psi_{\theta+\frac{\pi}{2}} = \frac{\partial v}{\partial x} \sin^2\theta - \frac{\partial u}{\partial y} \cos^2\theta - \left(\frac{\partial v}{\partial y} - \frac{\partial u}{\partial x} \right) \sin\theta \cos\theta \quad (9.25)
$$

The shear strain γ_θ is the difference between $\psi_\theta - \psi\theta + \frac{\pi}{2}$. Thus,

$$
\gamma_\theta = \left(\frac{\partial v}{\partial x} + \frac{\partial u}{\partial y} \right) \left(\cos^2\theta - \sin^2\theta \right) + 2 \left(\frac{\partial v}{\partial y} - \frac{\partial u}{\partial x} \right) \sin\theta \cos\theta
$$

$$
(9.26)
$$

$$
\begin{aligned}
\frac{1}{2}\gamma_\theta &= (\epsilon_y - \epsilon_x) \sin\theta \cos\theta + \frac{\gamma_{xy}}{2} \left(\cos^2\theta - \sin^2\theta \right) \\
&= \frac{(\epsilon_y - \epsilon_x)}{2} \sin 2\theta + \frac{\gamma_{xy}}{2} \cos 2\theta \quad (9.27)
\end{aligned}
$$

A comparison of Eqs. (9.24) and (9.27) with Eqs. (9.3) and (9.4) reveals that the two equations are identical if we replace $\sigma_{x'}$ by ϵ_θ, $\tau_{x'y'}$ by $\frac{\gamma_\theta}{2}$, σ_x by ϵ_x, σ_x by ϵ_x, and τ_{xy} by $\frac{\gamma_{xy}}{2}$.

Our interest here, was to express the strain components associated with x'-y' axes in term of the angle θ and the strain components ϵ_x, ϵ_y, γ_{xy} associated with x-y axes. The normal strain $\epsilon_{x'}$ associated with x'-axis is given by Eq. (9.23). Making use of the trigonometric relations, Eq. (9.23) can be written in an alternate form as

$$\epsilon_{x'} = \frac{(\epsilon_y + \epsilon_x)}{2} + \frac{(\epsilon_y - \epsilon_x)}{2}\cos 2\theta + \frac{\gamma_{xy}}{2}\sin 2\theta \qquad (9.28)$$

Replacing θ by $\theta + 90°$, the normal strain along y'-axis is obtained as

$$\epsilon_{y'} = \frac{(\epsilon_y + \epsilon_x)}{2} - \frac{(\epsilon_y - \epsilon_x)}{2}\cos 2\theta - \frac{\gamma_{xy}}{2}\sin 2\theta \qquad (9.29)$$

The shearing strain $\gamma_{x'y'}$ is given by Eq. (9.27) in which γ_θ corresponds to $\gamma_{x'y'}$.

Adding Eqs. (9.28) and (9.29), the following relationship is obtained:

$$\epsilon_{x'} + \epsilon_{y'} = \epsilon_x + \epsilon_y \qquad (9.30)$$

This implies that the sum of normal strains is invariant with respect to rotation.

9.3.1 Principal Strains and Maximum Shear Strain

Equations (9.28), (9.29) and (9.27), which represent the transformation of strains, closely resemble the equations for the transformation of stresses (Sec. 9.2). In the case of normal strains, if we replace σ by ϵ, we get the transformation for strains. However, τ_{xy} and $\tau_{x'y'}$ should be replaced by half the corresponding shear strains. Once we do this, the other things automatically follow (Sec. 9.2). Figure 9.4 can be redrawn with horizontal axis representing ϵ and the vertical axis $\frac{\gamma}{2}$. Points A and B (Fig. 9.4) will then correspond to minimum and maximum normal strain, respectively. Point C, corresponds to the maximum shear strain. The magnitude, however, will be double this value. The values of θ_p of the parameter θ which corresponds to the points A and B is obtained by setting $\gamma_{x'y'} = 0$. Thus, from Eq. (9.27)

$$\tan 2\theta_p = \frac{\gamma_{xy}}{\epsilon_x - \epsilon_y} \qquad (9.31)$$

The corresponding strains ϵ_θ are principal strains. The value of θ_s of the parameter θ, which corresponds to maximum shearing strain, is obtained by differentiating Eq. (9.27) with respect to θ and setting $(d\gamma_{x'y'}/d\theta) = 0$. Thus, for $\theta = \theta_s$,

$$\tan 2\theta_s = \frac{\epsilon_x - \epsilon_y}{\gamma_{xy}} \qquad (9.32)$$

9.4 Mohr's Circle for Plane Stress

The circle (Fig. 9.4) employed to describe some of the properties of stresses was suggested by the German Engineer Otto Mohr and is known as *Mohr's circle for plane stress*. One can use the Mohr's circle to solve a large number of problems in the area of plane stress without making use of analytical equations. This is a powerful method for a graphical solution. The use of Eqs. (9.3) to (9.17) sometimes result in difficulties because of many possible combinations of signs for the terms σ_x, σ_y, τ_{xy} and θ as also two possible roots of the square root. The graphical method helps to overcome some of these problems. The Mohr's circle results in a rapid and exact computation of the following:

1. The maximum and minimum normal stresses with sign and magnitude.

2. The maximum shear stress.

3. The orientation of the planes of maximum and minimum normal stress and maximum shear stress.

4. The magnitude and sign of normal stress that exists with maximum shear stress.

5. The state of stress at any orientation.

Consider a square element of a material subjected to a plane stress (Fig. 9.2(a)). Let σ_x, σ_y, τ_{xy} be the stresses acting on the

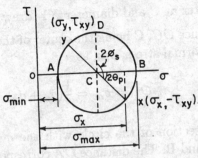

Fig. 9.7 Mohr's circle for 2D state of stress.

element. Mohr's circle for stress is drawn on a set of perpendicular axes with normal stress, σ, horizontally and shear stress, τ, vertically. The following convention is followed for drawing the Mohr's circle:

1. Positive normal stress (tensile) are drawn to the right of the origin.

2. Negative normal stresses (compressive) are drawn to left of the origin.

3. Shear stress that tend to rotate the element clock-wise are plotted above the horizontal axis or along the positive τ axis.

4. Shear stress that tend to rotate the element anticlockwise are plotted below the horizontal axis or along the negative axis.

5. The angle in the Mohr's circle is twice the angle in the actual physical system.

The stress element shown in Fig. 9.2(a) is used to draw the Mohr's circle in the following way:

1. Point X is plotted with coordinates σ_x and $-\tau_{xy}$, as σ_x is positive (tensile) and τ_{xy} produces an anticlockwise moment on X face. Point Y is plotted with coordinates σ_y and τ_{xy}, as τ_{xy} on the Y face produces a clockwise moment (Fig. 9.7).

2. Joining X and Y by a straight line, we define a point C, which is the intersection of this line with σ axis.

3. With center at C and diameter XY draw a circle.

4. The distance OC, C being the center of the circle, gives the average normal stress, σ_{av}, as

$$OC = \sigma_{av} = \frac{1}{2}(\sigma_x + \sigma_y) \qquad (9.33)$$

5. The intersection of the circle with the σ axis locates the points A and B. The distance OA corresponds to σ_{min} and OB corresponds to σ_{max}; which are the principal stresses.

6. The orientation of plane on which the principal stresses act is given by

$$\tan XCA = \frac{2\tau_{xy}}{(\sigma_x - \sigma_y)} \qquad (9.34)$$

Angle XCA is $2\theta_p$. Hence the orientation of the principal plane is half of the angle measured on Mohr's circle. Further if $\sigma_x > \sigma_y$ and $\tau_{xy} > 0$, as in the case considered here, the rotation which bring CX to CB is counter-clockwise.

7. Magnitude of maximum shear stress is given by the radius of the circle and is given by line CD.

8. The plane of the maximum shear stress is at 45° to the principal plane. The orientation of maximum shear plane is determined by finding the angle from Fig. 9.7. The angle is

$$2\phi_s = 90° + 2\theta_p \qquad (9.35)$$

Example 9.1

For the state of plane stress on an element (Fig. 9.8), construct the Mohr's circle and determine (i) the principal planes, (ii) the principal stresses and (iii) the maximum shearing stress and the associated normal stress. Draw also the complete principal stress element and the maximum shear stress element.

Solution

The various steps discussed in Section 9.4 will be used to obtain the solution for the problem.

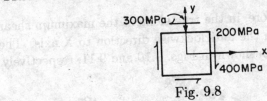

Fig. 9.8

Center of the circle C is at σ_{av} which can be evaluated as (Fig. 9.9).

$$\sigma_{av} = \frac{1}{2}(\sigma_x + \sigma_y) = \frac{1}{2}[400 + (-300)] = 50 \text{ MPa} \qquad (9.36)$$

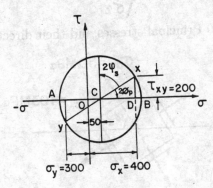

Fig. 9.9

The radius of the circle,

$$CX = \sqrt{(400 - 50)^2 + (-200)^2} = 403 \text{ MPa}$$

Also,

$$\sigma_1 = 403 + 50 = 453 \text{ MPa}$$
$$\sigma_2 = 50 - 403 = -353 \text{ MPa}$$
$$2\phi_p = \tan^{-1}\frac{XD}{CD} = \tan^{-1}\frac{200}{350} = 9.47°$$
$$\phi_p = 14.87°$$

Thus, X-axis has to be rotated by 14.87° in clockwise direction to coincide with the principal stress axis.

The maximum shear stress is 403 MPa and

$$2\phi_s = 90° - 2\phi_p = 60.36°$$

from X-axis. Therefore, in the true element the maximum shear stress is at 30.18° in the anticlockwise direction to X axis. The stress elements are as shown in Figs. 9.10 and 9.11, respectively.

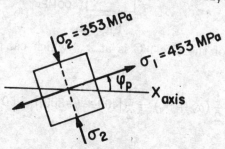

Fig. 9.10 Principal stresses and their directions.

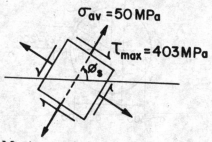

Fig. 9.11 Maximum shear stress and its direction.

Example 9.2

For the state of plane stress (Fig. 9.12), construct the Mohr's circle and determine (i) the principal planes (ii) the principal stresses and (iii) the maximum shearing stress and associated normal stress. Draw the complete principal stress element and the maximum shear stress element. Given: $\sigma_x = 60$ MPa, $\sigma_y = -40$ MPa and $\tau_{xy} = $ MPa

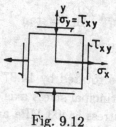

Fig. 9.12

Solution

For the given state of stress, since τ_{xy} on the x face produces an anticlockwise moment, it will be shown below the horizontal axis i.e. σ-axis (Fig. 9.13).

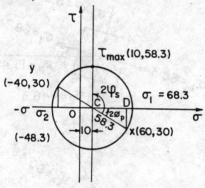

Fig. 9.13

The center of the circle C is at σ_{av} which is evaluated as

$$\sigma_{av} = \frac{1}{2}(\sigma_x + \sigma_y) = \frac{1}{2}[60 + (-40)] = 10 \text{ MPa} \qquad (9.37)$$

Radius of the circle R,

$$CX = \sqrt{(60-10)^2 + 30^2} = 58.3 \text{ MPa}$$

Also,

$$\sigma_1 = 58.3 + 10 = 68.3 \text{ MPa}$$
$$\sigma_2 = -58.3 + 10 = -48.3 \text{ MPa}$$
$$2\phi_p = \tan^{-1}\frac{XD}{CD} = \tan^{-1}\frac{30}{50} = 30.96°$$
$$\phi_p = 15.48°$$

That is, X-axis has to be rotated by 15.48° in the counter clockwise direction to coincide with the principal stress axis.
The maximum shear stress,

$$\tau_{max} = R = 58.3 \text{ MPa}$$

and

$$2\phi_s = 90° - 2\phi_p = 120.96°$$

$$\phi_s = 60.48°$$

from X-axis. The stress elements are shown in Figs. 9.14 and
9.15.

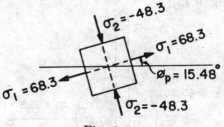

Fig. 9.14

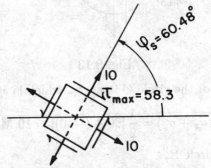

Fig. 9.15

Example 9.3

A state of plane stress exist on the element. A tensile stress
$\sigma_x = 80$ MPa is acting on the x face alongwith unknown shearing
stress τ_{xy} (Fig. 9.16).

Determine (i) the magnitude of the shearing stress τ_{xy} for
which $\sigma_x = 100$ MPa, (ii) the corresponding maximum shear
stress, (iii) the orientation of the principal plane with respect to
x-axis and (iv) the orientation of the maximum shear plane with
respect to x-axis.

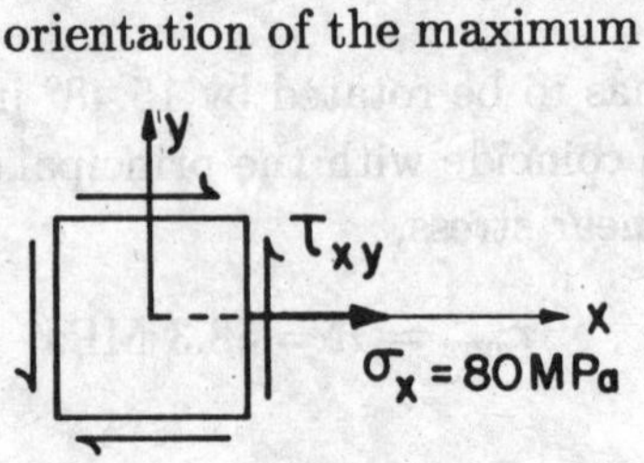

Fig. 9.16

Solution

We assume that the shearing stresses are acting in the direction shown in Fig. 9.16. Thus, the shearing stresses acting on the x face produce an anticlockwise moment and we plot the point x with coordinate 80 MPa and τ_{xy} below the horizontal axis. $\sigma_y = 0$ and τ_{xy} tends to rotate the element clockwise and hence will be shown above the horizontal axis. Also,

$$\sigma_{av} = \frac{1}{2}(\sigma_x + \sigma_y) = \frac{1}{2}(80 + 0) = 40 \text{ MPa}$$

The radius R of the circle is determined from the consideration that maximum normal stress. Now,

$$\begin{aligned}
\sigma_{max} &= 100 \text{ MPa} \\
\sigma_{max} &= \sigma_{av} + R \\
100 &= 40 + R, \\
R &= 60 \text{ MPa}
\end{aligned}$$

Draw the circle with center C and radius $R = 60$, as shown in Fig. 9.17.

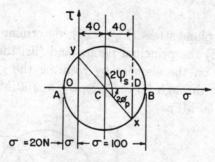

Fig. 9.17

The circle cuts σ axis at A and B, which locates σ_{min} and σ_{max}. Locate point D on σ axis such that $OD = 80$ MPa. Draw a vertical which cuts the circle at X. DX gives the magnitude of shearing stress τ_{xy}.

The shearing stress can be obtained as follows:

$$\cos 2\theta_p = \frac{CD}{CX} = \frac{40}{60} = 0.666$$

$$2\theta_p = 48.24°, \theta_p = 24.12°$$

$$\tau_{xy} = DX = CX \sin 2\theta_p = 60 \sin 48.24° = 44.76 \text{ MPa}$$

The maximum shearing stress,

$$\tau_{max} = R = 60 \text{ MPa}$$
$$2\phi_s = 90° + 2\theta_p = 90° + 48.24° = 138.24°$$
$$\phi_s = 69.12°$$

Figures 9.18 and 9.19 show the orientation of principal planes and maximum shear plane, respectively.

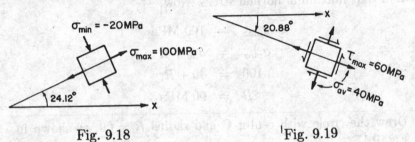

Fig. 9.18   Fig. 9.19

Example 9.4

For the state of plane stress (Fig. 9.20), determine (i) the principal stresses, (ii) the principal planes and (iii) the stress components exerted on the element by rotating the given element counterclockwise through 30°. Given $\sigma_x = 100$ MPa, $\sigma_y = 60$ MPa and $\tau_{xy} = 50$ MPa.

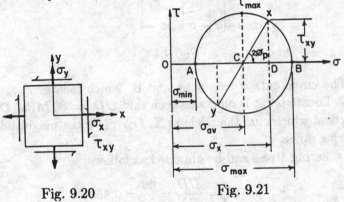

Fig. 9.20   Fig. 9.21

Solution

From the given data, on the X-face, normal stress σ_x is tensile and τ_{xy} produces a clockwise rotation of the element. Therefore, we plot 100 units to the right of the origin along the horizontal axis and 50 units above the horizontal axis (Fig. 9.21).

This gives

$$\sigma_{av} = \frac{1}{2}(\sigma_x + \sigma_y) = \frac{1}{2}(100 + 60) = 80 \text{ MPa}$$

The radius

$$R = \sqrt{CD^2 + DX^2} = \sqrt{20^2 + 50^2} = 53.85 \text{ MPa}$$

(i) Principal stress:

$$\sigma_{max} = OC + CB = 80 + 53.85 = 133.85 \text{ MPa}$$

$$\sigma_{min} = OC - AC = 80 - 53.85 = 26.15 \text{ MPa}$$

(ii) Principal plane:

$$\tan 2\theta_p = \frac{XD}{CD} = \frac{50}{20} = 2.5$$

$$2\theta_p = 68.2°, \qquad \theta_p = 34.1°$$

Figure 9.22(a) shows the stresses acting on the principal plane.

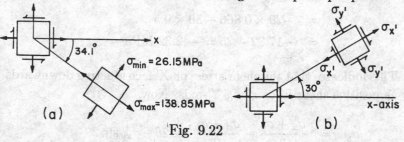

(a)

(b)

Fig. 9.22

(iii) Stress components on element rotated counterclockwise through 30° (Fig. 9.22(b)):

$$\phi = 180° - 60° - 68.2° = 51.8°$$

$$\sigma_{x'} = OE = OC - CE = \sigma_{av} - R \cos \phi = 80 - 53.85 \cos 51.8° = 46.7 \text{ MP}$$

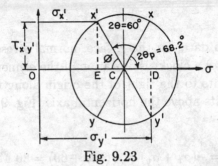

Fig. 9.23

$$\tau_{x'y'} = R\sin\phi = 53.85\sin 51.8° = 42.32 \text{ MPa}$$

$$\sigma_{y'} = OC + CD = \sigma_{av} + R\cos\phi = 80 + 53.58\cos 51.8° = 113.13 \text{ MPa}$$

Figure 9.23 indicates the stresses acting on the element.

The transformed stresses can also be obtained by employing Eqs. (9.5) to (9.7) which gives

$$
\begin{aligned}
\sigma_{x'} &= \frac{\sigma_x + \sigma_y}{2} + \frac{\sigma_x - \sigma_y}{2}\cos 2\theta + \tau_{xy}\sin 2\theta \\
&= \frac{100 + 60}{2} + \frac{100 - 60}{2}\cos 60° - 50\sin 60° \\
&= 80 + 20 \times 0.5 - 50 \times 0.866 \\
&= 90 - 43.3 = 46.7 \text{ MPa} \\
\tau_{x'y'} &= \frac{\sigma_x - \sigma_y}{2}\sin 2\theta + \tau_{xy}\cos 2\theta \\
&= \frac{60 - 100}{2}\sin 60° - 50\cos 60° \\
&= -20 \times 0.866 - 50 \times 0.5 \\
&= -17.32 - 25 = -42.32 \text{ MPa}
\end{aligned}
$$

This indicates that the shear stress on X' face is acting downwards as positive shear force along Y'-axis. Now,

$$
\begin{aligned}
\sigma_{y'} &= \frac{\sigma_x + \sigma_y}{2} - \frac{\sigma_x - \sigma_y}{2}\cos 2\theta - \tau_{xy}\sin 2\theta \\
&= \frac{100 + 60}{2} - \frac{100 - 60}{2}\cos 60° - (-50)\sin 60° \\
&= 80 - 20 \times 0.5 + 50 \times 0.866 = 113.13 \text{ MPa}
\end{aligned}
$$

This is identical to what is obtained from the Mohr's circle.

9.5   Mohr's Circle for Plane Strain

The transformation equations for plane strain are of the same form as plane stress and the concept of Mohr's circle can easily be extended for obtaining principal strains and strains on any other planes. The axes of the Mohr's circle for strain are represented by ϵ and $\frac{\gamma_{xy}}{2}$. As in the cases of shearing strain, if γ_{xy} is positive on the X-face it is plotted below the horizontal axis. Thus, if the shear deformation causes a given side to rotate clockwise, the corresponding point on the Mohr's circle for the plane is plotted above the horizontal axis, and if the deformation causes the side to rotate anticlockwise, the corresponding point is plotted below the horizontal axis.

The procedure to draw the Mohr's circle for plane strain is as follows (Fig. 9.24):

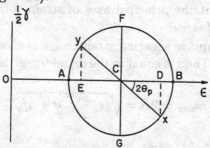

Fig. 9.24 Mohr's circle for plane strain.

Points X and Y are located on the ϵ and $\gamma/2$ axes on the basis of the given strain components ϵ_x, ϵ_y and γ_{xy}. Join X and Y with a straight line to locate point C, the center of the Mohr's circle. With C as center and CX as radius, draw the circle. This establishes the Mohr's circle for strain. The coordinates of point C are $(\epsilon_{av}, 0)$, and radius R of the Mohr's circle is

$$\epsilon_{av} = \frac{\epsilon_x + \epsilon_y}{2}$$

and

$$R = \sqrt{\frac{\epsilon_x + \epsilon_y}{2}^2 + \frac{\gamma_{xy}}{2}^2} \qquad (9.38)$$

The points A and B, where the Mohr's circle intersects the horizontal axis, correspond respectively to the minimum and maximum values of the normal strain, ϵ. These strains are

$$\epsilon_{min} = \frac{\epsilon_x + \epsilon_y}{2} - \sqrt{\frac{\epsilon_x + \epsilon_y}{2}^2 + \frac{\gamma_{xy}}{2}^2} \qquad (9.39)$$

$$\epsilon_{max} = \frac{\epsilon_x + \epsilon_y}{2} + \sqrt{\frac{\epsilon_x + \epsilon_y}{2}^2 + \frac{\gamma_{xy}}{2}^2} \qquad (9.40)$$

The direction θ_p along which the principal strains are acting is given by

$$\tan 2\theta_p = \frac{\gamma_{xy}}{\epsilon_x - \epsilon_y} \qquad (9.41)$$

For an isotropic material the shearing stress and strain are related through the material constant, G. Thus $\gamma_{xy} = 0$ when $\tau_{xy} = 0$, indicating that the principal axes of strain coincide with the principal axes of stress.

Maximum inplane shearing strain, γ_{max} is given by points F and G (Fig. 9.24) and is equal to the diameter of the circle. Thus,

$$\gamma_{max} = 2R = \sqrt{(\epsilon_x - \epsilon_y)^2 + \gamma_{xy}^2} \qquad (9.42)$$

Example 9.5

In structural element which is in a state of plane strain (Fig. 9.25(a)) the horizontal side of 10×10 mm square elongates by 4 μm (4×10^{-6} m), while the vertical side elongates by 1 μm (1×10^{-6} m) and the angle at the lower left corner increases by 0.4×10^{-3} radians (Fig. 9.25(b)). Determine (i) the principal directions and

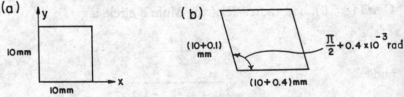

Fig. 9.25

principal normal strains and (ii) the maximum shearing strain and the associated normal strain.

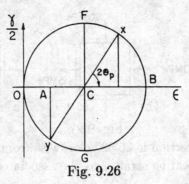

Fig. 9.26

Solution

(i) Principal directions and principal normal strains:
From the given data

$$\epsilon_x = \frac{4 \times 10^{-6}}{10 \times 10^{-3}} = 400 \ \mu m, \ \epsilon_y = \frac{1 \times 10^{-6}}{10 \times 10^{-3}} = 100 \ \mu m$$

$$\frac{\gamma_{xy}}{2} = 200 \ \mu m$$

Since the x-face associated with ϵ_x rotates clockwise, point X (Fig. 9.26) is located with $\epsilon_x = 400\mu m$ and $\frac{\gamma_{xy}}{2} = 200\mu m$ above the horizontal axes. Point Y is located below the horizontal plane as y-face rotates anticlockwise.

Points X and Y are joined by a straight line. This cuts the ϵ axis at point C, which locates the center of the Mohr's circle. With C as center and CX as radius, a circle is drawn which cuts the ϵ at points A and B. OA and OB gives the minimum and maximum normal strains ϵ_{min} and ϵ_{max}. Now,

$$OC = \frac{\epsilon_x + \epsilon_y}{2} = \frac{400 + 100}{2} = 250 \ \mu m$$

$$R = \sqrt{150^2 + 200^2} = 250 \ \mu m$$

$$\epsilon_{min} = \epsilon_{av} - R = 250 - 250 = 0$$

$$\epsilon_{max} = \epsilon_{av} + R = 250 + 250 = 500 \ \mu m$$

$$\tan 2\theta_p = \frac{gsxy}{\epsilon_x - \epsilon_y} = \frac{400}{400 - 100} = 1.33$$

$$2\theta_p = 53.06°$$

$$\theta_p = 26.53°$$

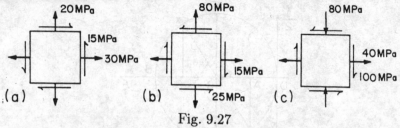

Fig. 9.27

The principal direction is 26.53° clockwise from X.

(ii) Maximum shearing strain and the associated normal strain:

$$\frac{\gamma_{xy}}{2} = R = 250\mu \;,\; \gamma_{max} = 500 \; \mu m$$

The strain ϵ at this point $= \epsilon_{av} = 250 \; \mu m$.

Problems

9.1 For a the state of stress shown in Figs. 9.27(a), (b) and (c) determine (i) the principal planes, (ii) the principal stresses and (iii) the maximum shear stress and associated normal stress.

9.2 The grains of a wooden member forms an angle of 20° with the vertical as shown in Figs. 9.28(a) and 9.28(b). For the state of stress shown, determine (i) the inplane shear stress parallel to the grain and (ii) the normal stress perpendicular to the grain.

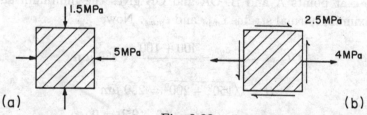

(a)   (b)

Fig. 9.28

9.3 For a state of stress shown in Figs. 9.29(a) and 9.29(b) determine the stresses on a plane obtained by rotating the vertical face anticlockwise through 15°.

9.4 For a plane stress element shown in Fig. 9.30, determine the range of values of τ_{xy} for which the largest inplane shear stress is equal to or less than 80 MPa.

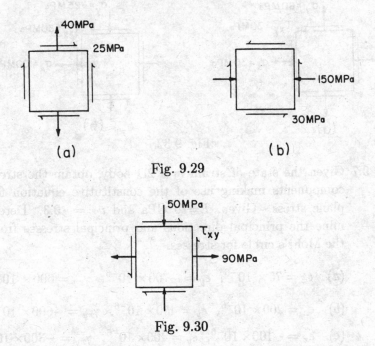

Fig. 9.29

Fig. 9.30

9.5 Certain stress components are known for each of the two orientations shown in Figs. 9.31(a) and 9.31(b). Determine the following stress components: τ_{xy}, $\tau_{x'y'}$ and $\sigma_{y'}$.

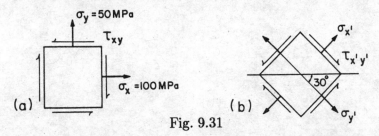

Fig. 9.31

9.6 For a given state of stress (Figs. 9.32(a) and 9.32(b), determine the direction and magnitude of maximum and minimum normal strain, using the material parameters $E=210$ GPa and $\nu = 0.3$. Use the 2D plane stress equations to determine the strains, and then using Mohr's circle for strains, determine the principal strain. Also determine the principal strains from the principal stress obtained from Mohr's circle for plane stress.

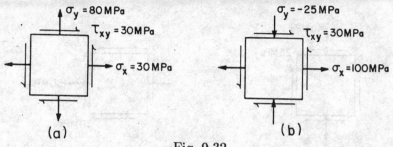

Fig. 9.32

9.7 Given the state of strain in a 2D body, obtain the stress components making use of the constitutive equation for plane stress. Given, $E=70$ GPa and $\nu = 0.3$. Determine the principal directions and principal stresses from the Mohr's circle for stresses.

(a) $\epsilon_x = 75 \times 10^{-6}$, $\epsilon_y = -700 \times 10^{-6}$, $\gamma_{xy} = 600 \times 10^{-6}$

(b) $\epsilon_x = 200 \times 10^{-6}$, $\epsilon_y = 450 \times 10^{-6}$, $\gamma_{xy} = -600 \times 10^{-6}$

(c) $\epsilon_x = -100 \times 10^{-6}$, $\epsilon_y = 400 \times 10^{-6}$, $\gamma_{xy} = -300 \times 10^{-6}$

10. BENDING AND DEFLECTION OF BEAMS

10.1 Introduction

A structural member subjected to a transverse load is known as *beam*. It is essentially a one-dimensional structure, as the cross-sectional dimensions are very much smaller compared to the length. We shall assume that the beam material is elastic and made up of isotropic material. Under the action of transverse load, the beam undergoes bending, resulting in bending stresses. The transverse load also causes shear stresses in the beam. Thus at any cross-section, both bending and shear stresses act. In order to calculate these stresses, it is necessary to determine the magnitude of shear forces and bending moment at that section. Bending results in transverse deflection of the beam. In this chapter, we shall discuss the methods to compute bending and shearing stresses as well as the transverse deflection of the beam.

10.2 Beam and Beam Supports

A beam is a structural member, essentially one-dimensional and is subjected to couples and transverse forces, which lie in the plane passing through the longitudinal axis of the member. The transverse forces acting on the beam are represented by arrows acting downwards or upwards. They may be acting at a point or over a small part of the length of the beam. A second kind of loading is the uniformly distributed load or varying load along the length of the beam, wherein the load is usually described as force per unit length of the beam.

A beam may be classified on the basis of the supports on which the beam is resting. Some examples of the ideal boundary conditions are given below.

(i) Simply supported beam: Also called as hinged beam is supported at the two ends in such a way that it can rotate freely but constrained against transverse deflection. There is a pin (or hinge) at one end while the other end is supported on rollers (Fig. 10.1), so that any changes (due to temperature variation) in the length of the beam can be accommodated. Usually for small transverse deflections the change in the length of a beam is negligible as compared to its length.

(ii) Cantilever beam: In the cantilever beam, one end of the beam is clamped such that at this end neither can the beam rotate nor have any transverse deflection. The other end is free to have any rotation and transverse deflection (Fig. 10.2).

(iii) Clamped-clamped beam: In this situation, both ends of the beam are clamped and are restrained against rotation and transverse deflection (Fig. 10.3).

(iv) Overhanging beam: An overhanging beam is one in which part of the loaded beam extends outside the supports. The overhang may be on either side of the support or may be on only one side (Fig. 10.4).

(v) Propped cantilever: In this case, one end of the beam is clamped while there is a hinged (or roller) support either at the end or at any intermediate point (Fig. 10.5).

(vi) Continuous beam: A beam which is supported at more than two points is termed as a continuous beam. The minimum number of spans are therefore two (Fig. 10.6).

The beams having end supports (Figs. 10.1, 10.2 and 10.4) are called *statically determinate* beams as all the support reactions can be computed using the equations of equilibrium studied in statics (Chapter 2) or mechanics. Rest of the supports mentioned above result in a *statically indeterminate* beam. As the name suggests, equations of equilibrium are not sufficient to obtain all

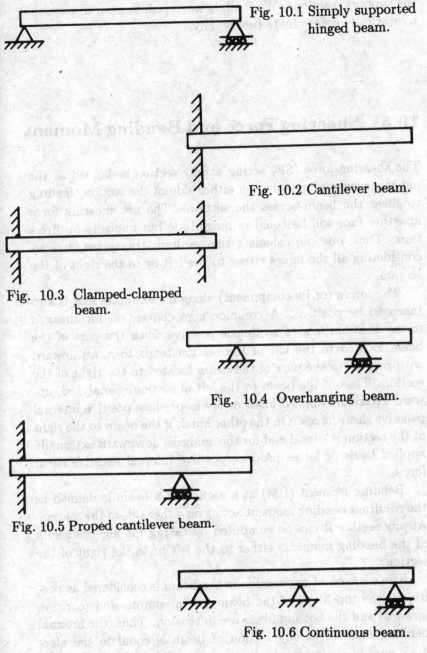

Fig. 10.1 Simply supported
hinged beam.

Fig. 10.2 Cantilever beam.

Fig. 10.3 Clamped-clamped
beam.

Fig. 10.4 Overhanging beam.

Fig. 10.5 Proped cantilever beam.

Fig. 10.6 Continuous beam.

the support reactions. We shall, however, confine our discussions to statically determinate beams only.

10.3   Shearing Force and Bending Moment

The shearing force (SF) acting at any section is defined as the net transverse force acting on either side of the section, tending to shear the beam across the section. The net shearing force at either face will be equal in magnitude but opposite in direction. Thus, one can calculate the resultant transverse force by considering all the forces either to the left or to the right of the section.

The forces (or its component) along a coordinate axis is also taken to be positive. A common sign convention for shear is to say that shear at a section is positive when the part of the beam located to the left of the section tends to move upward with respect to the part of the beam located to the right of the section. Thus, if the beam to the left of section is analyzed, upward externally applied loads or forces produce positive internal resistive shear forces. On the other hand, if the beam to the right of the section is considered for the analysis, downward externally applied loads or forces produce positive internal resistive shear forces.

Bending moment (BM) at a section of a beam is defined as the resultant bending moment acting on either side of the section. At any section it can be computed by taking the algebraic sum of the bending moments either to the left or to the right of the section.

The moment at any section in the beam is considered as positive if the top fibers of the beam are in compression (concave upward) and the bottom fibers are in tension. Thus, the internal bending moment at any section of beam is equal to the algebraic sum of the moments due to externally applied forces and concentrated moments (if any) about the centroid of the section.

10.4 Shear Force and Bending Moment Diagrams

The first step in analyzing a beam under a given loading is to show all the externally applied loads and support reactions on a free body diagram (FBD). It is very important to construct a correct free body diagram for the given system. The following procedures may be followed for getting the reactions (forces and moments) for a statically determinate beams:

1. Draw the free body diagram indicating all the applied loads (forces and moments) and support reactions (forces and moments).

2. Use the equilibrium equations,

$$\Sigma F = 0 \quad \text{and} \quad \Sigma M = 0 \qquad (10.1)$$

to arrive at a system of simultaneous, algebraic equations interms of reactive forces and moments. For the elastic beam, we have two force equations and one moment equation in the plane of loading. The problem is statically determinate if we are able to evaluate all the external reactive forces and moments.

Once the reactions are known, it is possible to draw the shear force and bending moment diagrams.

3. From equilibrium of forces, determine the shear force at any section within the beam.

4. Knowing the shear force at each section, the shear force diagram for the beam can be drawn.

5. Draw the free body diagram for a section of the beam indicating all reactions (forces and moments), applied transverse loading and moments and the reactive moment at the section. From the equilibrium of the moments, obtain the reactive moment at the section.

6. Knowing the reactive moment at each section, bending moment diagram for the entire beam can be drawn.

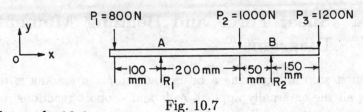

Fig. 10.7

Example 10.1

Compute the support reaction for the overhanging beam as shown in Fig. 10.7.

Solution

Consider the coordinate system shown passing through the centroid of the section. P_1, P_2 and P_3 are the applied external loads and R_1 and R_2 are the support reactions. The positive directions of X and Y are shown in the figure. Vertical equilibrium of forces gives

$$\sum F_y = 0; \quad R_1 + R_2 - 800 - 1000 - 1200 = 0$$

or $$R_1 + R_2 = 3000 \text{ N}$$

We can take the moment of all the forces (applied and reaction forces) either at points A or B about an axis passing through A and normal to XY plane. The choice of the point depends on the number of unknowns. Generally, we consider a point where the number of unknowns are minimum.

Take point A as the reference point. Then $\sum M_z$ at A = 0. Considering clockwise moment to be positive,

$$1000 \times 200 + 1200 \times 400 - 800 \times 100 - R_2 \times 250 = 0$$

or $$R_2 = 2400 \text{ N}$$

Therefore, from equilibrium of forces,

$$R_1 = 3000 - 2400 = 600 \text{ N}$$

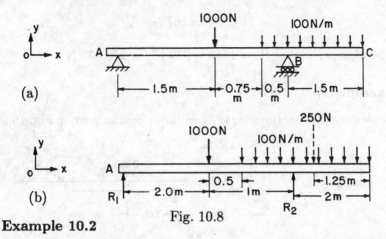

(a)

(b)

Fig. 10.8

Example 10.2

Determine the external reactions for the bean loaded as shown in Fig. 10.8(a).

Solution

Consider the coordinate system shown passing through the centroid of the section. The beam is supported externally by a pin joint at A and roller support at B. The support at A resists both horizontal and vertical motions, whereas at B it resists only vertical motion. For the given loading, no horizontal motion is possible. Hence, no horizontal reactions are developed. The free body diagram for the beam is, therefore, as shown in Fig. 10.8(b).

Consider the equilibrium of the entire beam. From vertical equilibrium,

$$\sum F_y = 0; \quad R_1 + R_2 - 1000 - 100 \times 2.5 = 0$$

or $$R_1 + R_2 = 1200 \text{ N}$$

For moment equilibrium, we consider the uniformly distributed load of 100 N/m to be lumped at the center of gravity of the load. The resultant vertical load being 250 N, applied at a distance of 1.25 m from the right hand end.

Taking moments at point A about an axis Z, normal to the XY plane, gives

$$\sum M_z = 0; 1000 \times 2 - 250 \times 3.75 - R_2 \times 3 = 0$$

$$R_2 = 979.167 \text{ N}$$

Therefore, $R_1 = 1250 - 979.167 = 270.833 \text{ N}$

Example 10.3

Determine the reactions for the cantilever beam shown in Fig. 10.9(a).

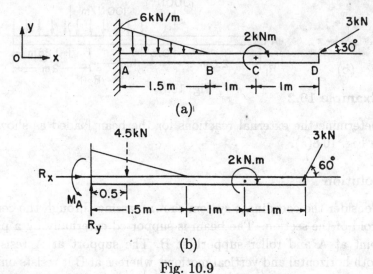

(a)

(b)

Fig. 10.9

Solution

The free body diagram of the beam is first drawn and the un-known reactions are found from equilibrium equations (Fig. 10.9(b)).

The linearly varying load is replaced by an equivalent load acting at the center of gravity of the load (i.e., at 0.5 m from the left support).

The total load $= \frac{1}{2} \times 6 \times 1.5 = 4.5 \text{ N}$

$\sum F_x = 0;$ $R_x - 3\cos 60° = 0$

          $R_x = 1.5 \text{ kN}$

$\sum F_y = 0;$ $R_y - 4.5 - 3\sin 60° = 0$

          $R_y = 7.098 \text{ kN}$

$\sum M_z = 0;$ $M_A - 4.5 \times 0.5 - 2 - 3\sin 60° \times 2.5 = 0$

          $M_A = 10.995 \text{ kN.m}$

Example 10.4

Calculate the shear forces and bending moment at sections C and D of the beam shown in Fig. 10.10(a).

Solution

The free body diagram of the beam is shown in Fig. 10.10(b). From equilibrium of the entire beam,

$$R_{Ax} = 0; \quad R_{Ay} + R_{By} = 70 \text{ kN}$$

Taking moment at A about an axis perpendicular to XY plane gives

$$R_{By} \times 6 = 100 \times 2 + 70 \times 4$$
$$R_{By} = 80 \text{ kN}$$

Hence, $$R_{Ay} = 90 \text{ kN}$$

To determine the internal forces and moments at section C, we take a section at C. Since at C, there is a concentrated load, the internal forces will change depending on whether the section is taken towards the right or left of the concentrated load.

We take a Section 1-1 just to left of the load at C (Fig.10.10(c)) and consider the equilibrium to the left of the section 1-1. The shear force and bending moment are both assumed to act in the positive direction according to the sign convention mentioned in Section 10.3.

From equilibrium,

$$\sum F_y = 0; \quad 90 - V_{CL} = 0;$$

or $$V_{CL} = 90 \text{ kN}$$

and $$M_{CL} = 90 \times 2 = 180 \text{ kN.m}$$

To find the internal forces just to the right of the concentrated load at C, we take a section 2-2 (Fig. 10.10(d))..

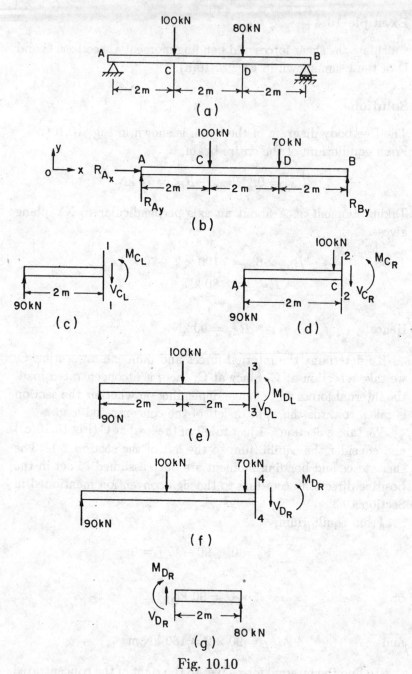

Fig. 10.10

Again consider the equilibrium of the beam to the left of the section 2-2. We now include the concentrated load at C in the free body diagram.

Equilibrium of vertical forces gives

$$95 - 100 - V_{CR} = 0$$

or
$$V_{CR} = -5 \text{ kN}$$

and
$$M_{CR} = 90 \times 2 - 100 \times 0 = 180 \text{ kN.m}$$

Thus, it is seen that there is a sudden decrease in the value of the internal shear force across the load at C. However, the bending moment remains constant at either side of C.

An upward load at the section, in general, causes a sudden increase in the shear force, while a downward load causes a sudden decrease in the shear force.

To determine the internal forces at D, consider a section 3-3 to the left of the load at D. Free body diagram of the beam to the left of the sections is shown in Fig. 10.10(e).

From equilibrium of vertical forces,

$$\sum F_y = 0; \quad 90 - 100 - V_{DL} = 0$$

$$\text{or} \quad V_{DL} = -5 \text{ kN}$$

$$\text{and} \quad M_{DL} = 90 \times 4 - 100 \times 2 = 160 \text{ kN.m}$$

Now consider a section just to the right of the load point D. From equilibrium of vertical forces (Fig. 10.10(f)),

$$\sum F_y = 0; \quad 90 - 100 - 70 - V_{DR} = 0$$

or
$$[V_{DR} = -80 \text{ kN}$$

$$M_{DR} = 90 \times 4 - 100 \times 2 - 70 \times 0 = 160 \text{ kN.m}$$

We can also obtain the values by considering the equilibrium of the beam to the right of section 4-4 (Fig. 10.10(g)). From the vertical equilibrium of forces,

$$V_{DR} = 80 \text{ kN} \quad \text{and} \quad M_{DR} = 80 \times 2 = 160 \text{ kN.m}$$

Example 10.5

Draw the shear force diagram for a simply supported (hinged-hinged) beam shown in Fig. 10.11(a).

Solution

The free body diagram for the beam is shown in Fig. 10.11(b).

Reactions R_{Ax}, R_{Ay} and R_{Fy} are obtained by considering the equilibrium of the beam under the applied external load and reactions. From horizontal equilibrium, $R_{Ax} = 0$, since there is no horizontal load.

From vertical equilibrium,

$$R_{Ay} + R_{Fy} = 3 + 4 + 1 + 3 = 11 \text{ kN}$$

Taking moment of all the forces at A about an axis normal to XY plane gives

$$R_{Fy} \times 1.8 = 3 \times 0.4 + 4 \times 0.8 + 1 \times 1.2 + 3 \times 1.5$$

or $R_{Fy} = 5.61$ kN

Therefore, $R_{Ay} = 5.39$ kN

Figure 10.11(c) shows the external loads and the support reactions. The shear force diagram (Fig. 10.11d) is drawn following the steps given below:

1. At point A, the shear force is positive (as per the sign convention) and has a value of 5.39 kN.

2. This value remains constant till we reach the load point B. To the left of point B, the shear force has a value of 5.39 kN and to the right of the load it has a value (5.39 − 3.0) = 2.39 kN.

3. This value remains constant till we reach the load point C. Just to the right of the load, the shear force becomes (2.39 − 4.0) = − 1.61 kN.

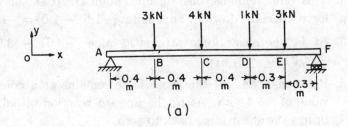

(a)

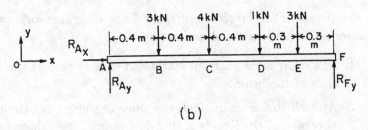

(b)

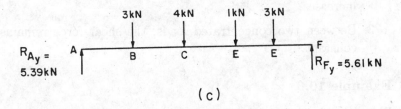

(c)

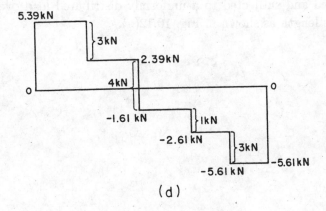

(d)

Fig. 10.11

4. This value remains constant upto point D. At D, the shear force will come down by an amount $(-1.61 - 1.0) = -2.61$ kN.

5. At E, the shear force attains a value $(-2.61 - 3.0) = -5.61$ kN.

6. Between E and F, the shear force remains at a constant value of -5.61 kN. At F, the upward reaction of 5.61 kN brings the shear force back to zero.

Some general characteristics of shear force diagrams for beams subjected to discrete loads are:

1. The shear force diagram starts and ends at zero at the two ends of the beam.

2. At each concentrated load the value of shear force changes abruptly. If the load is acting downwards, then shear force reduces, and if the load is directed upwards, the shear force increases.

3. Between two concentrated loads, the shear force remains constant.

Example 10.6

Draw the shear force diagram for a beam which is simply supported and subjected to a uniformly distributed load over part of its length as shown in Fig. 10.12(a).

Solution

The free body diagram for the beam is shown in Fig. 10.12(b). From the equilibrium of vertical forces,

$$R_{Ay} + R_{By} - 4 \times 1000 = 0$$

$$R_{Ay} + R_{By} = 4000 \text{ N}$$

For considering moment equilibrium of the beam, we lump the distributed load as a point load at the center of gravity of the load. Taking the moment at A,

$$R_{By} \times 7 = 8000 \text{ N.m} \quad R_{By} = 1142.86 \text{ N}$$

Hence, $\qquad\qquad R_{Ay} = 2857.14 \text{ N}$

To find the shear force at any point in the region A to B, we get sections at some specific distances from the end A.

Consider a section at a distance of 2 m from the end A as shown in Fig. 10.12(c).

From equilibrium of vertical forces,

$$V + 2000 = 2857.14$$

$$V = 857.14 \text{ N}$$

Taking a segment at a distance of 3 m from A (Fig. 10.12(d)) gives

$$V + 3000 = 2857.14$$

or $\qquad\qquad V = -142.86 \text{ N}$

If we take a section at a distance of 2.857 m from end A, it is seen that $V = 0$, since the applied load balances the reaction at A.

Now take a section at a distance of 5 m from the end A as shown in Fig. 10.12(e). The free body diagram for this case gives

$$V + 4000 = 2857.14$$

$$V = -1142.86 \text{ N}$$

The shear force diagram for the given loading is shown in Fig. 10.12(f). Note:

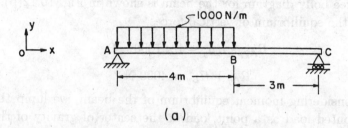

(a)

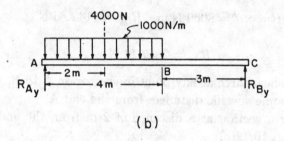

(b)

(c)   (d)

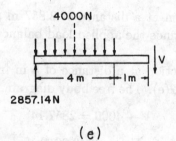

(e)

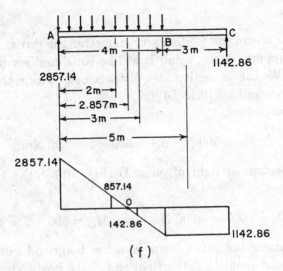

Fig. 10.12

1. For the portion of the beam carrying the uniformly distributed load, shear force curve is a straight line.

2. The slope of the straight line curve is equal to the rate of loading on the beam ie., the load per unit length.

3. The change in shear between any two points is equal to the area under the load diagram between these points.

Example 10.7

For a simply supported beam shown in Fig. 10.13(a), draw the bending moment diagram.

Solution

As the structure and loading are symmetric, the vertical reactions at the support points A and B will be equal and is equal to 500 N each. We take a section at a distance of 1 m form the support A as shown in Fig. 10.13(b).

From equilibrium,

$$V_C = 500 \text{ N}; \quad M_C = 500 \times 1 = 500 \text{ Nm}$$

Take a section at right of point D (Fig. 10.13(c)). Then from equilibrium,

$$V_D = 500 - 1000 = -500 \text{ N} \quad \text{and} \quad M_D = 500 \times 2 = 1000 \text{ Nm}$$

If we consider the entire beam as a free body and sum the moments about point B at the right end of the beam, then

$$M_B = 500 \times 2 - 1000 \times 1 = 0$$

Figure 10.13(d) shows the shear force and bending moment variations.

In general, (a) the bending moments at the ends of a simply supported beam are zero and (b) the change in bending moment between any two points on a beam is equal to area under the shear force curve between the two points.

Example 10.8

For the beam shown in Fig. 10.12(a), draw the bending moment diagram.

Solution

Consider once again, the free body diagram of a section of the beam of 2 m from the end A.

We can determine the bending moment in the beam (Fig. 10.14) by summing moments about the points due to all forces to the left of the sections. Thus,

$$M = 2857.14 \times 2 - 2000 \times 1 = 3714.28 \text{ N.m}$$

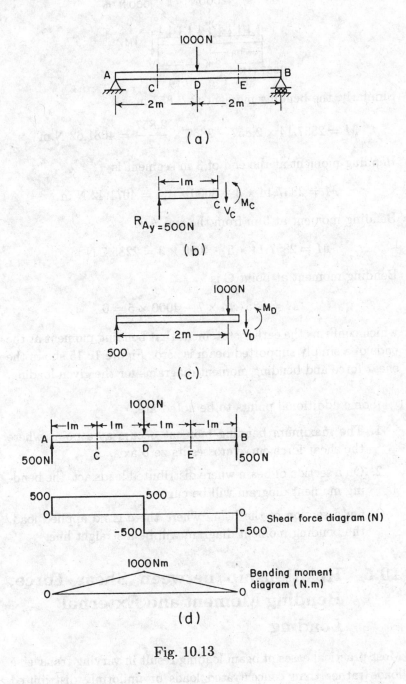

Fig. 10.13

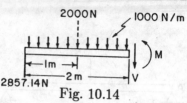

Fig. 10.14

Similarly, the bending moment at 2.857 m is

$$M = 2857.14 \times 2.857 - 2857 \times \frac{2.857}{2} = 4081.62 \text{ N.m}$$

Bending moment at the end of 3 m segment is

$$M = 2857.14 \times 3 - 3000 \times 1.5 = 4071.42 \text{ N.m}$$

Bending moment at 5 m from the end A is

$$M = 2857.14 \times 5 - 4000 \times 3 = 2285.7 \text{ N.m}$$

Bending moment at point C is

$$M = 2857.14 \times 7 - 4000 \times 5 = 0$$

which confirms the earlier statement that bending moment at the ends of a simply supported beam is zero. Figure 10.15 shows the shear force and bending moment diagrams for the given loading.

Some additional points to be noted are:

1. The maximum bending moment occurs at a point where the shear force curve crosses its zero axis.

2. On a section of beam where distributed loads act, the bending moment diagram will be curved.

3. On a section of the beam where there is no applied load, the bending moment diagram will be a straight line.

10.5 Relationship between Shear Force, Bending Moment and External Loading

Most practical cases of beam loading result in varying transverse loads rather than concentrated loads or uniformly distributed

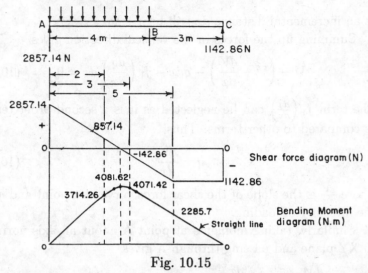

Fig. 10.15

loads we have discussed so far. Some examples are (i) loads on a bridge deck at any instant of time, (ii) platform on which gravel is dumped, (iii) lift loads on an aircraft wing, etc. For such loads, the bending moment and shear force diagrams can easily be constructed, if we can develop an analytical relationship between bending moment (BM), shear force (SF) and the transverse loading.

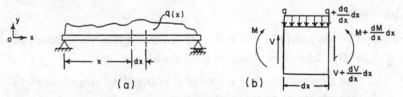

Fig. 10.16 (a) A simply supported beam subjected to transverse load and (b) forces and moments acting on a beam element of length dx.

Figure 10.16(a) shows a beam with a transverse load. The loading intensity at a distance x from the left hand end is q, which in general could be a function of x. For a uniform load it will be a constant.

Figure 10.16(b) shows a section of the beam of length dx (shown enlarged) with shear force and bending moment at either side of the section. As the load is also varying, the applied load at an incremental distance dx will be $q + dq$.

Summing up the forces in the vertical direction gives

$$V - \left(V + \frac{dV}{dx}\right) - qdx - f_1\left(\frac{dq}{dx}\right)dx = 0 \qquad (10.2)$$

The term $f_1\left(\frac{dq}{dx}\right)$ can be neglected as it is a second order term as compared to other terms. Thus,

$$q = \frac{dV}{dx} \qquad (10.3)$$

where $\frac{dV}{dx}$ is the slope of the shear force curve at a point and q is the intensity of loading.

Similarly, taking moments at point A about an axis normal to XY plane and passing through A gives

$$M + \frac{dM}{dx}dx + qdx\frac{dx}{2} + f_2(dq)(dx)^2 - M - Vdx = 0 \qquad (10.4)$$

Neglecting terms of the $O(dx)^2$,

$$\frac{dM}{dx} = V \qquad (10.5)$$

where $\frac{dM}{dx}$ is the slope of the BM curve at any point.

Based on Eqs. (10.3) and (10.5), following observations can be made:

1. The slope of the SF curve at a point gives the magnitude of the transverse load at that point.

2. The slope of the BM curve at a point gives the value of the shear force at that point. Thus,

$$\int_{V_1}^{V_2} dV = V_2 - V_1 = \int_{x_1}^{x_2} qdx \qquad (10.6)$$

The difference between shear force at any two points is equal to the area of the transverse load diagram between the same two points. Thus,

$$\int_{M_1}^{M_2} Mdx = M_2 - M_1 = \int_{x_1}^{x_2} Vdx \qquad (10.7)$$

The difference in BM between any two points is equal to the area under the SF diagram between the same points.

We shall now illustrate the above obeservations through some examples.

Example 10.9

For the overhang beam shown in Fig. 10.17(a), draw the SF and BM diagrams.

Solution

Figure 10.17(b) shows the free body diagram for the loaded beam. From equilibrium of vertical forces,

$$R_{Ay} + R_{Cy} = 24$$

Taking moment at A gives

$$R_{Cy} \times 16 = 16 \times 8 + 8 \times 18$$

$$R_{Cy} = 17 \text{ kN}$$

Therefore, $\qquad\qquad R_{Ay} = 7 \text{ kN}$

The shear force equations are

$$x = 0 \text{ to } 8 \text{ m}; \quad V = R_{Ay} = 7 \text{ kN}$$

$$x = 8 \text{ to } 16 \text{ m}; \quad V = R_{Ay} - 16 = -9 \text{ kN}$$

$$x = 16 \text{ to } 24 \text{ m}; \quad V = 7 - 16 + 17 - 2(x - 16) = 40 - 2x$$

For writing BM equations, it is convenient to choose three segments and for each segment the running coordinate x is defined seperately.

For region AB, $x_1 = 0$ to 8 m,

$$
\begin{aligned}
M &= M_A = 0 \quad \text{at } x_1 = 0 \\
M_{x_1} &= \int_0^{x_1} V \, dx_1 + M_A \\
&= 7 \int_0^{x_1} dx_1 + M_A = 7x_1 + 0 \\
M_{x_1} &= 56 \text{ kN.m at } x_1 = 8 \text{ m}
\end{aligned}
$$

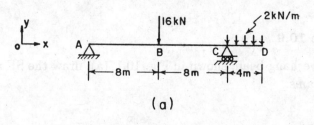

(a)

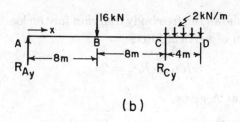

(b)

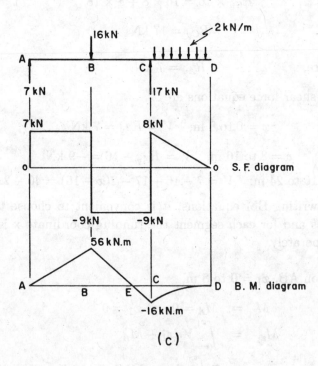

(c)

Fig. 10.17

Thus BM varies linearly with a positive slope. At point B, the maginitude is 56 kN.m.

For region BC, $x_2 = 0$ to 8 m,

$$M = M_B = 56 \text{ kN.m} \quad \text{at } x_2 = 0$$

and $\quad M_{x_2} = \int_0^{x_2} V dx_2 + M_B = -9 \int_0^{x_2} dx_2 + 56 = -9x_2 + 0$

Therefore, $\quad M_{x_2} = -72 + 56 = -16 \text{ kN.m at} \quad x_2 = 8 \text{ m}$

In this region, the BM changes sign and is zero at

$$x_2 = \frac{56}{9} = 6.22 \text{ m} \quad \text{from } B$$

For region CD, $x_3 = 0$ to 4 m,

$$M_{x_3} = \int_0^{x_3} V dx_1 + M_C = \int_0^{x_3} (8 - 2x_3) dx_3 + 56 = 8x_3 - x_3^2 - 16$$

$$M_{x_3} = 32 - 16 - 16 = 0 \text{ at} \quad x_3 = 4 \text{ m}$$
$$\frac{dM_{x_3}}{dx} = 8 - 2x_3 \quad \text{(positive slope)}$$

Fig. 10.17(c) shows the SF and BM diagrams.

Example 10.10

Draw the SF and BM diagrams for the beam shown in Fig. 10.18(a).

Solution

Figure 10.18(b) gives the FBD for the loaded beam.
From equilibrium of vertical forces,

$$R_{Ay} + R_{Fy} - 10 - 10 = 0$$

$$R_{Ay} + R_{Fy} = 20 \text{ kN}$$

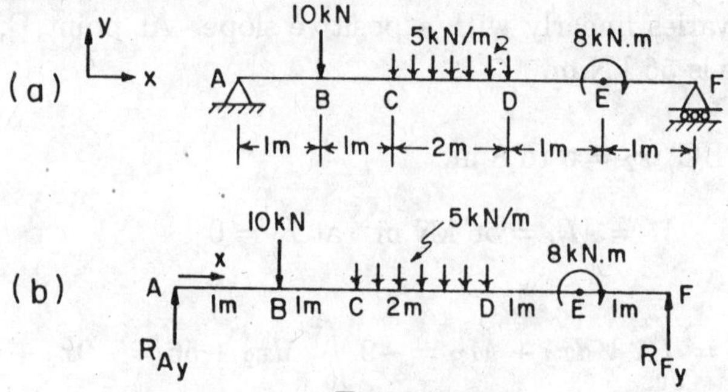

Fig. 10.18

$$R_{Fy} \times 6 = 10 \times 1 + 10 \times 3 + 8$$

$$R_{Fy} = 8 \text{ kN}$$

Therefore, $\qquad R_{Ay} = 12 \text{ kN}$

The SF equations are

$$x = 0 \text{ to } 1 \text{ m}; \quad V = R_{Ay} = 12 \text{ kN}$$

$$x = 1 \text{ to } 2 \text{ m}; \quad V = R_{Ay} - 10 = 2 \text{ kN}$$

$$x = 2 \text{ to } 4 \text{ m}; \quad V = 2 - 5(x - 2) = 12 - 5x$$

These represent straight line variation, i.e.,

$$V = 2 \text{ kN at } x = 2 \text{ m}$$

$$V = -8 \text{ kN at } x = 4 \text{ m}$$

$$V = 12 - 20 = -8 \text{ kN at } x = 4 \text{ to } 6 \text{ m}$$

The SF becomes zero at $x = \frac{12}{5} = 2.4$ m from A.

For BM equations, we once again consider each segment seperately and the running coordinate x is defined for seperate regions. For region AB, $x_1 = 0$ to 1 m,

$$M = M_A = 0 \quad \text{at } x_1 = 0$$

$$M_{x_1} = \int_0^{x_1} V \, dx_1 + M_A = 12 \int_0^{x_1} dx_1 + 0 = 12x_1$$

$$M_{x_1} = 12 \text{ kN.m at } x_1 = 1 \text{ m}$$

The BM varies linearly with a positive slope. At point B, the magnitude is 12 kN.m.

For region B to C, $x_2 = 0$ to 1 m,

$$M_B = 12 \text{ kN.m} \quad \text{at } x_2 = 0$$

$$M_{x_2} = \int_0^{x_2} V dx_2 + M_B$$

$$= 2 \int_0^{x_2} dx_2 + 12 = 2x_2 + 12$$

$$M_{x_2} = 14 \text{ kN.m at } x_1 = 1 \text{ m}$$

The BM in this region also varies linearly with a positive slope. At point C, the magnitude is 14 kN.m.

For region CD, $x_3 = 0$ to 2 m,

$$M_{x_3} = \int_0^{x_3} V dx_3 + M_C$$

$$= \int_0^{x_2} (2 - 5x_3) dx_3 + 14 = 2x_3 - \frac{5}{3}x_3^2 + 14$$

$$M_{x_3} = 4 - 10 + 14 = 8 \text{ kN.m at } x_1 = 1 \text{ m}$$

$$M_{x_3} = 2 \times 0.4 - \frac{5 \times 0.4^2}{2} + 14 = 14.4 \text{ kN.m at } x_3 = 0.4 \text{ m}$$

This is the maximum value of the BM. At this point the SF is zero.

For region DE, $x_4 = 0$ to 1 m,

$$M_{x_4} = \int_0^{x_4} V dx_4 + M_D$$

$$= -8 \int_0^{x_4} dx_4 + 8 = -8x_4 + 8$$

$$M_{x_4} = 0 \text{ at } x_4 = 1 \text{ m}$$

At E, there is a concentrated moment of 8 kN.m acting. This will cause a jump in the BM at this point.

For region EF, $x_5 = 0$ to 1 m,

$$
\begin{aligned}
M_{x_5} &= \int_0^{x_5} V\,dx_2 + M_E \\
&= -8\int_0^{x_5} dx_2 + 8 = -8x_5 + 8 \\
M_{x_5} &= 0 \text{ at } x_5 = 1 \text{ m}
\end{aligned}
$$

This agrees with the condition that at F, the BM must be zero as the end is simply supported.

Example 10.11

Draw the SF and BM diagrams for the cantilever beam shown in Fig. 10.19(a). Figure 10.19(b) gives the FBD.

Solution

This type of end condition (or boundary condition) for the beam is slightly different from that of simply supported beam. The end A·being clamped, it can resist bending moment unlike the case of hinged beam.

As the loading is in the plane and is vertical, from equilibrium $R_{Ax} = 0$. From equilibrium of vertical forces,

$$
R_{Ay} = (20 \times 2 + 5) = 45 \text{ kN}
$$

Taking moment about point A along an axis normal to XY plane gives

$$
M_A = 20 \times 2 \times 1 + 5 \times 3 = 55 \text{ kN.m}
$$

From the convention so far adopted M_A is negative, being anti-clockwise.

The SF diagram (Fig. 10.19(c)) shows that the shear force decreases linearly starting from a value of 45 kN at the end A to 5 kN just to the right of point C. From C to B it remains constant as there is no external load in this region. At B, the shear force curve comes to zero.

The BM diagram (Fig. 10.19(c)) starts with a value of -55 kN.m at point A. The bending moment in the region A to B is

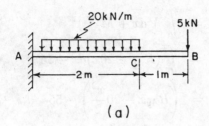

(a)

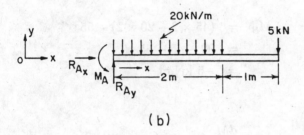

(b)

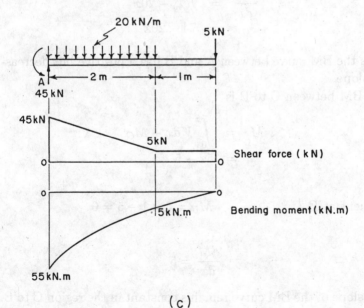

(c)

Fig. 10.19

given by

$$M = \int_0^x V\,dx + M_A$$

$$= \int_0^x (R_{Ay} - 20x)\,dx - 55$$

$$= \left(R_{Ay}x - 20\frac{x^2}{2} \right) - 55$$

Moment at B gives

$$M_B = (45 \times 2 - 20 \times 2) - 55$$

$$= 90 - 40 - 55$$

$$= -5 \text{ kN.m}$$

and

$$\frac{dM}{dx} = R_{Ay} - 20x$$

$$= 74 - 20x$$

Thus the BM curve between A and B has a positive but decreasing slope.

The BM between C to B is

$$M = \int_0^x V\,dx + M_C$$

$$= 5x - 5$$

Moment at B, $M_B = 5 \times 1 - 5 = 0$

and $$\frac{dM}{dx} = 5$$

The slope of the BM curve remains constant in the region C to B. The bending moment at the free end B is zero. Figure 10.19(c) shows the SF and BM diagrams.

Appendix 10.1 gives the maximum values of shear force and bending moment for some typical beams.

10.6 Deflection of Beams

As a consequence of application of transverse load, the elastic beam bends and undergoes displacement. The displacement of the various points of the beam from their original position (unloaded position) is called *deflection*. For the purposes of this book, we shall assume that the maximum deflection of the beam is much smaller compared to the smallest dimension of the beam (generally the depth). This ensures that the stresses any where inside the beam will not exceed the elastic limit.

In the design of machine elements, it is necessary to know the transverse deflection of the member at a certain section. If deflection is excessive, it may not line up properly in relation to other parts, thereby hindering its performance.

We shall consider the bending of a beam which is symmetrical about the plane of bending. The following assumptions are made for deriving the moment-curvature relations:

1. Material of the beam is isotropic and elastic.

2. Plane sections remain plane after deformation. This implies that the shear deformation can be neglected, though shear stress may exist at any cross-section.

3. Maximum deflection at any point of the beam is very much less than the small dimension of the beam.

4. The beam can be represented by a single line (neutral axis) for the measurement of deflection.

5. Stress-strain relation for the beam material is linear.

6. The neutral axis of the beam is the centroidal axis of the cross-section.

Consider a rectangular cross-section beam of length L (Fig. 10.20(a)) Fig. 10.20(a) and Fig. 10.20(b) subjected to a distributed transverse load $q = f(x)$. When the transverse load is applied to the beam, it deflects. The deflected position of the beam is shown by mean of dashed lines (Fig. 10.20(b)). Consider a section of the beam at a distance, ΔL apart. Figure 10.20(b) shows the exploded view of a small element of length ΔL after the application of load. Before application of the load the fiber AA' of the

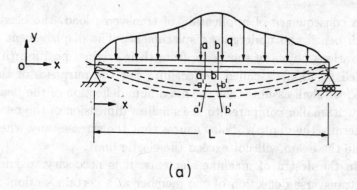

(a)

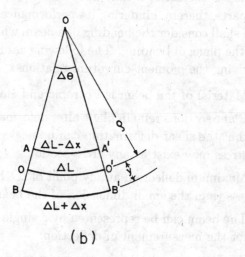

(b)

Fig. 10.20

cross-section at a distance y above the neutral axis, bottom fiber BB' at a similar location on the opposite side and neutral fiber OO' are of the same length ΔL. After application of the load, the three lengths do not remain the same. If ρ is the radius of curvature of the beam, then

$$OO' = \Delta L = \rho \Delta\theta \qquad (10.8)$$

Similarly,

$$\text{Similarly} \quad AA' = \Delta L - \Delta x = (P - y)\Delta\theta$$

$$BB' = \Delta L + \Delta x = (\rho + y)\Delta\theta \qquad (10.9)$$

From Eqs. (10.8) and (10.9),

$$\frac{\Delta L}{\rho} = \frac{\Delta L + \Delta x}{\rho + y} \qquad (10.10)$$

Simplification gives,

$$\frac{y}{\rho} = \frac{\Delta x}{\Delta L} = \frac{\text{increase in fiber length}}{\text{original fiber length}} \qquad (10.11)$$

Since

$$\frac{\Delta x}{\Delta L} = \epsilon_x$$

as shown in Chapter 7,

$$\epsilon_x = \frac{y}{\rho} \qquad (10.12)$$

Similarly, if we consider AA',

$$\epsilon_x = -\frac{y}{\rho} \qquad (10.13)$$

Equations (10.12) and (10.13) imply that the strain varies linearly along the depth of the beam. It is zero at the neutral axis (NA), tensile on the bottom side and compressive on the upper side for the loading shown.

The only non-zero stress in the beam is σ_x. Hence, from constitutive equation,

$$\sigma_x = E\epsilon_x \qquad (10.14)$$

where E is the elastic modulus of the material. Substituting in Eqs. (10.12) and (10.13),

$$\sigma_x = \frac{Ey}{\rho} \qquad (10.15)$$

for the bottom fibre and

$$\sigma_x = -\frac{Ey}{\rho} \qquad (10.16)$$

for the top fibre. Therefore, in general

$$\sigma_x = \frac{Ey}{\rho} \qquad (10.17)$$

and the nature of stress is decided by the response of the beam to the applied moment. The term $\frac{1}{\rho}$ is the curvature of the neutral axis.

10.6.1　Location of Neutral Axis

Since there is no net axial force acting on any cross-section of the beam, the summation of the forces along the axial direction must be zero. That is,

$$\int_{A_b} \sigma_x dA_b = 0 \qquad (10.18)$$

where A_b is the area of cross-section of the beam. Substituting for σ_x from Eq. (10.17),

$$\int_{A_b} \frac{Ey}{\rho} dA_b = 0 \qquad (10.19)$$

Since E and ρ are constants, Eq. (10.19) reduces to

$$\int y dA_b = 0$$

or

$$A_b \bar{y} = 0 \qquad (10.20)$$

where $\bar{y}$ is the distance of the centroid of area A_b from the plane from which y is measured. Since $A_b \neq 0$,

$$\bar{y} = 0 \qquad (10.21)$$

Therefore, NA passes through the centroid of the area.

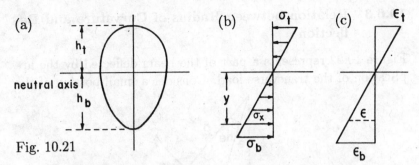

Fig. 10.21

10.6.2 Relation for Flexural Stress and Moment-Curvature

Fibers along the length of the beam are compressed or elongated due to the presence of flexural stress along the cross-section of the beam. From Fig. 10.20(a), for the given loading, the fibers above the neutral axis are compressed, while fibers below the neutral axis are stretched. Over an elemental area dA_b, if the flexural stress is σ_x, then the force on the area is

$$dF = \sigma_x dA_b \tag{10.22}$$

$$M = \int_{A_b} dF y = \int_{A_b} \sigma y dA_b \tag{10.23}$$

Figure 10.21(a) shows the exploded view of the symmetric cross-section of depth h. From Fig. 10.21(b),

$$\frac{\sigma_b}{h_b} = \frac{\sigma_t}{h_t} = \frac{\sigma_x}{y} \tag{10.24}$$

or

$$\sigma_x = \frac{\sigma_b}{h_b} y$$

Therefore,

$$M = \frac{\sigma_b}{h_b} \int_{A_b} y^2 dA_b = \frac{\sigma_x}{y} I \tag{10.25}$$

where $\int_{A_b} y^2 dA_b = I = $ the second moment of area of the cross-section.

Substituting for σ_x from Eq. (10.17) into Eq. (10.25) gives

$$\frac{M}{EI} = \frac{1}{\rho} \tag{10.26}$$

Equation (19.26), relates the BM at any section of the beam to the radius of curvature ρ and flexural rigidity, EI of the beam.

10.6.3　Relation between Radius of Curvature and Deflection

Figure 10.22 represents a part of the beam deflected by the application of the transverse load. Consider a small portion of the

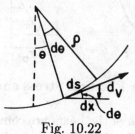

Fig. 10.22

beam of length ds at a distance x from the support and vertical distance v with respect to the same point. The slope of the beam is given by $\frac{dv}{ds}$. If θ is the angle which the tangent to the curve makes with the X axis, then

$$\frac{dv}{dx} = \tan\theta; \quad \frac{ds}{dx} = \sec\theta; \text{ and } \quad \frac{d\theta}{ds} = \frac{1}{\rho}$$

Therefore,

$$\frac{d^2v}{dx^2} = \frac{d}{dx}(\tan\theta) = \frac{d}{d\theta}(\tan\theta)\frac{d\theta}{dx} = \sec^2\theta\frac{d\theta}{dx}$$

$$= \sec^2\theta\frac{d\theta}{ds}\frac{ds}{dx} = \sec^3\theta\left(\frac{1}{\rho}\right)$$

or

$$\frac{1}{\rho} = \frac{\frac{d^2v}{dx^2}}{\sec^3\theta} \tag{10.27}$$

From trigonometric relations $\sec^2\theta = 1 + \tan^2\theta = 1 + \left(\frac{dv}{dx}\right)^2$,

$$\frac{1}{\rho} = \frac{\frac{d^2v}{dx^2}}{\left[1 + \left(\frac{dv}{dx}\right)^2\right]^{\frac{3}{2}}} \tag{10.28}$$

If $\frac{dv}{dx} \ll 1$, then

$$\frac{1}{\rho} = \frac{d^2v}{dx^2} \tag{10.29}$$

Hence, from Eq. (10.26),

$$\frac{M}{EI} = \frac{d^2v}{dx^2}$$

or

$$M = EI\frac{d^2v}{dx^2} \qquad (10.30)$$

Equation (10.30) is the moment-curvature relation for small deflection of a symmetric, isotropic and elastic beam. This equation can be used to obtain the deflection of a beam. The curvature of the beam can be positive or negative depending on (i) the sign convention used for the bending moment, M and (ii) the choice of the coordinates axis. If we follow the sign convention for the bending moment so far employed and also the coordinate system XY, then

$$M = -EI\frac{d^2v}{dx^2} \qquad (10.31)$$

Differentiating Eq. (10.31) twice with respect to x for a constant EI gives

$$\frac{d^2M}{dx^2} = -EI\frac{d^4v}{dx^4} \qquad (10.32)$$

Making use of Eqs. (10.3) and (10.5), Eq. (10.32) can be written as

$$EI\frac{d^4v}{dx^4} = -q \qquad (q \text{ acting upwards}) \qquad (10.33)$$

For $q(x, y)$ acting downwards,

$$EI\frac{d^4v}{dx^4} = q \qquad (10.34)$$

Equations (10.33) and (10.34) give the equations governing the transverse deflection of beams when load $q(x, y)$ is acting upwards or downwards, respectively. Equation (10.31) can also be used provided moment M, as a function of axial length, is known.

10.7 Boundary Conditions

Equation (10.31) is a second order, ordinary, linear differential equation with constant coefficients for a uniform cross-section beam and requires only two constants to be evaluated for determining the transverse deflection at any point of the beam. However, Eq. (10.34) requires the evaluation of four independent constants for the determination of transverse deflection. To evaluate these constants it would be necessary to know the geometric (static) and/or natural (dynamic) boundary conditions at the ends of the beam. The ideal boundary or end conditions are (i) simply supported end, (ii) clamped end 'and (iii) free end.

1. At a simply supported or hinged end:

   (a) The geometric condition is

   $$\text{Lateral deflection}, v = 0 \qquad (10.35)$$

   (b) The natural (dynamic) condition is
   $$\text{Bending moment}, M \left(= EI\frac{d^2v}{dx^2}\right) = 0$$

2. At a clamped or built-in end:

   (a) The geometric conditions are

   $$\text{Lateral deflection}, \quad v = 0 \qquad (10.36)$$
   $$\text{Bending slope}, \quad \frac{dv}{dx} = 0$$

3. At a free end:

   (a) The natural (dynamic) condition is

   $$\text{Bending moment}, \quad M \left(= EI\frac{d^2v}{dx^2}\right) = (10.37)$$

   $$\text{Shear force}, \quad V \left(= EI\frac{d^3v}{dx^3}\right) = 0$$

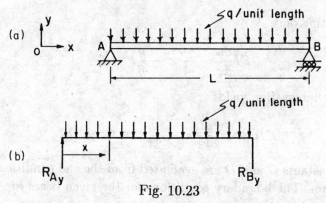

Fig. 10.23

We shall now consider some examples of evaluating transverse deflection by employing governing differential Eqs. (10.31) or (10.34).

Appendix A13 placed at the end of Chapter 13 gives the detailed derivation for neutral axis and moment of inertia.

Example 10.12

Obtain an expression for the transverse deflection of a constant cross-section simply supported beam subjected to a uniformly distributed load q per unit length shown in Fig. 10.23(a).

Solution

Figure 10.23(b) gives the FBD of the beam. The reactions at the ends R_{Ay} and R_{By} are obtained as

$$R_{Ay} = R_{By} = \frac{qL}{2}$$

The BM at any x is

$$M(x) = \frac{qL}{2}x - \frac{q}{2}x^2$$

From the equation governing the moment

$$EI\frac{d^2v}{dx^2} = \frac{qL}{2}x - \frac{q}{2}x^2$$

Integrating once

$$EI\frac{dv}{dx} = \frac{qL}{4}x^2 - \frac{q}{6}x^3 + C$$

Further integration yields

$$EIv(x) = \frac{qL}{12}x^3 - \frac{q}{24}x^4 + Cx + D$$

The constants C and D are evaluated from the end conditions of the beam. The boundary conditions for the given beam are

(a)   at $x = 0$, $v = 0$
(b)   at $x = L$, $v = 0$

Substituting these, gives

$$C = \frac{qL^3}{24}; \qquad D = 0$$

Therefore,

Therefore $\dfrac{dv}{dx} = \dfrac{q}{24EI}\left(6Lx^2 - 4x^3 - L^3\right)$

and $v(x) = \dfrac{q}{24EI}\left(2Lx^3 - x^4 - L^3x\right)$

$v(x)$ gives the deflection at any location within the domain. As the structure and the loading is symmetric, the maximum deflection at the mid-point $(x = L/2)$ of the beam is

$$v_{max} = \frac{q}{24EI}\left(\frac{-5}{16}L^4\right) = -\frac{5qL^4}{384EI}$$

Deflection is negative, since y is positive upwards.

Example 10.13

Obtain the expression for deflection of a constant cross-section cantilever beam loaded at the tip (Fig. 10.24(a)).

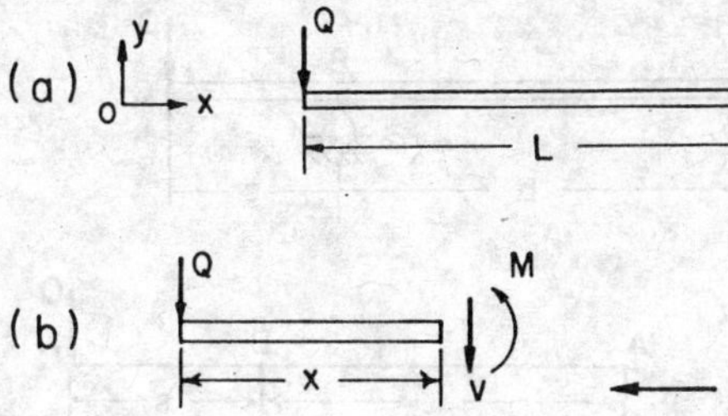

Fig. 10.24

Solution

Figure 10.24(b) shows the FBD of an element a distance x from the tip.

The bending moment at a distance x is $M = -Qx$ Therefore, $EI\frac{d^2v}{dx^2} = -Qx$ Integrating once gives $EI\frac{dv}{dx} = -Q\frac{x^2}{2} + C$ Integrating once again gives $EIv(x) = -Q\frac{x^3}{6} + Cx + D$ The boundary conditions for the beam are:

(a) $v = 0$; $x = 0$
(b) $\frac{dv}{dx} = 0$; $x = L$

Using the second condition, $C = QL^2/2$
Making use of the first condition, $D = -QL^3/3$

Hence, $EIv(x) = -Q\frac{x^3}{6} + \frac{QL^2}{2}x - \frac{QL^3}{3}$

Deflection at the tip of the beam is obtained by putting $x = 0$ in the equation for deflection. This gives

$$v = -\frac{QL^3}{3EI}$$

The negative sign indicates that the deflection is downward.

The slope of the beam at $x = 0$ is given by

$$\frac{dv}{dx} = \frac{QL^2}{2EI}$$

Example 10.14

For the overhang uniform beam shown in Fig. 10.25(a), obtain the expression for deflection $v(x)$ for the portion AB of the beam.

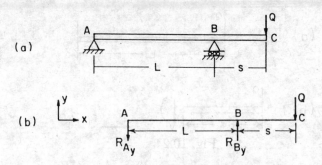

Fig. 10.25

Solution

Figure 10.25(b) shows the FBD for the beam system.
Taking moment at A,

$$R_{By} \times L = Q(L + s)$$

$$R_{By} = Q\left[1 + \frac{s}{L}\right]$$

$$R_{By} = R_{Ay} + Q$$

Hence, $\qquad R_{Ay} = Q\left[1 + \frac{s}{L}\right] - Q = Q\frac{s}{L}$

The bending moment M within the region AB is given as

$$M = -Q\frac{s}{L}x; \qquad 0 < x < L$$

The differential equation for the elastic curve is then,

$$EI\frac{d^2v}{dx^2} = -Q\frac{s}{L}x$$

Integrating twice,

$$EI\frac{dv}{dx} = -\frac{Qs}{L}\frac{x^2}{2} + D$$

$$EIv(x) = -\frac{Qs}{L}\frac{x^3}{6} + Dx + E$$

The conditions at the support are

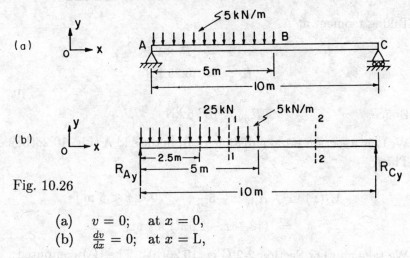

(a)

(b)

Fig. 10.26

(a) $v = 0$; at $x = 0$,
(b) $\frac{dv}{dx} = 0$; at $x = L$,

Making use of these conditions,

$$E = 0; \quad D = \frac{QsL}{6}$$

Therefore,

$$v(x) = \frac{1}{EI}\left(-\frac{Qs}{L}\frac{x^3}{6} + \frac{QsL}{6}x\right)$$

$$\frac{dv}{dx} = \frac{1}{EI}\left(\frac{Qs}{L}\frac{x^2}{2} + \frac{QsL}{6}\right)$$

Example 10.15

Obtain the general expression for deflections for the constant cross-section simply supported beam subjected to a uniform distributed load over half the beam length as shown in Fig. 10.26(a).

Solution

From the FBD (Fig. 10.26(b)), the support reactions are calculated.
From vertical equilibrium,

$$R_{Ay} + R_{Cy} = 25 \text{ kN}$$

Taking moment at A,

$$R_{Cy} \times 10 = 25 \times 2.5$$
$$R_{Cy} = 6.25 \text{ kN}$$

Hence, $R_{Ay} = 18.75 \text{ kN}$

We take a section 1-1 at a distance x from the end A (Fig. 10.26(b)). Then,

$$M_1(x) = R_{Ay} x - 5\frac{x^2}{2}; \qquad 0 < x < 5 \text{ m}$$
$$= (18.75x - 2.5x^2) \text{ kN.m}$$

We take another Section 2-2 (Fig. 10.26(b)). The BM computed from the right end is

$$M_2(x) = 6.259(10 - x) \text{ kN.m}; \qquad 5 < x < 10 \text{ m}$$

For the region AB,

$$EI\frac{d^2v}{dx^2} = M_1(x)$$
$$EI\frac{d^2v}{dx^2} = 18.75x - 2.5x^2$$
$$EI\frac{dv}{dx} = 18.75\frac{x^2}{2} - 2.5\frac{x^3}{3} + C_1$$
$$EIv(x) = 18.75\frac{x^3}{6} - 2.5\frac{x^4}{12} + C_1x + D_1$$

For the region BC,

$$EI\frac{d^2v}{dx^2} = 62.5 - 6.25x$$
$$EI\frac{dv}{dx} = 62.5x - 6.25\frac{x^2}{2} + C_2$$
$$EIv(x) = 62.5\frac{x^2}{2} - 6.25\frac{x^3}{6} + C_2x + D_2$$

The four constants of integration C_1, D_1, C_2 and D_2 have to be evaluated from the boundary conditions at A and C and continutity conditions at B. These are

1. $v = 0$ at $x = 0$
2. $v = 0$ at $x = 10$ m
3. the slope at B computed from both the equations must be equal
4. the deflection at B computed from both the equations must be equal

On application of these conditions,

$$D_1 = 0$$

$$62.5 \times \frac{10^2}{2} - 6.25 \times \frac{10^3}{6} + C_2 \times 10 + D_2 = 0$$

$$18.75 \times \frac{5^2}{2} - 2.5 \times \frac{5^3}{3} + C_1 = 62.5 \times 5 - 6.25 \times \frac{5^2}{2} + C_2$$

$$18.75 \times \frac{5^3}{6} - 2.5 \times \frac{5^4}{12} + C_1 \times 5 + D_1 = 62.5 \times \frac{5^2}{2} - 6.25 \times \frac{5^3}{6} + C_2 \times 5 + D_2$$

or

$$D_1 = 0$$
$$10C_2 + D_2 = -1041.667$$
$$C_1 + C_2 = 104.167$$
$$5C_1 - 5C_2 + D_2 = 390.625$$

Solving the system of simultaneous algebraic equations gives

$$C_1 = 149.739$$
$$C_2 = -45.572$$
$$D_1 = 0$$
$$D_2 = 585.947$$

Therefore, the beam deflections in the two regions are

$$v(x) = \frac{1}{EI} \left(\frac{18.75x^3}{6} - \frac{2.5x^4}{12} + 149.739x \right) ; \quad 0 < x < 5 \text{ m}$$

$$v(x) = \frac{1}{EI} \left(62.5\frac{x^2}{2} - 6.25\frac{x^3}{6} - 45.572x + 585.947 \right) ; \quad 5 < x < 10 \text{ m}$$

So far, we considered the differential equation approach for evaluating the deflections of beams. We shall now consider some examples making use of moment-area method.

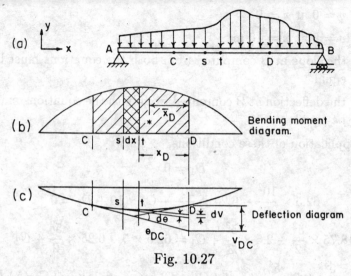

Fig. 10.27

10.8   Moment-Area Method

This method, which is semigraphical, is useful for deflections of beams, where the loading is complex or cross-section is varying along the length. One such example is shaft supported on bearings, where the cross-section varies along the length. In structural applications, varying cross-section beams are often used: larger sections are used where the bending moment is high, while at lower bending moment, the section required is small. This results in optimal utilization of the beam material. The moment-area method uses the quantity $\frac{M}{EI}$, the bending moment at a section divided by the flexural rigidity or bending rigidity at the same section.

Consider a uniform rectangular beam subjected to an arbitrary distributed load (Fig. 10.27(a)).

Take any two arbitrary points s and t on the elastic curve at a distance dx apart and let the radius of curvature of the deflection curve at these points be ρ. Then,

$$dx = \rho d\theta \qquad (10.38)$$

From Eq. (10.26),

$$\frac{1}{\rho} = \frac{M}{EI}$$

Substituting for ρ in Eq. (10.38) gives

$$d\theta = \frac{M}{EI}dx = \frac{1}{EI}dA_M \qquad (10.39)$$

where dA_M is the differential area of the element in the moment diagram between s and t (Fig. 10.27(b)). The angle

$$\theta_{DC} = \int_C^D d\theta = \int_C^D \frac{1}{EI}dA_M = \frac{1}{EI}\int_C^D dA_M = \frac{1}{EI}[A_M]_C^D \qquad (10.40)$$

In Eq. (10.40), $[A_M]_C^D$ represents the area of the moment diagram between the points C and D. Equation (10.40) can be stated as follows:

Theorem 1: *The angle between the tangents at any two points C and D of the deflection curve is equal to the area of the bending moment diagram between the vertical lines through C and D divided by the flexural rigidity of the beam (for a uniform beam).* For a non-uniform cross-sectional beam, the second moment of area I, has to obtained at each cross-section for integration.

The differential displacement between the points s and t can be obtained (for small slopes) by the expression (Fig. 10.27(c))

$$dv = x_D d\theta \qquad (10.41)$$

Integration of Eq. (10.41) gives

$$v_{DC} = \int_C^D dv = \int_C^D x_D d\theta \qquad (10.42)$$

Substituting for $d\theta$ from Eq. (10.39) gives

$$v_{DC} = \int_C^D \frac{x_D}{EI}dA_M \qquad (10.43)$$

Using the definition of the centroid of an area, Eq. (10.43), can be rewritten in a simpler form as

$$v_{DC} = \frac{1}{EI}\bar{x}_D[A_M]_C^D \qquad (10.44)$$

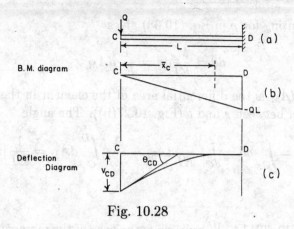

Fig. 10.28

where and $\bar{x}_D$ is the horizontal distance from the centroid of this area to point D. Equation (10.44) can be stated as follows:

Theorem 2: *The tangential deviation between any two points C and D of the deflection curve (vertical distance from D to the tangent at C) is equal to the area of the bending moment diagram between the vertical lines through C and D multiplied by the distance from the centroid of the area to point D divided by EI of the beam (for uniform cross-section only).* For non-uniform cross-section, one has to use the value of I at every section during integration.

Example 10.16

We shall consider once again the problem of a cantilever beam with a tip load discussed in Example 10.13 (Fig. 10.28(a)). We shall now obtain the maximum deflection and maximum slope of the tip.

Solution

The bending moment diagram is a triangle as shown in Fig. 10.28(b). The maximum value being QL, at the support. The area of the BM diagram between CD is

$$[A_M]_C^D = \frac{L(-PL)}{2} = -\tfrac{1}{2}PL^2$$

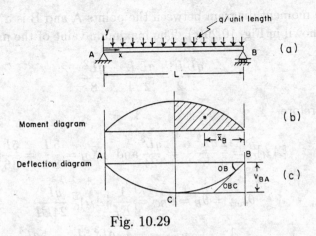

Fig. 10.29

and the centroid of this area from C is the distance of $\frac{2L}{3}$ from the tip. Now,

$$\theta_{max} = \theta_{CD} = \frac{1}{EI}[A_M]_C^D = \frac{1}{EI}\frac{PL^2}{2}$$

$$v_{max} = v_{CD} = \frac{1}{EI}[A_M]_C^D\,\bar{x}_C = \frac{1}{EI}\frac{PL^2}{2}\frac{2L}{3} = -\frac{PL^3}{3EI}$$

The negative sign indicates that point C is below point D (Fig. 10.28(c)).

Example 10.17

Obtain the maximum deflection and slope of a uniformly loaded constant cross-section simply supported beam of Example 10.12 (Fig. 10.29(a)).

Solution

The structure and loading is symmetric with respect to the vertical axis passing through $x = \frac{L}{2}$, the center C of the beam. Hence, the deflection must be symmetric with respect to C. Therefore, the tangent at point C is horizontal and

$$\theta_B = \theta_{BC}$$

The moment diagram between the points A and B is a parabola as shown in Fig. 10.29(c). The maximum value of the moment is

$$\frac{qL}{2}\frac{L}{2} - \frac{qL}{2}\frac{L}{4} = \frac{qL^2}{8}$$

Therefore,

$$[A_M]_C^B = \frac{2}{3}\frac{qL^2}{8}\frac{L}{2} = \frac{qL^3}{24} \text{ and } \quad \overline{X}_B = \frac{5}{8}\frac{L}{2} = \frac{5L}{16}$$

$$\theta_{max} = \theta_B = \theta_{BC} = \frac{1}{EI}[A_M]_C^B \frac{qL^3}{24EI}$$

$$v_C = \frac{1}{EI}[A_M]_C^B \overline{X}_B = \frac{1}{EI}\frac{qL^3}{24}\frac{5L}{16} = \frac{5qL^4}{384EI}$$

Example 10.18

For the uniform rectangular beam loaded (Fig. 10.30(a)), obtain the deflection at A using Moment-Area method.

Solution

Figure 10.30(b) gives the FBD of the beam. Here,

$$R_{Cy} + R_{By} = \frac{qL}{4}$$

$$R_{Cy} \times L = \frac{qL}{4}\left(\frac{L}{8} + L\right) = \frac{9qL^2}{32}$$

Therefore,

$$R_{Cy} = \frac{9qL}{32}, \qquad R_{By} = \frac{qL}{32}$$

Figure 10.30(c) shows the BM diagram.

The $\frac{M}{EI}$ diagram will be similar, since EI is constant along the length of the beam. The BM curve is a parabola up to C, later on it is a straight line. The maximum magnitude at C is $\frac{qL^2}{32}$ and

$$A_{M1} = \frac{1}{3}\left(-\frac{qL^2}{32EI}\right)\frac{L}{4} = -\frac{qL^3}{384EI}$$

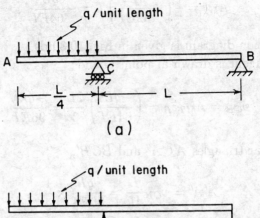

(a)

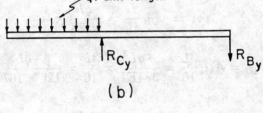

(b)

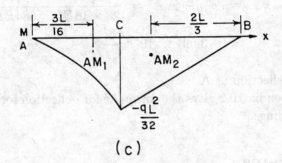

(c)

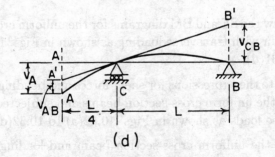

(d)

Fig. 10.30

$$A_{M2} = \tfrac{1}{2}\left(-\frac{qL^2}{32EI}\right)L = -\frac{qL^3}{64EI}$$

The deflection diagram is given in Fig. 10.30(d).

The tangent is drawn at point C and

$$v_{CB} = A_{M2}\bar{x}_B = \left(\frac{-qL^3}{64EI}\right)\frac{2L}{3} = \frac{-qL^4}{96EI}$$

From similar triangles $A'CA''$ and BCB',

$$A'A'' = v_{CB}\frac{L}{4L} = \frac{-qL^4}{96EI}\times\frac{1}{4}$$

$$v_{AB} = A_{M1}\frac{3L}{16} = \frac{-qL^3}{384EI}\times\frac{3L}{16} = \frac{-qL^4}{128\times16EI}$$

$$v_A = v_{AB} + A'A'' = \frac{-qL^4}{128\times16EI} - \frac{-qL^4}{384EI}$$

$$= -3.091\times10^{-3}\frac{qL^4}{EI}$$

is the deflection at A.

Appendix 10.2 gives the formulae for deflection for some typical loadings.

Problems

10.1 Draw the SF and BM diagrams for the uniform cross-section beam with transverse loading as shown in Figs. 10.31(a) to 10.31(d).

10.2 Write the expressions for shear forces and bending moments for the uniform cross-section beams and subjected to transverse loads as shown in Figs. 10.32(a) to 10.32(d).

10.3 For the uniform cross-section beam and loading shown in Figs. 10.33(a) to 10.33(d), determine (i) the magnitude of the maximum bending moment, (ii) the location and (iii) the maximum normal stress due to bending.

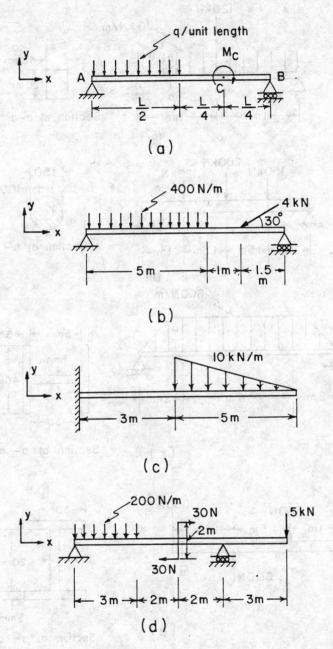

Fig. 10.31

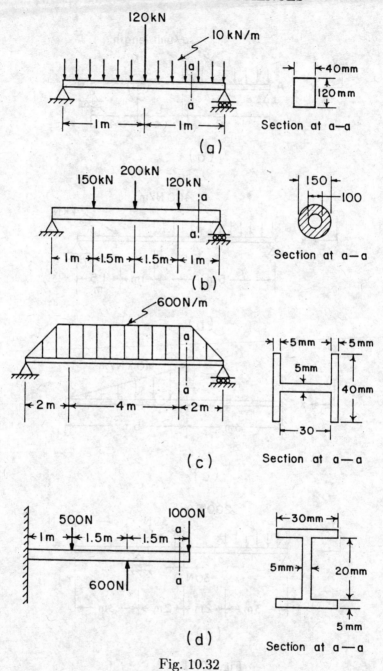

Fig. 10.32

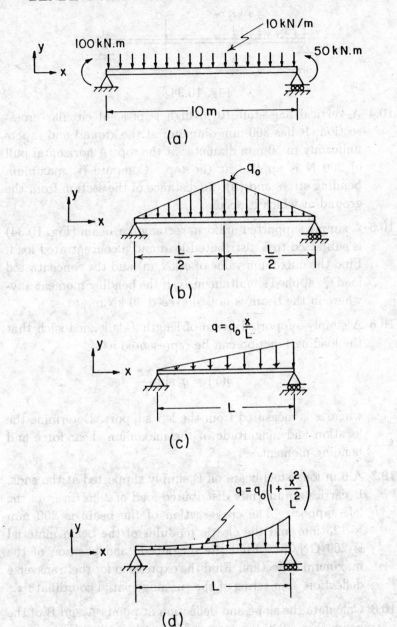

Fig. 10.33

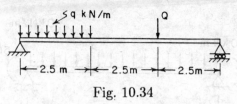

Fig. 10.34

10.4 A vertical flag staff 10 m high is of solid circular cross-section. It has 800 mm diameter at the ground and tapers uniformly to 50 mm diameter at the top. A horizontal pull of 200 N is applied at the top. Compute (i) maximum bending stress and (ii) the distance of the section from the ground at which it occurs.

10.5 A simply supported uniform rectangular beam (Fig. 10.34) is subjected to a distributed load and a concentrated load. Find the maximum value of q(kN/m) and the concentrated load Q, applied simultaneously, if the bending moment anywhere in the beam is not to exceed 30 kN.m.

10.6 A simply supported beam of length L is loaded such that the loading function can be represented as

$$q(x) = q_o \sin \frac{\pi x}{L}$$

where x is measured from the left support. Determine the location and magnitude of the maximum shear force and bending moment.

10.7 A 6 m long steel beam off is simply supported at the ends. It carries a uniformly distributed load of 2 kN/m from the left support. The cross-section of the beam is 200 mm × 125 mm and the elastic modulus of the beam material is 200 GN/m². Find the magnitude and position of the maximum deflection. Find the expression for the transverse deflection, v in terms of the running spatial coordinate x.

10.8 Calculate the slope and deflection at points A and B of the beam (Fig. 10.35) in terms of flexural rigidity, EI by using the Moment-Area method.

10.9 Determine the slope and deflection of the free end of the cantilever beam of rectangular cross-section, when it is loaded

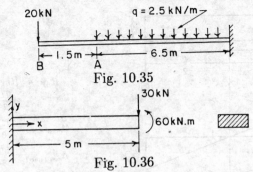

Fig. 10.35

Fig. 10.36

as shown in Fig. 10.36. Assume the flexural rigidity, $EI = 40$ MN.m^2.

10.10 A stepped cantilever beam is loaded as shown in Fig. 10.37. Calculate the transverse deflection of the point B and C employing the Moment-Area method.

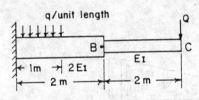

Fig. 10.37

Appendix A10.1

Bending Moment and Shear Force Diagrams for Uniform Cross-section Beams (After Das and Hassler, 1988)

(a) Simple Beam - Uniformly Distributed Load (Fig. A10.1(a)

$$R_{AY} = R_{BY} = V_A = V_B = \frac{qL}{2}$$

$$M_{\max(\text{center})} = \frac{qL^2}{8}; \quad \Delta_{\max(\text{center})} = \frac{5qL^4}{384EI}$$

(b) Simple Beam - Triangular Load (Fig. A10.1(b)

$$Q = \frac{qL}{2}; \quad R_{AY} = V_A = \frac{Q}{3}; \quad R_{BY} = V_B = \frac{2Q}{3}$$

$$M_{\max} = 0.1283QL; \quad \Delta_{\max(\text{at } 0.519 \text{ L})} = \frac{0.01304QL^3}{EI}$$

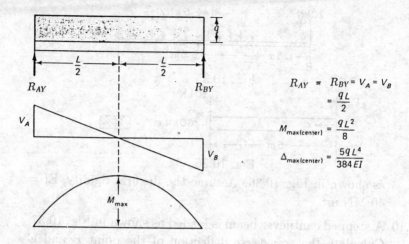

$$R_{AY} = R_{BY} = V_A = V_B$$
$$= \frac{qL}{2}$$
$$M_{max(center)} = \frac{qL^2}{8}$$
$$\Delta_{max(center)} = \frac{5qL^4}{384EI}$$

Simple Beam - Uniformly Distributed Load

Fig. A10.1(a)

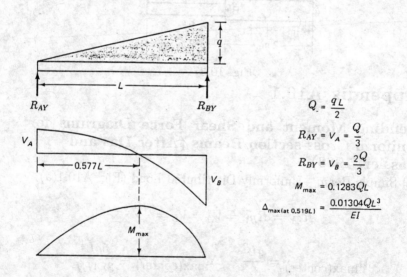

$$Q = \frac{qL}{2}$$
$$R_{AY} = V_A = \frac{Q}{3}$$
$$R_{BY} = V_B = \frac{2Q}{3}$$
$$M_{max} = 0.1283QL$$
$$\Delta_{max(at\ 0.519L)} = \frac{0.01304QL^3}{EI}$$

Simple Beam - Triangular Load

Fig. A10.1(b)

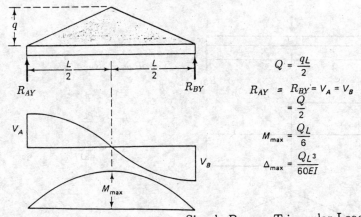

$$Q = \frac{qL}{2}$$

$$R_{AY} = R_{BY} = V_A = V_B$$

$$= \frac{Q}{2}$$

$$M_{max} = \frac{QL}{6}$$

$$\Delta_{max} = \frac{QL^3}{60EI}$$

Simple Beam - Triangular Load

Fig. A10.1(c)

(c) Simple Beam - Triangular Load (Fig. A10.1(c)

$$Q = \frac{qL}{2}; \quad R_{AY} = R_{BY} = V_A = V_B = \frac{Q}{2}$$

$$M_{max} = \frac{QL}{6}; \quad \Delta_{max} = \frac{QL^3}{60EI}$$

(d) Simple Beam - Concentrated Load at Center (Fig. A10.1(d)

$$R_{AY} = R_{BY} = \frac{P}{2}; \quad V_A = V_B = \frac{P}{2}$$

$$M_{max} = \frac{PL}{4}; \quad \Delta_{max} = \frac{PL^3}{48EI}$$

(e) Simple Beam - Concentrated Load at any Point (Fig. A10.1(e)

$$a + b = L; \quad R_{AY} = V_A = \frac{Pa}{L}; \quad R_{BY} = V_B = \frac{Pb}{L}$$

$$M_{max} = \frac{Pab}{L}; \quad \Delta_{max} \text{ (at } x = \sqrt{\frac{a(a+2b)}{3}} \text{ when } a > b)$$

$$= \frac{Pab(a + 2b)\sqrt{3a(a + 2b)}}{27EIL}$$

(f) Fixed end Beam - Uniform Load (Fig. A10.1(f)

$$R_{AY} = R_{BY} = V_A - V_B = \frac{qL}{2}; \quad M_{max(ends)} = \frac{QL^2}{12}$$

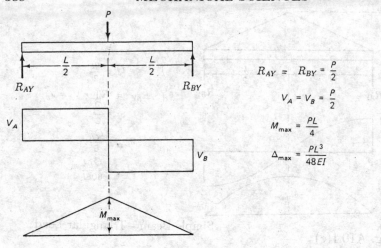

$$R_{AY} = R_{BY} = \frac{P}{2}$$

$$V_A = V_B = \frac{P}{2}$$

$$M_{max} = \frac{PL}{4}$$

$$\Delta_{max} = \frac{PL^3}{48EI}$$

Simple Beam - Concentrated Load at Center

Fig. A10.1(d)

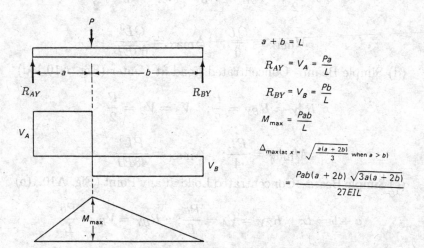

$$a + b = L$$

$$R_{AY} = V_A = \frac{Pa}{L}$$

$$R_{BY} = V_B = \frac{Pb}{L}$$

$$M_{max} = \frac{Pab}{L}$$

$$\Delta_{max\,(at\ x\ =\ \sqrt{\frac{a(a + 2b)}{3}}\ when\ a > b)}$$

$$= \frac{Pab(a + 2b)\sqrt{3a(a + 2b)}}{27EIL}$$

Simple Beam - Concentrated Load at any Point

Fig. A10.1(e)

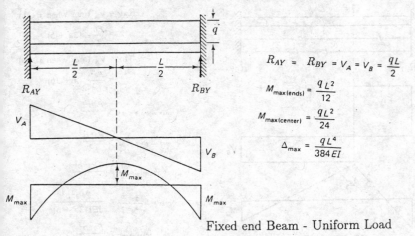

$$R_{AY} = R_{BY} = V_A = V_B = \frac{qL}{2}$$

$$M_{max(ends)} = \frac{qL^2}{12}$$

$$M_{max(center)} = \frac{qL^2}{24}$$

$$\Delta_{max} = \frac{qL^4}{384EI}$$

Fixed end Beam - Uniform Load

Fig. A10.1(f)

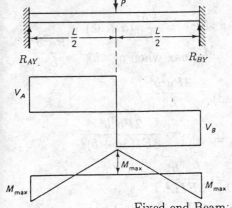

$$R_{AY} = V_A = R_{BY} = V_B = \frac{P}{2}$$

$$M_{max} = \frac{PL}{8}$$

$$\Delta_{max} = \frac{PL^3}{192EI}$$

Fixed end Beam - Concentrated Load at Center

Fig. A10.1(g)

$$M_{max(center)} = \frac{QL^2}{24}; \quad \Delta_{max} = \frac{QL^4}{384EI}$$

(g) Fixed end Beam - Concentrated Load at Center (Fig. A10.1(g)

$$R_{AY} = V_A = R_{BY} = V_B = \frac{P}{2}; \quad M_{max} = \frac{PL}{8}$$

$$\Delta_{max} = \frac{PL^3}{192EI}$$

(h) Fixed end Beam - Concentrated Load at any Point (Fig. A10.1(h)

$$R_{AY} = V_A \text{ (max when a<b)}; \quad = \frac{Pb^2}{L^3}(3a + b)$$

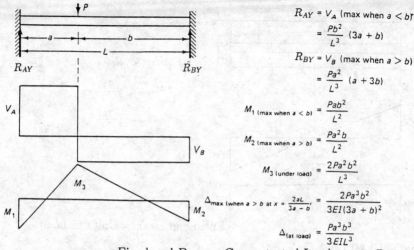

$$R_{AY} = V_A \text{ (max when } a < b)$$
$$= \frac{Pb^2}{L^3}(3a+b)$$
$$R_{BY} = V_B \text{ (max when } a > b)$$
$$= \frac{Pa^2}{L^3}(a+3b)$$
$$M_1 \text{ (max when } a < b) = \frac{Pab^2}{L^2}$$
$$M_2 \text{ (max when } a > b) = \frac{Pa^2b}{L^2}$$
$$M_3 \text{ (under load)} = \frac{2Pa^2b^2}{L^3}$$
$$\Delta_{\text{max (when } a > b \text{ at } x = \frac{2aL}{3a-b})} = \frac{2Pa^3b^2}{3EI(3a+b)^2}$$
$$\Delta_{\text{(at load)}} = \frac{Pa^3b^3}{3EIL^3}$$

Fig. A10.1(h)   Fixed end Beam - Concentrated Load at any Point

$$R_{BY} = V_B \text{ (max when } a>b); \quad = \frac{Pa^2}{L^3}(a+3b)$$
$$M_1(\text{max when } a < b) = \frac{Pab^2}{L^2}; \quad M_2(\text{max when } a > b) = \frac{Pa^2b}{L^2}$$

$$M_3(\text{under load}) = \frac{2Pa^2b^2}{L^3}$$

$$\Delta_{\text{max}} \left(\text{when } a > b \text{ at } x = \tfrac{2aL}{3a+b}\right) = \frac{2Pa^3b^2}{3EI(3a+b)^2}$$

$$\Delta_{\text{(at load)}} = \frac{Pa^3b^3}{3EIL^3}$$

(i) Cantilever Beam - Uniform Load (Fig. A10.1(i)

$$R_{BY} = V_B = qL; \quad M_{\text{max(ends)}} = \frac{QL^2}{2}$$

$$\Delta_{\text{max(at free end)}} = \frac{QL^4}{8EI}$$

(j) Cantilever Beam - Concentrated Load at any Point (Fig. A10.1(j)

$$R_{BY} = V_B = P; \quad M_{\text{max}} = Pb$$

$$\Delta_{\text{max(at free end)}} = \frac{Pb^2(3L-b)}{6EI}$$

(k) Cantilever Beam - Concentrated Load at end (Fig. A10.1(k)

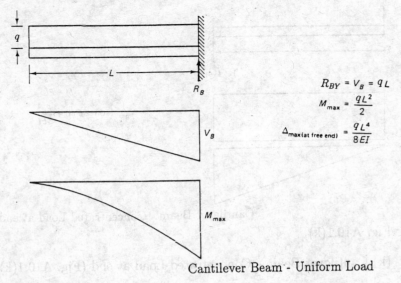

$$R_{BY} = V_B = qL$$

$$M_{max} = \frac{qL^2}{2}$$

$$\Delta_{max\,(at\,free\,end)} = \frac{qL^4}{8EI}$$

Cantilever Beam - Uniform Load

Fig. A10.1(i)

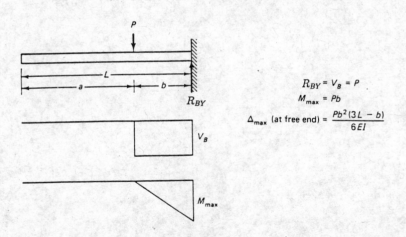

$$R_{BY} = V_B = P$$

$$M_{max} = Pb$$

$$\Delta_{max}\ (at\ free\ end) = \frac{Pb^2(3L - b)}{6EI}$$

Cantilever Beam - Concentrated Load at any Point

Fig. A10.1(j)

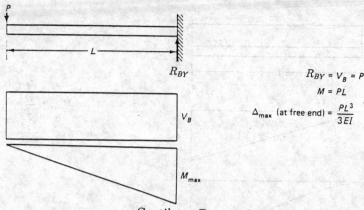

$R_{BY} = V_B = P$

$M = PL$

Δ_{max} (at free end) $= \dfrac{PL^3}{3EI}$

Cantilever Beam - Concentrated Load at end

Fig. A10.1(k)

(k) Cantilever Beam - Concentrated Load at end (Fig. A10.1(k)

$$R_{BY} = V_B = PM = PL; \quad \Delta_{\text{max(at free end)}} = \frac{PL^3}{3EI}$$

Appendix A 10.2

Formulas for Deflection of Beams (After Timoshenko and MacCullough, 1961)

Beams	Slope at free end.	Deflection at any section in terms of x: Δ is positive downward.	Maximum deflection.
1. Cantilever Beam — Concentrated load P at the free end.			
	$\theta = \dfrac{Pl^2}{2EI}$	$\Delta = \dfrac{Px^2}{6EI}(3l-x)$	$\Delta_{max} = \dfrac{Pl^3}{3EI}$
2. Cantilever Beam — Concentrated load P at any point.			
	$\theta = \dfrac{Pa^2}{2EI}$	$\Delta = \dfrac{Px^2}{6EI}(3a-x)$ for $0<x<a$. $\Delta = \dfrac{Pa^2}{6EI}(3x-a)$ for $a<x<l$	$\Delta_{max} = \dfrac{Pa^2}{6EI}(3l-a)$
3. Cantilever Beam — Uniformly distributed load of q kg. per unit length.			
	$\theta = \dfrac{ql^3}{6EI}$	$\Delta = \dfrac{qx^2}{24EI}(x^2+6l^2-4lx)$	$\Delta_{max} = \dfrac{ql^4}{8EI}$
4. Cantilever Beam — Uniformly varying load: maximum intensity w lbs. per unit length.			
	$\theta = \dfrac{ql^3}{24EI}$	$\Delta = \dfrac{qx^2}{120\,lEI}(10l^3-10l^2x+5lx^2-x^3)$	$\Delta_{max} = \dfrac{ql^4}{30EI}$
5. Cantilever Beam — Couple M applied at the free end.			
	$\theta = \dfrac{ML}{EI}$	$\Delta = \dfrac{Mx^2}{2EI}$	$\Delta_{max} = \dfrac{Ml^2}{2EI}$
6. Beam freely supported at ends — Concentrated load P at the center.			
	$\theta_1 = \theta_2 = \dfrac{Pl^2}{16EI}$	$\Delta = \dfrac{Px}{12EI}\left(\dfrac{3l^2}{4}-x^2\right)$ for $0<x<\dfrac{l}{2}$	$\Delta_{max} = \dfrac{Pl^3}{48EI}$

Formulas for Deflection of Beams

Beams	Slope at ends.	Deflection at any section in terms of x; Δ is positive downward.	Maximum and center deflections.
7. Beam freely supported at the ends — Concentrated load at any point.			
$R_{AY}=\dfrac{Pb}{l}$ $R_{BY}=\dfrac{Pa}{l}$	Left End. $\theta_1=\dfrac{Pb(l^2-b^2)}{6lEI}$ Right End. $\theta_2=\dfrac{Pab(2l-b)}{6lEI}$	To the left of load P: $\Delta=\dfrac{Pbx}{6lEI}(l^2-x^2-b^2)$ To the right of load P: $\Delta=\dfrac{Pb}{6lEI}\left[\dfrac{l}{b}(x-a)^3+(l^2-b^2)x-x^3\right]$	$\Delta_{max}=\dfrac{Pb(l^2-b^2)^{3/2}}{9\sqrt{3}\,lEI}$ at $x=\sqrt{\dfrac{l^2-b^2}{3}}$ At center, if $a>b$: $\Delta=\dfrac{Pb}{48EI}(3l^2-4b^2)$
8. Beam freely supported at the ends — Uniformly distributed load of q kg. per unit length.			
$R_{AY}=\dfrac{ql}{2}$ $R_{BY}=\dfrac{ql}{2}$	$\theta_1=\theta_2=\dfrac{ql^3}{24EI}$	$\Delta=\dfrac{qx}{24EI}(l^3-2lx^2+x^3)$	$\Delta_{max}=\dfrac{5ql^4}{384EI}$
9. Beam freely supported at the ends — Couple M at the right end.			
$R_{AY}=\dfrac{M}{l}$ $R_{BY}=\dfrac{M}{l}$	$\theta_1=\dfrac{Ml}{6EI}$ $\theta_2=\dfrac{Ml}{3EI}$	$\Delta=\dfrac{Mlx}{6EI}\left(1-\dfrac{x^2}{l^2}\right)$	$\Delta_{max}=\dfrac{Ml^2}{9\sqrt{3}EI}$ at $x=\dfrac{l}{\sqrt{3}}$ At center $\Delta=\dfrac{Ml^2}{16EI}$
10. Beam freely supported at the ends — Couple M at the left end.			
$R_{AY}=\dfrac{M}{l}$ $R_{BY}=\dfrac{M}{l}$	$\theta_1=\dfrac{Ml}{3EI}$ $\theta_2=\dfrac{Ml}{6EI}$	$\Delta=\dfrac{Mx}{6lEI}(l-x)(2l-x)$	$\Delta_{max}=\dfrac{Ml^2}{9\sqrt{3}EI}$ at $x=\left(1-\dfrac{1}{\sqrt{3}}\right)l$ At center $\Delta=\dfrac{Ml^2}{16EI}$

11. Torsion of Circular Tubes

11.1 Introduction

In structural design, the load carrying members are often subjected to twisting loads in addition to axial and bending. Analogous to bending (or flexural) rigidity, EI, the torsional rigidity, GJ (or GI_p for circular cross-section), becomes relevant as the rigidity of the structure depends on this parameter. The higher the value of GJ greater is the resistance to torsional deformation. Any load applied off from the axis and transverse to the member produces torsional or twisting moment. Some common examples of this mode of load transmission are door knobs, screw driver, drill bits, shafts for transmission of torque in gas turbines, generators, motors etc.

Shafts can be solid or hollow depending on the design. For a given cross-sectional area, hollow shafts are found to be economical as compared to solid shafts. Further, the distribution of shear stresses across the cross-section is very much different in solid and hollow shafts. In many practical situations, open section members are subjected to torsional loads. The torsional rigidity of these members are much lower than solid sections and hence these are susceptible to large torsional deformations. Even then such sections cannot be avoided in actual designs. In this chapter, we shall confine ourselves to study the torsional response of solid and hollow circular cross-sections only.

11.2   Torsion of Solid Circular Shaft

Consider a uniform solid circular shaft of length L and radius R (Fig. 11.1). The far end of the shaft is constrained against

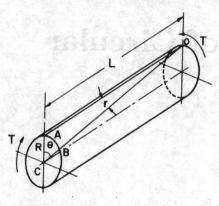

Fig. 11.1 Solid circular shaft.

rotation, while the near end is subjected to a torque T. As a consequence, an internal resistive torque T is generated, which is the result of shear stresses developed in the shaft.

The following assumptions are made for the development of torsional relations:

1. The shaft is of uniform cross-section along its length.
2. Torque is applied only at the free end and hence remains constant over the entire length.
3. The material behavior is linearly elastic.

The line OA on the surface of the shaft, which was parallel to the axis of the shaft before application of the torque, takes the new position OB after application of the torque. The angle γ being the shear strain of the material at the surface. Since the applied load is pure torque, the only non-zero stress is the shear stress τ.

Assuming γ to be small,

$$AB = \gamma L \quad \text{or} \quad \gamma = \frac{AB}{L} \qquad (11.1)$$

From the constitutive equation

$$\gamma = \frac{\tau_s}{G} \qquad (11.2)$$

where τ_s is the shear stress at the surface.

The radius CA also gets rotated through an angle θ and moves to a new position CB due to the applied torque. Therefore,

$$AB = R\theta \tag{11.3}$$

From Eqs.(11.1), (11.2) and (11.3)

$$\frac{L\tau_s}{G} = R\theta \quad \text{or} \quad \frac{\tau_s}{R} = \frac{G\theta}{L} \tag{11.4}$$

Because the material is linearly elastic, the shear stress τ at any radius r, can be written as

$$\frac{\tau}{r} = \frac{\tau_s}{R} \tag{11.5}$$

or

$$\tau = \frac{\tau_s}{R}r \tag{11.6}$$

Therefore,

$$\frac{\tau}{r} = \frac{G\theta}{L} \tag{11.7}$$

In Eq. (11.7), θ is the angle of twist at a distance L from the fixed end. Sometimes Eq. (11.7) is also written as

$$\frac{\tau}{r} = G\theta' \tag{11.8}$$

where θ' is the twist per unit length.

11.3 Twisting Moment and Shear Stress

The shear stress τ at any radius r acts uniformly on a small ring shaped area (Fig. 11.2). The total tangential force on the ring is

$$dF = \tau \, 2 \, \pi \, r \, dr \tag{11.9}$$

It is assumed that over this small element area, the shear stress remains constant. The torque dT developed by this force about the axis of the shaft is

$$dT = dF \, r = \tau \, 2\pi \, r^3 \, dr$$

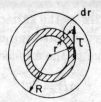

Fig. 11.2 Stresses acting on a ring element

Substituting for τ from Eq. (11.7)

$$dT = \frac{G\theta}{L} 2\pi r^3\, dr \qquad (11.10)$$

Equation (11.10) is the internal resistive torque developed on a small area dA. The total torque on the entire area is the sum of the individual elemental torque. Hence,

$$T = \int_0^R dT = \int_0^R \frac{G\theta}{L} 2\pi r^3 dr = \frac{G\theta}{L}\frac{\pi R^4}{2} \qquad (11.11)$$

or

$$T = \frac{G\theta}{L}J \qquad (11.12)$$

where J is the polar moment of inertia of cross-section about point C.

From Eq. (11.4)

$$\frac{\tau_s}{R} = \frac{G\theta}{L}$$

Hence Eq. (11.12) reduces to

$$T = \tau_s\, \frac{J}{L} \qquad (11.13)$$

Combining Eqs. (11.7) and (11.12),

$$\frac{T}{J} = \frac{\tau}{r} = \frac{G\theta}{L} \qquad (11.14)$$

Equation (12.13) can be used to compute the maximum shear stress on a circular shaft subjected to torque T and Eq. (11.14) can be used to compute shear stresses within the shaft.

From Eq. (11.14),

$$\frac{T}{\theta} = \frac{G\,J}{L} \qquad (11.15)$$

The torque per unit twist $\left(\frac{T}{\theta}\right)$ is called the *Torsional Stiffness of the shaft.* Equation (11.15) can be rewritten as

$$\frac{T}{\left(\frac{\theta}{L}\right)} = GJ \qquad (11.16)$$

which is torque divided by the angle of twist per unit length (or the rate of twist) and is referred as *torsional rigidity.* This is analogous to bending rigidity, EI, for beams.

11.3.1 Polar Moment of Inertia for Hollow Circular Cross-section

An approach similar to the one carried out for solid circular shaft can be used in this case as well. Consider a hollow shaft as shown in Fig. 11.3. The polar moment of Inertia, J, for this is given by

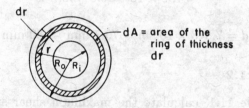

Fig. 11.3 Hollow circular cross-section.

$$J = \int r^2 dA$$

From the figure,

$$dA = 2\pi r dr; \ R_i \leq r \leq R_o$$

Thus,

$$\begin{aligned}
J &= \int_{R_i}^{R_o} 2\pi r^3 dr \\
&= \frac{2\pi}{4}\left[r^4\right]_{R_i}^{R_o} = \frac{\pi}{2}\left[R_o^4 - R_i^4\right] = \frac{\pi}{32}\left[D_o^4 - D_i^4\right] (11.17)
\end{aligned}$$

where $R_o = D_o/2$ and $R_i = D_i/2$

In the case of a hollow tube, there exist a finite value of shear stress at the inner boundary.

Example 11.1

Calculate the diameter of the solid shaft required to transmit a torque of 4500 Nm. The twist of the shaft is not to exceed $1°$ over a length of 2 m. The shear modulus G of the shaft material is 133 GN/m^2.

Solution

Given $\theta = 1°$, i.e., $= 0.01745$ radians. From the torsion formula for solid shaft,

$$T = \frac{G\theta}{L}J = \frac{G\theta}{L}\frac{\pi}{32}d^4$$

Therefore,

$$d^4 = \frac{4500 \times 2 \times 32}{130 \times 10^9 \times 0.01745 \times \pi}$$

or $d = 7.973 \times 10^3 \ m = 79.7$ mm ≈ 80 mm

Example 11.2

For Example 11.1, calculate the maximum shear stress in the shaft.

Solution

As the shear stress is directly proportional to the radius of the shaft, the shear stress is maximum at the outer most surface. Therefore,

$$\tau_s = \frac{G\theta}{L}R = \frac{133 \times 10^9 \times \pi}{2 \times 180}40 \times 10^{-3} = 46.42 \ MN/m^2$$

Example 11.3

A propeller drive shaft is made up of a hollow tube having outside diameter of 50 mm and inside diameter of 35 mm. Find the stresses at both outer and inner surfaces if it transmits a torque of 1.8 kNm.

Solution

The polar moment of inertia J for the section is

$$J = \frac{\pi}{32} \left(D_o^4 - D_i^4 \right) = \frac{\pi}{32} \left(50^4 - 35^4 \right) = 0.4663 \times 10^6 \text{ mm}^4$$

At the outer surface,

$$\tau_s = \frac{TR_o}{J} = \frac{1.8 \times 10^3 \times 25 \times 10^3}{0.4663 \times 10^6} = 96.5 \text{ N/mm}^2$$

which is the maximum stress at the surface.
    At the inner surface,

$$\tau_s = \frac{TR_i}{J} = \frac{1.8 \times 10^3 \times 17 \times 10^3}{0.4663 \times 10^6} = 67.55 \text{ N/mm}^2$$

Example 11.4

The hollow shaft in Example 11.3, is to be replaced by a solid shaft subjected to the same torque, $T = 1.8$ kNm and having the same strength. Determine (i) the diameter of the solid shaft and (ii) the ratio of weight of solid shaft to the hollow shaft.

Solution

To have the same strength, the solid shaft must have maximum shear stress τ_s equal to 96.5 N/mm², when subjected to a torque T equal to 1.8 kNm. Therefore,

$$\tau_s = \frac{1.8 \times 10^3 \times d}{2 \times \frac{\pi d^4}{32}}$$

or

$$d^3 = \frac{1.8 \times 10^3 \times 16}{\pi \tau_s} = \frac{9.167 \times 10^3 \times 10^{-6}}{96.5}$$

Hence

$$d = 45.6 \text{ mm}$$

The cross-sectional area of solid shaft

$$A_s = \frac{\pi}{4} 45.6^2 = 1632 \text{ mm}^2$$

The cross-sectional area of hollow shaft of Example 11.3,

$$A_h = \frac{\pi}{4} \left[50^2 - 35^2 \right] = 1001 \text{ mm}^2$$

Therefore, $\qquad \dfrac{W_s}{W_h} = \dfrac{1632}{1001} = 1.63$

Thus, the solid shaft weighs more than one and half times the hollow shaft.

Example 11.5

Determine the angle of twist between two sections 300 mm apart when a torque of 20 Nm is applied. The diameter of the solid shaft is 15 mm. The shear modulus, $G = 80 \times 10^9$ N/m^2.

Solution

The angle of twist $\theta = \frac{TL}{GJ}$ and J for solid shaft $= \frac{\pi D^4}{32} = 4970$ mm^4.

Therefore,

$$\theta = \frac{20 \times 300 \times 10^9}{80 \times 10^9 \times 4970} = 0.0151 \text{ rad} = 0.865°$$

11.4 Shafts with Discrete Torques

In many practical situations, shafts are designed to carry torque loads which are applied at discrete locations on the shaft. Further, different sections of the shaft may be of different diameters for transmitting the torques. In this section, circular shafts subjected to discrete torque loads are considered.

Before stress and twisting deformation of the shafts are evaluated, it is required to determine the internal resistive torque at any section developed due to the application of the external torque. The method of section as employed while studying bending of beams is applied to determine the internal resistive torque.

Consider a shaft subjected to three balanced external torques shown in Fig. 11.4(a).

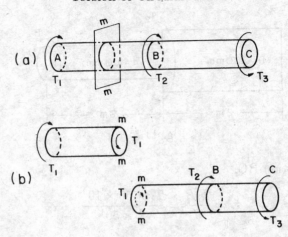

Fig. 11.4 (a) Shaft with discrete torques and (b) section of the shaft with discrete torque.

To determine the internal torque at a section between any two loaded points, a plane m-m is passed through the section, cutting the shaft and separating it into two parts as shown in Fig. 11.4(b). If we consider the equilibrium of the portion of the shaft to the left of section m-m, the internal torque is found to be T_1. The equilibrium of the shaft to the right of the section is also T_1, but in opposite direction. Since such a value represents the internal torque at the same section, the torque should have the same sign. Hence, in this case, both are considered positive.

Similarly, if one takes a section n-n between B and C, the torque acting on the section will be $T_1 + T_2$ in the anticlockwise direction if we consider left half of the section and clockwise if we consider the equilibrium of the right half of the section. Both are positive as per assumption.

Example 11.6

What should be the values of the torques T_1 and T_2 acting on the steel shaft so that the maximum shear stress value does not exceed 90 MPa at any point on the shaft. Given the diameter to be 200 mm and 100 mm respectively and each sections of 2 m as shown in Fig. 11.5.

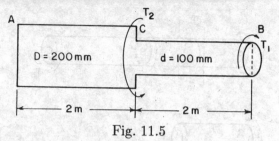

Fig. 11.5

Solution

For the portion BC,

$$\tau_s = \frac{T_1 R}{J} = \frac{T_1 \times 50 \times 10^9}{\frac{\pi}{2} \times 50^4}$$

or

$$T_1 = \frac{90 \times 10^6 \times \pi \times 125 \times 10^3}{2 \times 10^9} = 17.671 \ \text{kN-m}$$

In the portion AB, the torque is $T_2 - T_1$. Thus,

$$\tau_s = \frac{(T_2 - T_1) \times 100}{\frac{\pi}{2} \times 100^4} \times 10^9$$

or

$$T_2 - T_1 = \frac{90 \times 10^6 \times \pi \times 10^6}{2 \times 10^9} = 141.376 \ \text{kN-m}$$

Therefore, $\quad T_2 = (141.376 + 17.671) = 159.047 \ \text{kN-m}$

Example 11.7

A steel shaft attached with three discs is shown in Fig. 11.6. One

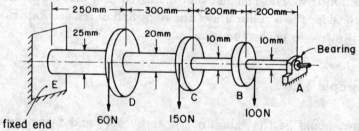

Fig. 11.6

end of the rod is constrained against rotation, while the other end is free to rotate inside the bearing. Determine the angle of rotation of the shaft at the bearing end. Assume shear modulus $G = 80$ GPa. The diameter of disc is 200 mm.

Solution

In view of the varying shaft length between the disc, the torque resisted by each segment is different. Further, diameters being different, twist will also be different. The first step is to determine the internal torque in each segment. This is done by cutting the element and drawing the free body diagram and balancing the torques.

We assume that there is no bending of the shaft due to the applied load.

Torque on the discs B,C and D are

$$T_B = 100 \times 100 = 10,000 \, \text{N mm} = 10 \, \text{Nm} \quad \text{(clockwise)}$$

$$T_C = 150 \times 100 = 15,000 \, \text{N mm} = 15 \, \text{Nm} \, \text{(anticlockwise)}$$

$$T_D = 60 \times 100 = 6,000 \, \text{N mm} = 6 \, \text{Nm} \quad \text{(anticlockwise)}$$

To determine the internal torque in member BC, cut a section to the right of C. We observe that the internal torque must be 10 Nm to balance the external torque. Therefore, $T_{BC} = 10$ Nm.

Similarly in member CD,

$$T_{CD} = T_C - T_B = 15 - 10 = 5 \, \text{Nm}$$

and $$T_{DE} = T_D + T_C - T_B = 6 + 5 = 11 \, \text{Nm}$$

With this information, it is now possible to determine the twist of the segments. Thus,

$$\theta_{AB} = \frac{T_{AB}L}{GJ_{AB}}$$

since $T_{AB} = 0$, θ_{AB} is also zero. Therefore,

$$\theta_{BC} = \frac{T_{BC}L}{GJ_{BC}}$$

$$T_{BC} = 10 \, \text{Nm}, \quad L = 200 \, \text{mm}, \quad G = 80 \times 10^9 \, \text{N/m}^2$$

$$J_{BC} = \frac{\pi}{2} R^2 = \frac{\pi}{2} 5^4 = 981.75 \, \text{mm}^4$$

Therefore,

$$\theta_{BC} = \frac{10 \times 200}{80 \times 10^9 \times 981.75} \times 10^9 = 0.025 \, \text{rad}$$

This implies that section B rotates 0.025 radians clockwise relative to section C.

Similarly,

$$\theta_{CD} = \frac{T_{CD}L}{GJ_{CD}}$$

$$T_{CD} = 5 \, \text{Nm}, \quad L = 300 \, \text{mm}, \quad G = 80 \times 10^9 \, \text{N/m}^2,$$

$$J_{CD} = \frac{\pi}{2} R^2 = \frac{\pi}{2} 10^4 = 1.5707 \times 10^4 \, \text{mm}^4$$

Therefore,

$$\theta_{CD} = \frac{5 \times 300}{80 \times 10^9 \times 1.5707 \times 10^4} \times 10^9 = 0.0012 \, \text{rad}$$

This indicates that section C rotates 0.0012 rad anticlockwise relative to section D.

In member DE,

$$\theta_{DE} = \frac{T_{DE}L}{GJ_{DE}}$$

$$T_{DE} = 11 \, \text{Nm}, L = 250 \, \text{mm}, G = 80 \times 10^9 \, \text{N/m}^2$$

$$J_{DE} = \frac{\pi}{2} R^2 = \frac{\pi}{2} \times 12.5^4 = 38349.5 \, \text{mm}^4$$

Therefore,

$$\theta_{DE} = \frac{11 \times 250 \times 10^9}{80 \times 10^9 \times 38349.5} = 0.00089 \, \text{rad}$$

indicating that section D rotated by 0.00089 radians anticlockwise relative to E.

The total angle of twist

$$\theta_{AE} = \theta_{AB} + \theta_{BC} + \theta_{CD} + \theta_{CE} = 0 + 0.025 - 0.0012 - 0.00089 = 0.0229 \, \text{rad}$$

Problems

11.1 Obtain the maximum shear stress in a solid circular shaft of 25 mm diameter when subjected to a torque of 300 Nm.

11.2 A steel tube is used to carry a torque of 500 Nm. The outside diameter is 50 mm and a wall thickness of 10 mm. Compute the torsional stress at the outer and inner surface. $G = 80 \times 10^9$ N/m^2.

11.3 Determine the maximum shear stress in the steel shaft shown in the Fig. 11.7.

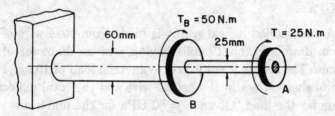

Fig. 11.7

11.4 If $G = 80 \times 10^9$ N/m^2, Compute the twist of the free end of the steel shaft shown in Fig. 11.8.

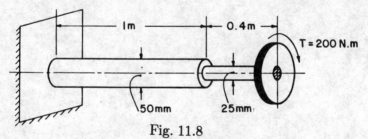

Fig. 11.8

11.5 A beryllium copper wire having a diameter of 1.5 mm and a length of 50 mm is used in a torsional instrument. Determine what angle of twist would result in the wire when it is stressed to 200 MPa.

11.6 The propeller shaft of a ship is 400 mm in diameter. The allowable shearing stress for the material of the shaft is 45 MN/m^2 and the angle of twist is not to exceed 1° in length of 5000 mm. Determine the maximum permissible torque that can be transmitted through the shaft if $G = 80 \times 10^9$ N/m^2.

11.7 Compare the weights of equal lengths of two shafts, one hollow and the other solid. For the hollow shaft the internal diameter is 0.8 times the outer diameter. The maximum shear stress in each of them is the same while subjected to the same torque.

11.8 A close-coiled helical spring (Fig. 11.6), is to be designed with each coil of 10 cm mean diameter from a wire of 0.6 cm diameter and of such a stiffness that it will elongate axially 2.5 cm for an axial pull of 20 kg. How many coils should you have?. The modulus of rigidity G, for the spring is 80 GPa.

11.9 A close coiled helical spring is made from steel wire of 0.6 cm diameter and 10 coils having a mean diameter of 7.5 cm. The spring is subjected to an axial load of 10 kg. Find the shear stress in the helical wire and the total deflection under the load. Given $G = 80$ GPa for the material.

12. THIN WALLED CYLINDERS AND HELICAL SPRINGS

12.1 Introduction

Pressurized cylinders or spheres such as petrol tankers, gas storage tanks, liquid fuel engines of rockets and missiles, pipelines etc., whose wall thickness is small as compared to the diameter of the vessels are generally referred to as thin walled pressure vessels. These pressure vessels are generally spheres or closed-end cylinders. Walls being thin, they offer very little resistance to bending. The internal forces exerted on the given portion of the wall are tangential to the surface of the wall. The internal pressure tends to burst the vessel because of tensile stresses developed in its walls. If the ratio of the mean radius to wall thickness is greater than 10, the stresses across the thickness are very nearly uniform. The mean radius is defined as the average of outer and inner radii. This chapter restricts to thin walled cylinders under internal and/or external pressures.

12.2 Thin Walled Cylinders

These structural elements are frequently employed as storage tanks, hydraulic and penumatic actuators and for piping of fluids under pressure. Consider a thin walled cylinder closed at both ends and subjected to an internal pressure p (Fig. 12.1).

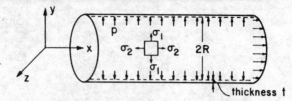

Fig. 12.1 Thin walled cyclinder with close ends under internal pressure.

Since the structure and loading is axisymmetric, no shearing stress exist on the element shown. The normal stresses σ_1 and σ_2 are therefore principal stresses. The stress σ_1 is termed as *hoop stress* or *tangential stress* and stress σ_2 as *longitudinal stress*.

To determine the hoop stress, we consider a section of length Δx at a distance x and bounded by two planes parallel to the yz plane as shown in Fig. 12.2.

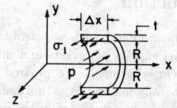

Fig. 12.2 Curved element of the cylinder.

The force parallel to z-axis due to the internal stress σ_1 acting on the walls of the cylinder is equal to $\sigma_1 \times$ elemental area. Since there are two surfaces, the total force along z is equal to $2\sigma_1 \times t \times \Delta x$.

The force due to the internal pressure will be p times the project area and is equal to $p \times 2R \times \Delta x$.

Therefore, for equilibrium along z, $\Sigma F_z = 0$. This gives

$$2\sigma_1 t \Delta x = p\, 2R \Delta x \qquad (12.1)$$

or

$$\sigma_1 = \frac{pR}{t} \qquad (12.2)$$

where R is the radius of the pressure vessel and t is the wall thickness.

To determine the longitudinal stress σ_2, consider a section perpericular to x-axis and the free body consisting the portion

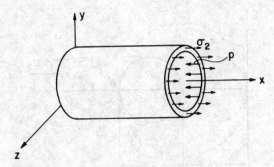

Fig. 12.3 Free body diagram of a sectin of the cylinder.

of the vessel with internal pressure to the left of the section as shown in Fig. 12.3.

The forces acting on this section are due to internal axial stress σ_2 acting on the walls of the cylinder and internal pressure p. The force due to σ_2 acting along x is given by

$$F_x = \sigma_2 2\pi Rt \tag{12.3}$$

The force due to internal pressure p is pressure multiplied by the internal area of the cylinder, i.e., equal to $p\pi R^2$. For equilibrium of the section,

$$p\pi R^2 - \sigma_2 2\pi Rt = 0 \tag{12.4}$$

or

$$\sigma_2 = \frac{pR}{2t} \tag{12.5}$$

Equations (12.2) and (12.5) show that the hoop stress is twice as large as the longitudinal stress, i.e.,

$$\sigma_1 = 2\sigma_2 \tag{12.6}$$

Figure 12.4 shows the Mohr's circle drawn with σ_1 and σ_2 which gives the maximum shear stress τ_{max} as

$$\tau_{max} = \frac{\sigma_1 - \sigma_2}{2} = \frac{2\sigma_2 - \sigma_2}{2} = \frac{\sigma_2}{2} = \frac{pr}{4t} \tag{12.7}$$

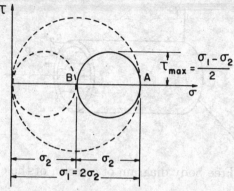

Fig. 12.4 Mohr's circle for the state of stress in a cylinder.

This shear stress is exerted on an element obtained by rotating the original element shown in Fig. 12.1 through 45°. The maximum shearing stress in the wall of the vessel, is at an angle of 45° about the longitudinal axis.

$$\tau_{max} = \frac{pr}{2t} \qquad (12.8)$$

This is because $\sigma_3 = 0$.

A better result can be obtained if we consider the mean radius,

$$R_m = R + \frac{t}{2} \qquad (12.9)$$

Equation (12.1), gives

$$\sigma_1 t \Delta x = p2(R + \frac{t}{2})\Delta x$$

or

$$\sigma_1 = \frac{pR}{t}(1 + \frac{t}{2R}) \qquad (12.10)$$

Equation (12.4) becomes

$$p\pi(R + \frac{t}{2})^2 = 2\sigma_2\pi t(R + \frac{t}{2})$$

or

$$\sigma_2 = \frac{pR}{2t}\left[1 + \frac{t}{2R}\right] \qquad (12.11)$$

Example 12.1

A cylindrical tank of 250 mm inside radius has a wall thickness of 15 mm. It is subjected to an internal pressure of 1000 kPa. Compute the hoop stress and the longitudinal stress in the walls of the cylinder. Calculate also the maximum inplane shear stress.

Solution

We first check whether the cylinder is thin walled or not.

$$D_m = D_o - t = 500 - 15 = 4985 \text{ mm}$$

$$\frac{D_m}{t} = \frac{485}{15} = 32.33 > 20$$

Hence, the cylinder is thin walled. Therefore from Eqs. (12.2) and (12.5),

$$\text{Hoop stress} = \frac{pR}{t} = \frac{1000 \times 10^3 \times 250}{15} = 16.667 \text{ MPa}$$

$$\text{Longitudinal stress} = \frac{pR}{2t} = 8.333 \text{ MPa}$$

From Eq. (12.7), the maximum inplane shear stress is

$$\tau_{max} = \frac{pr}{4t} = \frac{1000 \times 10^3 \times 250}{15 \times 4}$$
$$= 4.166 \text{ MPa}$$

12.3 Thin Walled Spheres

The objective here is to determine the stresses in the walls of the pressure vessel and compare with the specified design value to ensure that the structure is safe. It also gives us the information about factor of safety. Since the structure and loading is symmetric, it is sufficient to analyse only one-half of the structure shown in Fig. 12.5(a). The internal pressure acts perpendicular to the walls, uniformly over the entire internal surface. If we cut

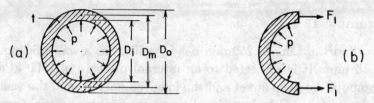

Fig. 12.5 (a) Cross-section of a thin walled sphere under internal pressure and (b) a section of a sphere under internal pressure.

the sphere shown in Fig. 12.5(b), the forces in the wall act horizontally. Because of the symmetry, no other component of the force exists.

Hence,

$$F = pA \tag{12.12}$$

or

$$F = p\frac{\pi D^2}{4} \tag{12.13}$$

where A is the projected area of the sphere on the plane through the mean diameter D_m. Force F can also be evaluated as

$$F = \sigma A_{annular} \tag{12.14}$$

or

$$A_{annular} = \frac{\pi}{4/(D_o^2 - D_i^2)} \tag{12.15}$$

Here D_i and D_o are the inner and outer radii of the spherical vessel. If thickness t of the sphere is very small compared to D, Eq. (12.14) can be approximated to

$$A_{annular} = \pi D_i t \tag{12.16}$$

Making use of Eqs. (12.12), (12.13) and (12.15),

$$\sigma = \frac{pD_i}{4t} \tag{12.17}$$

Since any section that passes through the center of the sphere yields the same result, this stress condition is called all round tension.

From symmetry considerations, the normal stresses acting on the four faces of the element must be equal. Hence,

$$\sigma_1 = \sigma_2 = \frac{pD}{4t}m \qquad (12.18)$$

Since the principle stresses σ_1 and σ_2 are equal, the Mohr's circle reduces to a point and the inplane maximum shear stress is zero. The maximum shear stress in the wall is not zero. It is equal to $\sigma_2/2$ or $\sigma_1/2$ acting at 45° out of plane.

Example 12.2

A spherical vessel with an inside diameter of 210 mm and a wall thickness of 10 mm is made of a material which has an allowable stress of 42 MPa. Determine the maximum allowable internal pressure that the vessel can withstand.

Solution

Equations (12.12), (12.13) and (12.14) gives

$$\sigma \frac{\pi}{4}(D_o^2 - D_i^2) = \frac{\pi}{4}pD_i^2$$

or

$$\sigma = \frac{pD_i^2}{(D_o^2 - D_i^2)}$$

$$D_i = D_o - 2t = 190 \text{ mm}$$

Substituting the values,

$$42 \times 10^6 = \frac{p \times 190^2}{(210^2 - 190^2)} = 4.512p$$

or

$$p = 9.31 \times 10^6 \text{ N/m}^2 = 9.31 \text{ MPa}$$

12.4 Helical Springs

The basic purpose of all kinds of springs is to absorb energy and restore it slowly or rapidly accordingly to the function of the particular spring under consideration. For example, in a clock a certain amount of work is done by the external moment in

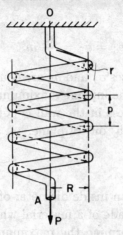

Fig. 12.6 Close Coiled Helical Spring.

winding up, i.e., deforming the spring. This work is stored in the form of strain energy and is regained when the spring is allowed to return to its original configuration. Two other most common use of springs is for absorbing shocks, such as the springs of buffers of railway rolling stock, and springs of wheels on all types of vehicles. Springs are also used to provide a means of restoring various mechanisms to their original configurations against the action of some extrnal force.

The properties of springs which are usually of interest to the engineers are: (i) the capacity for absorbing energy and (ii) the deformation produced by the given load.

Springs in practice belong usually to one of the two definite families. One in which the length of the rod or wire is made up into a coil of some kind called *coiled spring* and those consisting of one or more approximately flat plates called *leaf springs*.

12.4.1   Close-coiled Helical Springs

A close-coiled helical spring (Fig. 12.6) carries an axial load which causes torsion and bending moment in every section of the spring and alters the curvature of the coils.

There is also a shear stress in the coil due to axial load. When the coils are closely wound, the bending stress, in general, is negligible in comparison to the torsional stress. Because of the cross-sectional dimensions of the coil, in comparison to the length, the shear stress is not of consequence.

If there are N number of coils in the spring and the mean radius of the spring is R, then the length L of the wire is approximately given by

$$L = 2\pi RN \qquad (12.19)$$

The torque T developed due to the applied axial load P is

$$T = PR \qquad (12.20)$$

If δ is the axial displacement of the end A with respect to O, and β is the twist of the spring due to the torque T, then

$$\frac{1}{2}P\delta = \frac{1}{2}T\beta \qquad (12.21)$$

or

$$\delta = \frac{T}{P}\beta$$

Using Eq. (12.1) and replacing θ by β,

$$\beta = \frac{TL}{GJ} = \frac{PR \times 2\pi RN}{G \times \frac{\pi r^4}{2}}$$

or

$$\beta = \frac{4PR^2N}{Gr^4} \qquad (12.22)$$

in which

$$J = \pi\frac{r^4}{2}$$

where r is the radius of the wire.

Hence,

$$\delta = \frac{4TR^2N}{Gr^4} \qquad (12.23)$$

$$= \frac{4PR^3N}{Gr^4} \qquad (12.24)$$

The elastic energy E, stored in the spring is given by

$$E = \frac{1}{2}T\beta = \frac{1}{2}PR\beta \qquad (12.25)$$

Using Eq. (12.14) and replacing θ by β,

$$E = \frac{1}{2}PR\frac{\tau L}{Gr} = \frac{1}{2}(\tau\frac{J}{r})\frac{\tau L}{Gr} \qquad (12.26)$$

For circular cross-section, $J = (\pi r^4)/2$ Therefore, Eq. (12.25) reduces to

$$E = \frac{1}{4}\frac{\tau^2}{g}\pi r^2 L \qquad (12.27)$$

If the close coiled spring is subjected to an axial torque at the end A, then the bending moment at any cross-section is of constant magnitude. This bending moment will increase or decrease the curvature of the coils depending on the direction of the moment.

Assuming each coil to act as a beam of large curvature, the strain energy of the beam U_b given by

$$U_b = \frac{1}{2}\int \frac{M^2}{EI}dx \qquad (12.28)$$

where M is the bending moment at any section and I is the second moment of area.

Since M is constant,

$$U_b = \frac{1}{2}\frac{M^2}{EI}\int dx$$

$$U_b = \frac{1}{2}\int \frac{M^2 L}{EI} \qquad (12.29)$$

If ϕ is the angle of twist of the free end,

$$\frac{1}{2}M\phi = \frac{1}{2}\frac{M^2 L}{EI} \quad \text{or} \quad \phi = \frac{ML}{EI}$$

Substituting for L and I, the above equation gives

$$\phi = \frac{(2\pi RN)M}{E(\pi/4r^4)} = \frac{8RNM}{Er^4} \qquad (12.30)$$

Problem

12.1 A stainless steel cylindrical pressure vessel of 250 mm inside radius is subjected to an internal pressure of 3.0 MPa. If the allowable stress is 140 MPa, determine the wall thickness of the vessel.

12.2 Titanium alloy is used to fabricate a spherical tank having an outside diameter of 1150 mm. The working pressure inside the tank is maintained at 4 MPa. Determine the required thickness of the vessel if a design factor of 2.0 based on yield strength is desired. The yield strength of the titanium alloy is 1070 MPa.

12.3 A steel pipe has an inside diameter of 400 mm and wall thickness of 20 mm. If the ultimate strength of steel is 450 MPa, determine the bursting pressure of the pipe.

12.4 Determine the normal stress, inplane shear stress and out of plane shear stress in a basketball of 300 mm diameter and 3 mm wall thickness which has been inflated to a pressure of 100 kPa.

12.5 Determine the largest internal pressure that can be applied to a cylindrical tank of 1.8 m diameter and 15 mm wall thickness if the ultimate stress of steel used is 400 MPa. The factor of safety of 5.0 is to be maintained.

12.6 A 20 m spherical tank is used to store gas. The wall thickness of the tank is 10 mm and the working stress of the material is 100 MN/m. What is the maximum permissible gas pressure?

12.7 A cylindrical pressure vessel has the following dimensions: Diameter = 2.0 m, Thickness of the plate = 15 mm, Internal fluid pressure = 750 kPa. Determine the hoop, longitudinal and inplane shear stress in the wall.

12.8 A close-coiled helical spring shown in Fig. 12.6 is to be designed with each coil of 10 cm mean diameter from a wire of 0.6 cm diameter and of such a stiffness that it will elongate axially 2.5 cm for an axial pull of 20 kg. How many coils should have. The modulus of rigidity G, for the spring is 80 GPa.

12.9 A close coiled helical spring is made from steel wire of 0.6 cm diameter and 10 coils having a mean diameter of 7.5 cm. The spring is subjected to an axial load of 10 kg. Find the shear stress in the helical wire and the total deflection under the load. Given $G = 80$ GPa for the material.

13. ELASTIC STRAIN ENERGY: Concept and Applications

13.1 Introduction

Earlier chapters discussed the responses of structures to a variety of loading conditions in terms of stresses, strains and deformations. This was based on two basic concepts, namely 'stress' and 'strain'. We shall introduce now a third concept, that of 'strain energy'. When an external force acts on an elastic (deformable) structure, the structure undergoes deformation. The point of application of force also moves from its original position. Therefore, some work is done by the force. We know that for a conservative system, the total energy must be conserved. From the First Law of Thermodynamics, the work done by the external force must be equal to change in the internal energy, assuming that no energy is lost in the form of heat etc. Further, if we assume that at the time of application of load the internal energy was zero (undeformed stage), then the internal energy stored in the system due to the application of external force is called *strain energy*. One such example is the energy stored in a spring when it is compresed. This energy is released, when the force is removed.

For a perfectly elastic body, the work done by the force on the body is completely recovered when the external force is removed. This recoverable stored energy in the structure is termed as *elastic strain energy*. In this chapter, we shall discuss the strain energy stored in the structural member due to variety of loadings, assuming that the material behaviour is completely elastic.

13.2 Strain Energy of an Elastic Rod Subjected to Tensile/Compressive Load

Figure 13.1(a) shows a uniform rectangular rod of length L and cross-sectional area A. The rod is fixed at one end and at the other end a uniform load P is applied. The load P is gradually increased in steps making sure that at each increment of load, equilibrium has been attained.

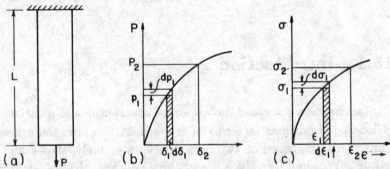

Fig. 13.1 (a) Uniform rectangular cross-section rod under axial load, (b) load-displacement curve and (c) stress-strain curve.

Figures 13.1(b) and (c) show the load versus displacement, and stress versus strain diagrams, respectively. When load P attains a value P_1, let the displacement of the free end of the rod with respect to the fixed end be δ_1. Therefore work done for deforming the rod by an amount $d\delta_1$ is

$$dW = P_1 d\delta_1 + \frac{1}{2} dP_1 d\delta_1 \qquad (13.1)$$

The contribution of the second term to the incremental work is small compared to the first term. Further, for very small increments in deformation δ_1, the contribution of the second term will be still smaller. Hence,

$$dW = P_1 d\delta_1 \qquad (13.2)$$

This corresponds to the hatched area in Fig. 13.1(b). The strain energy stored in the member is equal to the work done by the load P. Hence,

$$dU = P_1 d\delta_1 \tag{13.3}$$

The total work done by the load as the rod undergoes a deformation δ is

$$U = \int_o^\delta P_1 d\delta_1 \tag{13.4}$$

or

$$U = \frac{1}{2} P\delta \tag{13.5}$$

for a linear system since $\frac{P_1}{\delta_1} = \frac{P}{\delta}$. This corresponds to the area under the load-deformation diagram i.e., the area of the triangle.

Also,

$$P_1 = A_r \sigma_1 \tag{13.6}$$

where A_r is the area of the rod.

Further,

$$d\delta_1 = L d\epsilon_1 \tag{13.7}$$

Using Eqs. (13.6) and (13.7) in Eq. (13.3) gives

$$dU = A_r \sigma_1 L d\epsilon_1 \tag{13.8}$$

Using the constitutive equation for the material,

$$\sigma_1 = E\epsilon_1 \tag{13.9}$$

Equation (13.8) becomes,

$$dU = A_r \frac{L}{E} \sigma_1 d\sigma_1 \tag{13.10}$$

and total strain energy U for a stress level σ_2 is

$$U = \int_o^{\sigma_2} A_r \frac{L}{E} \sigma_1 d\sigma_1$$

or

$$U = \frac{A_r L \sigma_2^2}{2E} \tag{13.11}$$

Since $P_2 = \sigma_2 A_r$, Eq. (13.10) can be rewritten in terms of load as

$$U = \frac{LP_2^2}{2A_r E} \qquad (13.12)$$

Equation (13.10) shows that strain energy U depends on the dimensions of the member. In Eq. (13.12), the dimensions of the member explicitly appears in the expression for energy. This can be overcome by considering the strain energy per unit volume. Dividing the strain energy U by volume V of the rod gives

$$\frac{U}{V} = \bar{U} = \frac{\sigma_2^2}{2E} \text{ (for a linear system)} \qquad (13.13)$$

or from Eq. (13.8)

$$\bar{U} = \int_o^\epsilon \sigma_1 d\epsilon_1 \qquad (13.14)$$

$\bar{U}$ can also be written as

$$\bar{U} = \frac{1}{2}(\sigma_2)(\frac{\sigma_2}{E}) = \frac{1}{2}\epsilon_2\sigma_2 = \frac{1}{2}(\text{stress} \times \text{strain}) \qquad (13.15)$$

Therefore,

$$U = \int \frac{\sigma^2}{2E} dV \qquad (13.16)$$

$$= \int \frac{P^2}{2A_r^2 E} dV \qquad (13.17)$$

Setting $dV = A_r dx$

$$U = \int_o^L \frac{P^2}{2A_r E} dx \qquad (13.18)$$

Example 13.1

A circular rod ABC is made up of the same material but having different cross-sectional area as shown in Fig. 13.2. It is clamped at one end and at the other an axial compensive load P_1 is applied. Write the expression for the strain energy of the rod. Given $D_2 = nD_1$.

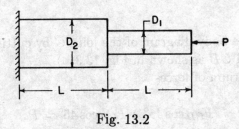

Fig. 13.2

Solution

We write the strain energy for the two portions separately and add them up, assuming that the material behaves linearly. Hence,

$$U = \frac{1}{2}\frac{P^2 L}{A_1 E} + \frac{1}{2}\frac{P^2 L}{A_2 E}$$

Now

$$A_2 = \frac{\pi}{4}D_1^2$$

$$A_1 = \frac{\pi}{4}D_2^2 = \frac{\pi}{4}n^2 D_1^2 = n^2 A_2$$

Therefore,

$$U = \frac{1}{2}\frac{P^2 L}{E}\left[\frac{1}{n^2 A_2} + \frac{1}{A_2}\right] = \frac{P^2 L}{2A_2 E}\left(\frac{1+n^2}{n^2}\right)$$

Example 13.2

A two bar truss member is subjected to a load P as shown in Fig. 13.3(a). The two rods are made of the same material and have the same cross-sectional area which is constant along the length. Obtain strain energy of the structure, if the areas of the member is A.

Fig. 13.3

Solution

Draw the free body diagram of the joint C by cutting the members CD and CB as shown in Fig. 13.3(b).

From equilibrium of forces

$$F_{CD} \cos 45° + F_{CB} \cos 45° = P$$

and

$$F_{CD} \sin 45° = F_{CB} \sin 45°$$

Solving the above two equations,

$$F_{CD} = F_{CB} = \frac{P}{2}$$

From the sign convention of forces

$$F_{CD} = +\frac{P}{2},$$

$$F_{CB} = -P/2$$

Now the strain energy,

$$U = \frac{F_{BC}^2 L}{2AE} + \frac{F_{CD}^2 L}{2AE} = \frac{0.25 P^2 L}{2AE} + \frac{0.25 P^2 L}{2AE} = \frac{P^2 L}{4AE}$$

13.3   Strain Energy for Bending of a Beam

Consider a beam, shown in Fig. 13.4, subjected to a transverse load. It is assumed that the beam is slender, therefore, we can neglect the effect of shear.

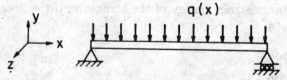

Fig. 13.4

Taking into account only the normal stresses produced by the moment M at any cross-section, the normal stresses σ_x in the cross-section at any distance y from the neutral axis is

$$\sigma_x = \frac{My}{I} \quad \text{(See Chapter 10)} \qquad (13.19)$$

Substituting for σ_x in Eq. (13.15), the strain energy is

$$U = \int_{vol} \frac{\sigma_x^2}{2E} dv = \int_{vol} \frac{M^2 y^2}{2EI^2} dv \qquad (13.20)$$

Setting $dV = dA.dx$, where dA represents an element of the cross-section area, Eq. (13.20) becomes

$$U = \int_0^L \frac{M^2}{2EI^2} \left[\int_{Area} y^2 dA \right] dx \qquad (13.21)$$

The quantity inside square brackets represents the second moment of area I of the cross-section about its neutral axis. Hence,

$$U = \int_0^L \frac{M^2}{2EI} dx \qquad (13.22)$$

Eq. (13.21) can be used for evaluating the strain energy if we know the bending moment M as a function of spatial coordinate x.

Using of Moment-curvature relation (Chapter 10), Eq. (13.21) can be rewritten as

$$U = \int_0^L \frac{EI}{2} \left(\frac{d^2 v}{dx^2} \right)^2 dx \qquad (13.23)$$

Bending moment M is written as

$$M = EI \frac{d^2 v}{dx^2} \qquad (13.24)$$

Example 13.3

Determine the strain energy of the rectangular beam, shown in Fig. 13.5, taking into account only the normal stresses.

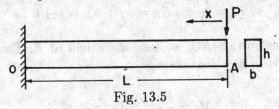

Fig. 13.5

Solution

Eq. 13.22 was used in writing the strain energy due to normal stresses. The bending moment at any distance x from the free end is $M = -Px$.

Therefore,

$$U = \int_o^L \frac{P^2 x^2}{2EI} dx$$

For the given cross-section, $I = \frac{1}{12}bh^3$, which is constant along the length. Substituting the value of I and integrating with respect to x gives

$$U = \frac{12P^2}{2Ebh^3} \cdot \frac{L^3}{3}$$

$$U = \frac{2P^2}{Ebh^3} \cdot L^2$$

Example 13.4

Obtain the expression for strain energy due to bending of the beam shown in Fig. 13.6(a). Given $I = 100 \times 10^{-6}$ m^4, $E = 200$ GPa, $L = 4$ m, $a = 0.5$ m, $b = 3.5$ m, $P = 100$ kN.

Solution

First draw the free body diagram of the system and obtain reactions at the supports A and B. The support is idealised as a simple (pin) for drawing the free body diagram (Fig. 13.6(b)). The bending moment diagram is shown in Fig. 13.6(c).

From the free-body diagram, the reactions are

$$R_a = \frac{Pb}{L}; \quad R_B = \frac{Pa}{L}; \quad L = a + b$$

For the portion AC, we take the origin at A. The bending moment M at any distance x is

$$M_1 = \frac{Pb}{L}x; \quad 0 < x < a$$

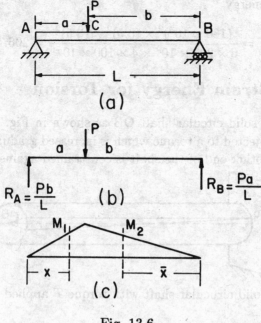

Fig. 13.6

For the portion BC, we take the origin at B. The bending moment M at any distance $\bar{x}$ is

$$M_2 = \frac{Pa}{L}\bar{x}; \quad 0 < \bar{x} < b$$

Since strain energy is a scalar quantity, we can add the strain energy contributions from the two portions of the beam to obtain the total strain energy of the beam. Thus,

$$
\begin{aligned}
U &= U_{AC} + U_{BC} = \int_o^a \frac{M_1^2}{2EI}dx + \int_o^b \frac{M_2^2}{2EI}d\bar{x} \\
&= \frac{1}{2EI}\int_o^a \frac{P^2b^2}{L^2}x^2dx + \frac{1}{2EI}\int_o^b \frac{P^2a^2}{L^2}\bar{x}^2d\bar{x} \\
&= \frac{1}{2EI}\left[\frac{P^2b^2}{3L^2}a^3 + \frac{P^2a^2}{3L^2}b^3\right] = \frac{P^2a^2b^2}{6EIL^2}(a+b) = \frac{P^2a^2b^2}{6EIL}
\end{aligned}
$$

Substituting the respective values in the above expression,

the strain energy,

$$U = \frac{(180 \times 10^3)^2 \times (0.5)^2 \times (3.5)^2}{6 \times 200 \times 10^9 \times 4 \times 100 \times 10^{-6}} = 206.7$$

13.4   Strain Energy for Torsion

Consider a solid circular shaft OB as shown in Fig. 13.7. The shaft is subjected to a torque which is increased gradually from 0 to T. The other end of the shaft is constrained against rotation.

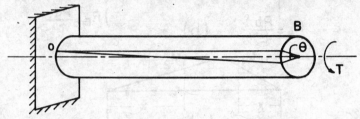

Fig. 13.7 Solid cirucular shaft with torque T applied at the free end.

The torsional relationship (Eq. 11.14) is

$$\frac{T}{J} = \frac{G\theta}{L} = \frac{\tau}{r} \qquad (13.25)$$

As the shaft is of constant cross-section, Eq. (13.25) indicates that the relationship between T and θ is linear. The shaft is subjected to a pure torque and, therefore, the only non-zero stress present is the shear stress τ. Therefore, strain energy due to torsion, per unit volume is

$$\bar{U} = \int_o^\gamma \tau.d\gamma \qquad (13.26)$$

Since $\tau = G\gamma$,

$$\bar{U} \;\; = \;\; \int_o^\gamma G\gamma dz = \frac{G\gamma^2}{2} = \frac{1}{2}\tau\gamma = \frac{\tau^2}{2G} \qquad (13.27)$$

Therefore total energy U_T is

$$U_T = \int_{vol} \frac{\tau^2}{2G} dV \qquad (13.28)$$

Using Eq. (13.25), Eq. (13.28) can be rewritten as

$$U_T = \int_{vol} \frac{T^2}{2GJ^2} r^2 \, dA \, dx$$

$$= \int_o^l \frac{T^2}{2GJ^2} \{ \int_A r^2 \, dA \} dx = \int_o^l \frac{T^2}{2GJ} dx \quad (13.29)$$

Equation (13.29) represents the torsional strain energy for a shaft. If T and J are constant across the lengths of the shaft, Eq. (13.29), can be integrated and rewritten as

$$U = \frac{T^2}{2GJ} L \quad (13.30)$$

or

$$U = \frac{G\tau\theta^2}{2L} \quad (13.31)$$

Example 13.5

A stepped shaft, shown in Fig. 13.8, is subjected to a torque T at the free end. Obtain the expression for strain energy of the shaft in terms of diameter, length and material of the shaft.

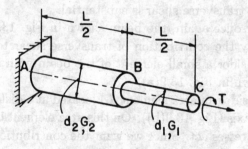

Fig. 13.8

Solution

We use Eq. (13.29) to calculate the strain energy of each of the two lengths of shaft. Since energy is a scalar quantity, we can add up the two energies to get the total energy of the shaft. Strain energy for the portion AB is

$$U_{AB} = \frac{T^2(L/2)}{2G_2 J_2}$$

where $J_2 = \frac{\pi}{32}d_2^4$ (solid shaft).

Strain energy for the portion BC is

$$U_{BC} = \frac{T^2(L/2)}{2G_1J_1}$$

where $J_1 = \frac{\pi}{32}d_1^4$ (solid shaft).

Here G_1 and G_2 are the shear modulus of the material of the two portions of the shaft.

Therefore,

$$U = U_{AB} + U_{BC} = \frac{T^2(L/2)}{2}[\frac{1}{G_2J_2} + \frac{1}{G_1J_1}]$$

13.5   Strain Energy for Transverse Shear

Section 13.3 described the problem of beam subjected to transverse load. For such a case, bending moment and shear force vary along the length. There we neglected the contribution of transverse shear to the energy of the system, by assuming that the beam is slender. However, if the beam is stubby, the contribution from transverse shear is substantial.

Consider once again, the beam shown in Fig. 13.4. We now consider only the contribution of transverse shear to the strain energy. Consider a small element of the beam with an enlarged view as shown in Fig. 13.9(a).

Also consider an area element dA located at a distance y from the neutral axis (Fig. 13.9(b)). On this area element both normal and shear stresses act. Since we want the contribution of τ, only τ is shown in Fig. 13.9(c). It is assumed that τ is uniformly distributed over the element dA.

The strain energy for this element dA of length dx is

$$d(dU_s) = \frac{\tau^2}{2G}dAdx \qquad (13.32)$$

or

$$dU_s = \int_A \frac{\tau^2}{2G}dAdx = \frac{dx}{2G}\int_A \tau^2 dA \qquad (13.33)$$

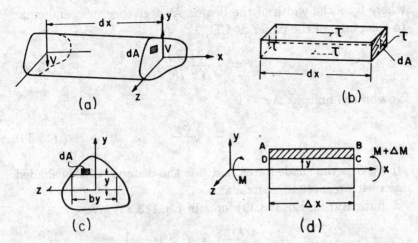

Fig. 13.9 (a) Enlarged view of section of the beam element, (b) cross-sectional view of the element, (c) shear stresses acting on the element and (d) bending moment acting on the element.

In Eq. (13.33) τ is expressed in terms of the applied transverse force. Consider once again the beam shown in Fig. 13.9(d). We shall include the bending moment across the section Δx.

The normal force on face BC, in general, is different from that on AD. The difference of the two will act as shear force over surface CD, since surface AB is free it cannot have any force.

The shear stress on face DC is

$$\tau = \frac{\text{Shear force across the force DC}}{\text{Area of surface DC}} \tag{13.34}$$

$$= \frac{\int_C^B \sigma_x dA - \int_D^A \sigma_x dA}{\text{Area of surface DC}} \tag{13.35}$$

Substituting for normal stress from Eq. (13.19),

$$\tau = \int_C^B -\frac{(M + \Delta M)y}{I} dA - \int_D^A \left(-\frac{My}{I}\right) dA$$

$$= \frac{\int_e^B -\frac{\Delta M y}{I} dA}{\text{Area of surface CB}} \tag{13.36}$$

Assuming that the beam cross-section is uniform along the length of the beam

$$\tau = -\frac{\Delta M}{I} \frac{\int_C^D y dA}{\Delta x b_y} \tag{13.37}$$

where b_y is the width of the beam. At a given y, perpendicular to the plane of the paper at CD,

$$\tau = -\frac{\Delta M}{\Delta x}\frac{S\bar{y}}{Ib_y} \qquad (13.38)$$

Now in the limit as $\Delta x \to 0$,

$$\tau = -Vy\frac{S\bar{y}}{Ib_y} \qquad (13.39)$$

where S is the shaded area and $\bar{y}$ is the distance of this shaded area with respect to neutral axis.

Substituting Eq. (13.39) for τ in Eq. (13.33) gives

$$dU_s = \frac{V_y^2 dx}{2GA}\left[\frac{A}{I^2}\int_A \frac{S^2\bar{y}^2}{b_y^2}dA\right] \qquad (13.40)$$

or

$$dU_s = \alpha_y \frac{V_y^2}{2GA}dx \qquad (13.41)$$

in which

$$\alpha_y = \frac{A}{I^2}\int_A \frac{S^2\bar{y}^2}{b_y^2}dA \qquad (13.42)$$

The strain energy due to transverse shear, for the entire beam is

$$U_s = \frac{1}{2}\int_o^\ell \alpha_y \frac{V_y^2}{2GA}dx \qquad (13.43)$$

Example 13.6

A uniform cross-section rectangular beam of length L is subjected to a transverse load P at the tip as shown in Fig. 13.10. Obtain the contributions due to bending energy and shear energy to the overall strain energy of the beam.

Solution

The total strain energy is due to bending moment and shear force at any section of the beam. Thus,

$$U = U_b + U_s = \int_o^L \frac{M^2}{2EI}dx + \alpha_y \int_o^L \frac{V_y^2 dx}{2GA}$$

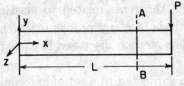
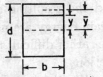

Enlarged view of section A B

Fig. 13.10

For the present case $M = -P(L - x)$ and $V_y = P$

Therefore,

$$U = \int_0^L \frac{P^2(L-x)^2}{2EI} dx + \alpha_y \frac{P^2 L}{2AG}$$

$$\alpha_y = \frac{A}{I^2} \int_A \frac{S^2 \bar{y}^2}{b_y^2} dx$$

where $\bar{y}$ is the centroidal distance of shaded area from neutral axis.

For the rectangular cross-section of width b and depth d,

$$A = bd \quad I = bd^3$$

$$S = \text{Shaded area} = b\left(\frac{d}{2} - y\right)$$

or

$$S = y + \frac{(d/2 - y)}{2} = \frac{1}{2}\left(\frac{d}{2} + y\right)$$

and

$$S\bar{y} = \frac{b}{2}\left(\frac{d}{2} + y\right)\left(\frac{d}{2} - y\right) = \frac{b}{2}\left[\left(\frac{d}{2}\right)^2 - y^2\right]$$

Here $b_y = b$ (constant across depth) and $dA = bdy$. Substituting these values in the expression for α_y, we get $\alpha_y = 6/5$, which on integration gives

$$U = \frac{P^2 L^3}{6EI} + \frac{6}{5}\left(\frac{P^2 L}{2AG}\right)$$

13.6 Work and Energy for Linear Elastic Systems under Several Loads

The strain energy of a linear elastic structure of an istropic material subjected to several loads may be expressed in terms of loads and the resulting deflections. The theorems related to elastic strain energy are useful in the analysis of many problems which are rather difficult to handle with the method discussed in earlier sections.

Consider a body in equilibrium subjected to a set of external forces, $\bar{F}_1$, $\bar{F}_2 \ldots \bar{F}_n$. Let δ_1, $\delta_2 \ldots \delta_n$ be the displacements corresponding to these forces. Consider the displacement δ_1 alone. It is quite obvious that all the forces contribute to this displacement. We imagine the forces to be acting one at a time and evaluate its component to the displacement δ_1. The displacement at point 1 due to a force $\bar{F}_1$ is written as $\alpha_{11}\bar{F}_1$, where α_{11} is the displacement at point 1 due to a unit force acting at 1. Similarly, the displacement at 1 due to F_2 acting at 2 is given by $\alpha_{12}\bar{F}_2$ and so on. Thus,

$$\delta_1 = \alpha_{11}\bar{F}_1 + \alpha_{12}\bar{F}_2 + \alpha_{13}\bar{F}_3 + \ldots \alpha_{1n}\bar{F}_n$$

Following similar argument, one can write

$$\delta_2 = \alpha_{21}\bar{F}_1 + \alpha_{22}\bar{F}_2 + \alpha_{23}\bar{F}_3 + \ldots \alpha_{2n}\bar{F}_n$$

$$\delta_3 = \alpha_{31}\bar{F}_1 + \alpha_{32}\bar{F}_2 + \alpha_{33}\bar{F}_3 + \ldots \alpha_{3n}\bar{F}_n$$

and

$$\delta_n = \alpha_{n1}\bar{F}_1 + \alpha_{n2}\bar{F}_2 + \alpha_{n3}\bar{F}_3 + \ldots \alpha_{nn}\bar{F}_n \qquad (13.44)$$

The above equations in a matrix form can be written as

$$\begin{Bmatrix} \delta_1 \\ \delta_2 \\ \cdot \\ \cdot \\ \cdot \\ \delta_n \end{Bmatrix} = \begin{bmatrix} \alpha_{11} & \alpha_{12} & \ldots & \alpha_{1n} \\ \alpha_{21} & \alpha_{22} & \ldots & \alpha_{2n} \\ \cdot & \cdot & & \cdot \\ \cdot & \cdot & & \cdot \\ \cdot & \cdot & & \cdot \\ \alpha_{n1} & \alpha_{n2} & \ldots & \alpha_{nn} \end{bmatrix} = \begin{Bmatrix} \bar{F}_1 \\ \bar{F}_2 \\ \cdot \\ \cdot \\ \cdot \\ \bar{F}_n \end{Bmatrix} \qquad (13.45)$$

Maxwell's Reciprocal Theorem states that *the displacement at any location i of a linear elastic system, due to a unit load at another location j, is equal to the displacement at j due to the unit load at i, provided, that the forces and displacements are measured in the same direction at each point.* In view of this, the coefficients α_{ij} are symmetric, i.e., $\alpha_{ij} = \alpha_{ji}$.

Let us now multiply first of Eq. (13.44) by F_1, second by F_2 and so on. Adding them gives

$$
\begin{aligned}
F_1\delta_1 + F_2\delta_2 + \ldots F_n\delta_n =\ & \alpha_{11}F_1^2 + \alpha_{12}F_1F_2 + \ldots \alpha_{1n}F_1F_n \\
& + \ \alpha_{21}F_2F_1 + \alpha_{22}F_2^2 + \ldots \alpha_{2n}F_2F_n \\
& + \ \ldots
\end{aligned}
\tag{13.46}
$$

If we now consider that the loads or forces are gradually applied, increasing from zero to the final value, the work done by the force F_1 is $\frac{1}{2}F_1\delta_1$; and so on. The total work done by the set of forces $F_1, F_2 \ldots F_n$ on the body, i.e., the strain energy U of the body is given by

$$
U = \frac{1}{2}F_1\delta_1 + \frac{1}{2}F_2\delta_2 + \ldots \frac{1}{2}F_n\delta_n = \frac{1}{2}\sum_{i=1}^{n}\sum_{j=1}^{n}\alpha_{ij}F_iF_j
\tag{13.47}
$$

In order to make the process clear, consider a simply supported elastic beam subjected to two discrete forces F_1 and F_2 as shown in Fig. 13.11.

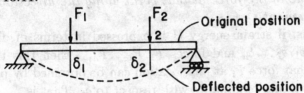

Fig. 13.11 Simply supported beam with two discrete forces.

The strain energy of the beam is equal to the work done by the forces F_1 and F_2 through the displacements δ_1 and δ_2 in the direction of the load. It is assumed that the forces F_1 and F_2 are applied one at a time. The force F_1 produces a displacement δ_{11} and δ_{21} at locations 1 and 2, respectively. Therefore,

$$
\delta_{11} = \alpha_{11}F_1, \quad \delta_{21} = \alpha_{21}F_1
\tag{13.48}
$$

Similarly, when force F_2 is applied, the displacements δ_{12} and δ_{22} at locations 1 and 2, respectively are

$$
\delta_{12} = \alpha_{12}F_2, \quad \delta_{22} = \alpha_{22}F_2
\tag{13.49}
$$

Combining Eqs. (13.47) and (13.48),

$$\delta_1 = \delta_{11} + \delta_{12} = \alpha_{11}F_1 + \alpha_{12}F_2 \tag{13.50}$$

$$\delta_2 = \delta_{21} + \delta_{22} = \alpha_{21}F_1 + \alpha_{22}F_2 \tag{13.51}$$

When forces F_1 and F_2 are applied together,

$$\begin{aligned}
U &= \frac{1}{2}(F_1\delta_1 + F_2\delta_2 \\
&= \frac{1}{2}(\alpha_{11}F_1^2 + \alpha_{12}F_1F_2 + \alpha_{21}F_2F_1 + \alpha_{22}F_2^2) \tag{13.52}
\end{aligned}$$

From Maxwell's reciprocal theorem, $\alpha_{12} = \alpha_{21}$, therefore,

$$U = \frac{1}{2}(\alpha_{11}F_1^2 + 2\alpha_{12}F_1F_2 + \alpha_{22}F_2^2) \tag{13.53}$$

13.7   Castigliano's First Theorem

The first theorem states that *the partial derivative of strain energy U, with respect to a particular displacement, gives the value of the force at that point along the direction of the displacement.*

Thus, if strain energy U is expressed in terms of displacements $\delta_1, \delta_2 \ldots \delta_n$ and forces $F_1, F_2 \ldots F_n$, , then this theorem states that force F_i at any location can be obtained by partially differentiating function U with respect to δ_i. That is,

$$F_i = \frac{\partial U}{\partial \delta_i} \tag{13.54}$$

The displacements and forces have their generalized meaning; δ, in general, represents displacement, rotation and angle of twist, while F represents force, bending moment and twisting moment.

13.8   Castigliano's Second Theorem

The second theorem states that *for a linearly elastic system which is in equilibrium under the action of applied forces, the partial derivative of the strain energy with respect to a particular independent force gives the corresponding displacement at the point of application of that particular force.* Therefore,

$$\delta_j = \frac{\partial U}{\partial F_j} \tag{13.55}$$

Here again, the forces and displacements have their generalized meaning. Thus,

$$\theta_j = \frac{\partial U}{\partial M_j} \tag{13.56}$$

$$\theta_j = \frac{\partial U}{\partial T_j} \tag{13.57}$$

Example 13.7

The uniform cantilever OA is subjected to an external load as shown in Fig. 13.12. Obtain the tip deflection of the beam by making use of the Castigliano's theorem.

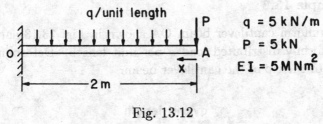

Fig. 13.12

Solution

We shall assume that the beam is slender and therefore, the contributions of the transverse shear energy to the total energy of the beam is negligible. Major contribution to the strain energy is due to bending alone. The deflection δ_{tip} is in the direction of the applied load P. If in a problem, the transverse load is not applied externally, we can introduce a fictitious load and carry out the entire analysis and at the end, put the fictitious load equal to zero to get the deflection at that point. Thus,

$$U_b = \int_o^L \frac{M^2}{2EI} dx$$

$$\delta_{tip} = \frac{\partial U_b}{\partial p} = \int_o^L \frac{M}{EI} \frac{\partial M}{\partial P} dx$$

The bending moment M at a distance x from A is given by

$$M = -(Px + \frac{1}{2}qx^2) \quad \text{and} \quad \frac{\partial M}{\partial P} = -x$$

Hence,

$$\delta_{tip} = \frac{L}{EI}\int_o^L \left(Px + \frac{1}{2}qx^2\right)xdx$$

$$= \frac{1}{EI}\left[\frac{Px^3}{3} + \frac{1}{2}q\frac{x^4}{4}\right]_o^L = \frac{1}{EI}\left[\frac{PL^3}{3} + \frac{qL^4}{8}\right]$$

Substituting the values,

$$\delta_{tip} = \frac{1}{5 \times 10^6}\left[\frac{5000 \times 8}{3} + \frac{5000 \times 16}{8}\right]$$

$$= 4.666 \text{ mm}$$

Example 13.8

The uniform cantilever beam OA shown in Fig. 13.13 supports a uniformly distributed load q per unit length. Determine the slope of the tip of the cantilever beam.

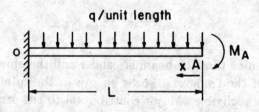

Fig. 13.13

Solution

To determine the slope at A, we have to have an external moment applied at that point. Since no moment is acting, we introduce a dummy clockwise moment at A as shown in Fig. 13.13. Then from Eq. (13.55),

$$\theta_A = \frac{\partial U}{\partial M_A}$$

Since

$$U_b = \frac{1}{2} \int_o^L \frac{M^2}{EI} dx$$

therefore,

$$\theta_A = \frac{\partial U_b}{\partial M_A} = \int_o^L \frac{M}{EI} \frac{\partial M}{\partial M_A} dx$$

The bending mement M at a distance x from A is

$$M = -M_A - \frac{1}{2} qx^2$$

Therefore,

$$\frac{\partial M}{\partial M_A} = -1$$

Substituting for M and $\partial M/\partial MA$,

$$\theta_A = \frac{\partial U}{\partial M_A} = \frac{1}{EI} \int_o^L -\left(M_A + \frac{1}{2} qx^2\right)(-1)dx$$

or

$$\theta_A = M_A L + \frac{1}{6} \frac{qL^3}{EI} \, .$$

We put $M_A = 0$, since in the actual system there is no moment applied at the point A. Therefore,

$$\theta_A = \frac{1}{6} \frac{qL^3}{EI}$$

Problems

13.1 The steel rod ABC shown in Fig. 13.14, has a yield strength of 250 MPa. The elastic modulus of the rod material is 200 GPa. Determine the maximum axial strain energy acquired by the rod without any permanent deformation for the loading shown.

13.2 Determine the strain energy of a structure made up of steel loaded as shown in Fig. 13.15. Given: $P = 500$ N, diameter of the rod = 20 mm, $\ell = 0.5$ m, $E = 200$ GPa and $E/G = 2.5$.

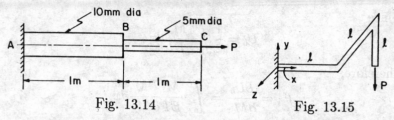

Fig. 13.14   Fig. 13.15

13.3 Taking into account only the normal stresses due to bending, obtain the expression for the strain energy of the beam AB, for loadings shown in Figs. 13.16(a), 13.16(b) and 13.16(c).

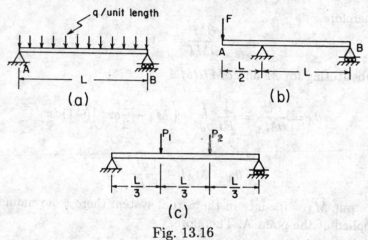

(a)   (b)

(c)

Fig. 13.16

13.4 Determine the strain energy of the stepped shaft subjected to a torque as shown in Fig. 13.17. G for the shaft material = 40 GPa.

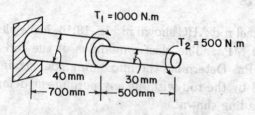

Fig. 13.17

13.5 For the beam and loading shown in Fig. 13.18, determine the transverse deflection in terms of EI at a distance of 1.5 m from the left end.

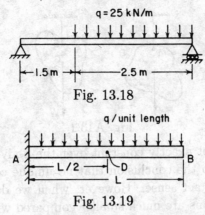

Fig. 13.18

Fig. 13.19

13.6 For the rectangular beam loaded as shown in Fig. 13.19, obtain the transverse deflection of the mid-point D.

APPENDIX A13

A13.1 Centre of Gravity

A rigid body is regarded as an aggregate of particles, each having a weight, or force of attraction, directed towards the centre of the earth. The resultant of this parallel system of gravitational forces in space is the weight of the body. The resultant axis of the parallel gravitational forces that act on the various particles is called the *gravity axis* of the body. Its location with respect to the body depends on the orientation of the body with respect to the earth. For example, when a body is suspended from point A (Fig. A13.1) the gravity axis is the dotted line passing through A. If the body is suspended from point B, the gravity axis is different and is shown as a line passing through B. If the body is rotated and suspended from another point, one can describe another gravity axis. Surprisingly, all such line pass through a common point, called the *centre of gravity*.

An exact analysis, however, would take into account the fact that the direction of the gravity forces for the various particles may not be exactly parallel since they converge towards the center of attraction of the earth. Further, since the particles are at different distances from the earth, the intensity of the earth's

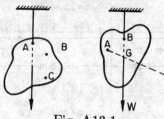

Fig. A13.1

force field is not exactly constant over the body. These considerations lead to the conclusion that no unique center of gravity exists in the exact sense. However, when we deal with bodies whose dimensions are much smaller compared with those of the earth, the assumption of a parallel force field due to gravitational attraction is reasonably accurate and the centre of gravity can then be uniquely determined. Further, in such cases, the *centre of mass* coincides with that of centre of gravity.

Mathematically, the location of the centre of gravity of any body can be determined using *varignon's theorem* of principle of moments. The next section deals with the procedure for evaluating the centre of gravity.

A13.1.1 Centroid of an Area (First Moment of an Area)

Consider an arbitrary shaped plate of area A, of constant thickness, in the XY-plane as shown in Fig. A13.2. Consider that the plate is made up of a large number of particles having weights ΔW_1, ΔW_2, ΔW_3, etc. The total weight W of the plate can then be written as

$$W = \Delta W_1 + \Delta W_2 + \Delta W_3 \qquad (A.1)$$

where W is an equivalent force acting at point C (x_g, y_g).

One can locate the coordinates (x_g, y_g) by taking moment of all the forces about X and Y axes passing through O. Thus,

$$\sum M_x = 0 \qquad (A.2)$$

$$\sum M_y = 0 \qquad (A.3)$$

Equation (A.2) gives

$$y_g = \frac{\Delta W_1 y_1 + \Delta W_2 y_2 + \Delta W_3 y_3 + \ldots}{\Delta W_1 + \Delta W_2 + \Delta W_3 + \ldots}$$

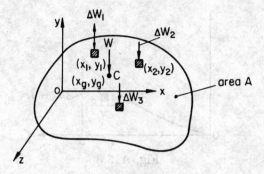

Fig. A13.2

$$= \frac{\Delta W_1 y_1 + \Delta W_2 y_2 + \Delta W_3 y_3 + \dots}{\Delta W} \qquad (A.4)$$

Similarly, Eq. (A.3) gives

$$x_g = \frac{\Delta W_1 x_1 + \Delta W_2 x_2 + \Delta W_3 x_3 + \dots}{\Delta W} \qquad (A.5)$$

If the elemental forces are very small, we can replace Eqs. (A.4) and (A.5) in terms of integrals rather than discrete values. Rewriting Eqs. (A.4) and (A.5) in terms of integrals gives

$$x_g = \frac{\int x\,dW}{\int dW} \qquad (A.5)$$

$$y_g = \frac{\int y\,dW}{\int dW} \qquad (A.6)$$

Whenever the density of a body is uniform throughout, it will be a constant factor in both the numerators and denominators of Eqs. (A.5) and (A.6), and will therefore cancel. The expressions that remain define a purely geometrical property of the body, since any reference to its physical properties has disappeared. The term *centroid* is used when the calculations concern a geometrical shape. If the density is uniform throughout the body, the positions of centroid and centre of mass or centre of gravity are identical.

For a homogeneous plate of constant thickness,

$$dW = \rho t\,dA \qquad (A.7)$$

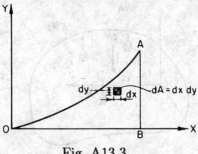

Fig. A13.3

Substituting for dW in Eqs. (A.5) and (A.6),

$$x_g = \frac{\int x dA}{\int dA} = \frac{\int x dA}{A} \qquad (A.8)$$

$$y_g = \frac{\int y dA}{\int dA} = \frac{\int y dA}{A} \qquad (A.9)$$

where A is the area of the plate in the XY-plane. Consider an arbitrary area A enclosed by a curve and X and Y axes as shown in Fig. A13.3.

To evaluate the location of x_g and y_g, we need to apply Eqs. (A.8) and (A.9). Substituting for dA, these equations can be written as

$$x_g = \frac{\int\int x dy dx}{\int\int dx dy} \qquad (A.10)$$

$$y_g = \frac{\int\int y dx dy}{\int\int dx dy} \qquad (A.11)$$

With this representation we have to work with double integrals, which is generally quite involved. This can be overcome by considering a very thin rectangular strip as shown in Figs. A13.4(a) and A13.4(b) for obtaining centre of gravity locations.

In Fig. A13.4(a), the elemental strip has an area $dA = y dx$. This elemental area has the centroidal distance $\bar{x} = x$, from y-axis. Therefore x_g for the entire area is

$$x_g = \frac{\int y dx . x}{\int y dx} \qquad (A.12)$$

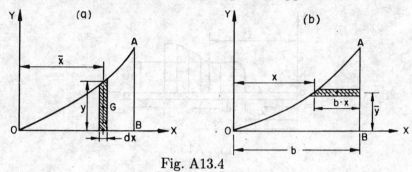

Fig. A13.4

In a similar manner, from Fig. A13.4(b),

$$y_g = \frac{\int (b-x)dy.y}{\int (b-y)dy} \qquad (A.13)$$

A13.1.2 Centre of Pressure

The pressure of wind on the side walls of a building, water pressure on the retaining wall of a dam are some examples of distributed loads. Usually, these forces act normal to the surface and for a plane surface, they constitute a parallel force system. The resultant force can be obtained by the algebraic addition of these forces. The line of action of the resultant force can be obtained, as in Section A.2, by using Varignon's theorem. The intersection of the line of action of the resultant of the distributed force system and the plane on which it acts is termed as the *centre of pressure*.

Frequently, in the analysis of beams one encounters distributed loads. In order to find the reactions at the supports, the distributed load can be replaced by an equivalent resultant force passing through the centre of pressure. However, for drawing bending moment diagrams or for the determination of stresses one should not replace the distributed load in the above mentioned manner as this will result in gross errors in the estimation of stresses.

Figure A13.5 shows the load distribution diagram of a floor beam.

The ordinate q of the diagram indicates the intensity of the load in force per unit length and in SI units it is N/m. The load

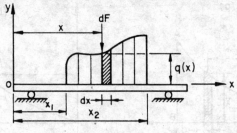

Fig. A13.5

dF on an element of the beam of length dx is

$$dF = -qdx\, j \qquad (A.14)$$

where j is a unit vector along y-direction. The total load is

$$F = -\int_{x_1}^{x_2} qdx\, j = -j\int_{x_1}^{x_2} qdx \qquad (A.15)$$

The moment of the resultant force about an axis through O is

$$M_0 = \int x \times dF = \int_{x_1}^{x_2} xi \times (-qdx)j = -k\int_{x_1}^{x_2} (xqdx) \quad (A.16)$$

Let x_p be the location of the *centre of pressure*, then by Varignon's theorem,

$$x_p \times F = M_0 \qquad (A.17)$$

Using Eqs. (A.16) and (A.17) we get,

$$x_p = \frac{M_0}{F}$$

From Eqs. (A.15) and (A.17), the magnitude of the resultant of the force distribution is the area under the load-distribution diagram and x_p is nothing but the centroid of the above area.

A.2   Moment of Inertia

While designing structure, we have to select the material and also the type of cross-section of the load bearing member. Interestingly the rails are made of I-section though difficult to manufacture. One comes across several other sections such as angle, box, hat and other special sections that suit a particular application. It is important to understand the logical basis for arriving at a particular choice of cross-section and how such an approach can be mathematically represented in the analysis. Looking from a different point of view, the area or mass is distributed in a particular way for each section. The nature of this distribution significantly affects the response of the structure to the applied loading. The purpose of this section is to arrive at a mathematical quantity which represents the distribution of mass or area. Selection of an appropriate cross-section for an application requires further study and is beyond the scope of this book.

If the load bearing member has to resist either tension or compression forces then only the total area of the cross-section is important. The area distribution does not affect the performance. On the other hand, when the force is distributed over the area, as in the case of a beam, the distribution of the area plays an important role in determining the response of the structure to the applied loading.

The integrals like $\int x^2 dA$ and $\int y^2 dA$ appear in the derivation of the stress distribution in the case of symmetrical bending. These are referred to as *moment of inertia* about the axis in question. However, a more fitting term is *second moment of area*.

The above mentioned integrals are similar in form to the representation of distribution of mass while evaluating the resultant moments of the inertia forces in the case of rotating bodies. Eüler coined the term *moment of inertia of mass* to refer to the effect of distribution of mass. An area has no mass and hence no inertia. However, in view of the similarity of the mathematical form, the term moment of inertia is used. To distinguish it from mass moment of inertia, the term moment of inertia is used.

When the moment axis is in the plane of the area, the second moment of the area is called the *rectangular moment of inertia*. If the moment axis is perpendicular to the plane of the area then it is called the *polar moment of inertia*. The second moment of inertia is only a function of the area distribution and can be used in several instances. The reference to beam problem is done as an illustration and its use is not restricted to such problems

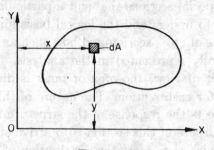

Fig. A13.6

only.

The moment of inertia of a cross-section is defined as the sum of the products obtained by multiplying the elemental area by the square of the perpendicular distance of the centre of gravity of the elemental area from the reference axis. Mathematically this can be defined as

$$I = \int (\text{distance from the reference axis})^2 \times dA \qquad (A.18)$$

Consider an arbitrary plane area in the XY-plane as shown in Fig. A13.6.

Using of Eq. (A.18), the moment of inertia of the area about the X-axis can be written as

$$I_x = \int y^2 dA \qquad (A.19)$$

Similarly, the second moment of area about the y-axis is

$$I_y = \int x^2 dA \qquad (A.20)$$

Here x and y are distances measured from the center of gravity of the elemental area.

Example A13-1

Consider the rectagular cross-section shown in Fig. A13.7 and evaluate I_x and I_y.

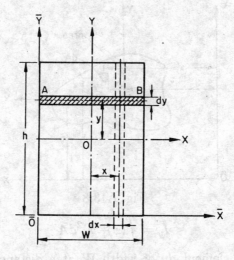

Fig. A13.7

Solution

The cross-section is doubly symmetric with O as the position of the centre of gravity. The reference axes X and Y are chosen to pass through O. We wish to determine I_x and I_y of the section.

Determination of I_x:

Consider an arbitrary area A with the center of gravity at O. OX and OY are the axes passing through the point O. Let $\bar{x}$ and $\bar{y}$ be another set of axes passing through O. These set of axes are parallel to the axes OX and OY as shown in Fig. A13.8.

The moment of inertia of the area A about $\bar{x}$ and $\bar{y}$ axes can be written as

$$I_{\bar{x}} = \int y'^2 dA \qquad (A.21)$$

$$I_{\bar{y}} = \int x'^2 dA \qquad (A.22)$$

From Fig. A13.8,

$$y' = y + \bar{y} \quad \text{and} \quad x' = x + \bar{x}$$

Substituting these values in Eqs. (A.21) and (A.22),

$$I_{\bar{x}} = \int (y + \bar{y})^2 dA \qquad (A.23)$$

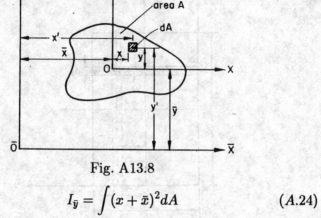

Fig. A13.8

$$I_{\bar{y}} = \int (x + \bar{x})^2 dA \qquad (A.24)$$

Consider an element dy of width W at a distance y from the x-axis as shown in Fig. A13.7. Then,

$$dA = W\,dy$$

and

$$I_x = \int_{-h/2}^{h/2} y^2 dy W = \frac{W}{3}\left[y^3\right]_{-h/2}^{h/2} = \frac{Wh^3}{12}$$

Determination of Iy:

Consider a strip of width dx and depth h at a distance x from the y-axis as shown in Fig. A13.7. Then,

$$I_y = \int_{-W/2}^{W/2} x^2 dx h = \frac{h}{3}[x^3]_{-W/2}^{W/2} = \frac{hW^3}{12}$$

A13.2.1 Parallel Axes Theorem

In many situations, the moment of inertia or the second moment of area about the axes passing through the center of gravity may be known. However, we may require the moment of inertia about an axes parallel to these axes. In such a situation, make use of parallel axes theorem, which gives

$$I_{\bar{x}} = \int y^2 dA + 2\bar{y}\int y dA + \int \bar{y}^2 dA$$

or

$$I_{\bar{x}} = I_x + \int \bar{y}^2 dA + 2\bar{y} \int y dA \qquad (A.25)$$

The last term of Eq. (A.25) is zero since OX is the centroidal axis. Therefore,

$$I_{\bar{x}} = I_x + \bar{y}^2 A \qquad (A.26)$$

In a similar manner it can be shown that

$$I_{\bar{y}} = I_y + \bar{x}^2 A \qquad (A.27)$$

where $\bar{x}$ and $\bar{y}$ are the distances between $\bar{x}$ and $\bar{y}$ axes and the centroidal axis passing through O.

Example A13-2

Making use of parallel axis theorem, obtain the second moment of area of $I_{\bar{x}}$ and $I_{\bar{y}}$ of a rectangle of width W and depth h.

Solution

Consider once again the rectangular section shown in Fig. A13.7. OX and OY represent the centroidal axis. Coinciding with the edges of the rectangle consider another set of axis $\bar{o}\bar{x}$ and $\bar{o}\bar{y}$.
Invoking Eqs. (A.26) and (A.29),

$$I_{\bar{x}} = \frac{1}{12}Wh^3 + Wh\left(\frac{h}{2}\right)^2 = \frac{1}{12}Wh^3 + \frac{1}{4}Wh^3 = \frac{1}{3}Wh^3$$

and

$$I_{\bar{y}} = \frac{1}{12}hW^3 + \left(\frac{W}{2}\right)^2 Wh = \frac{1}{12}hW^3 + \frac{1}{4}hW^3 = \frac{1}{3}hW^3$$

Example A13-3

Obtain the moment of inertia I_x of the channel section shown in Fig. A13.9.

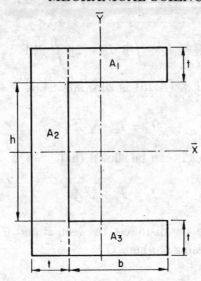

Fig. A13.9

Solution

The second moment of area I_x of the cross-section about X axis can be evaluated in the following three alternate ways:

(i) Divide the section into three rectangles as shown in Fig. A13.9 and find the second moment of area of the rectangles about its own centre of gravity (CG) axis and then transfer the values about CG of the section by employing parallel axis theorem.

(ii) Divide the cross-section as shown in Fig. A13.10 into three rectangles and calculate the second moment of area.

(iii) Consider a bigger solid rectangle made up of outer walls of the section and another smaller rectangle as shown in Fig. A13.11. The difference in the values of I_x for the two rectangles gives the value of I_x for the channel section.

Alternative (iii)

The location of CG of both the rectangles with respect to the vertical axis remains the same. However, the location with respect to X-axis is different. This does not concern us since we are interested only in I_x. Thus,

$$I_{\bar{x}} = \frac{1}{12}(b+t)(h+2t)^3 - \frac{1}{12}bh^3$$

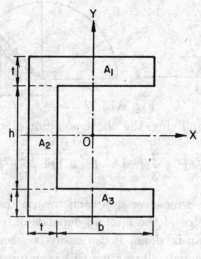

Fig. A13.10

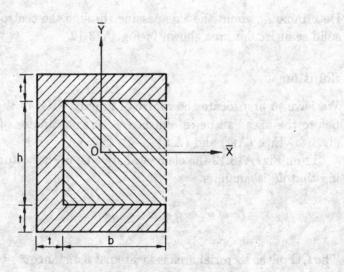

Fig. A13.11

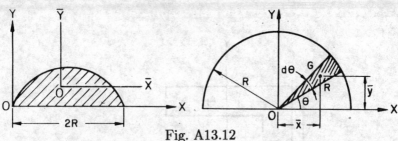

Fig. A13.12

Expanding and simplifying, the above equation gives

$$I_{\bar{x}} = \frac{1}{2}bh^2t + bht^2 + \frac{2}{3}bt^3 + \frac{1}{12}h^3t + ht^3 + \frac{1}{2}h^2t^2 + \frac{2}{3}t^4$$

Thus, we see that, whichever approach we use the value $I_{\bar{x}}$ is the same. The choice of the method depends on the case with which evaluation can be done. It can easily be seen that for the cross-section considered, alternative (iii) is quite convenient to apply.

Example A13-4

Determine $I_{\bar{y}}$, about the axis passing through the centroid of the solid semicircular area shown in Fig. A13.12.

Solution

We have to first locate the centre of gravity (CG) of the section before the axes can be established. The coordinates of CG are given by Eqs. (A.4) and (A.5).

From Fig. A13.12 the elemental area dA of the sector, assuming that $d\theta$ is small, is

$$dA = \frac{1}{2}R^2 d\theta$$

The CG of the sectorial area is located at a distance $\bar{x} = \frac{2}{3}R\cos\theta$, and $\bar{y} = \frac{2}{3}R\sin\theta$ for X and Y axes. Hence

$$x_g(\text{Total area}) = \frac{\int x dA}{\int dA} = \frac{\int_o^\pi \frac{2}{3}R\cos\theta \, \frac{1}{2}R^2 d\theta}{\int_o^\pi \frac{1}{2}R^2 d\theta} = \frac{\frac{1}{3}R^3 \sin\theta|_o^\pi}{\frac{1}{2}R^2[\theta]_o^\pi} = 0$$

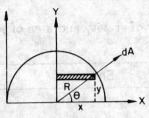

Fig. A13.13

which implies that X-axis passes through O. For vertical location,

$$y_g(\text{Total area}) = \frac{\int y dA}{\int dA} = \frac{\int_o^\pi \frac{2}{3} R \sin\theta \frac{1}{2} R^2 d\theta}{\int_o^\pi \frac{1}{2} R^2 d\theta}$$

$$= \frac{\frac{1}{3} R^3 (-\cos\theta)|_o^\pi}{\frac{1}{2} R^2 [\theta]_o^\pi} = \frac{4}{3} \frac{R}{\pi}$$

Thus, CG of the semicircular disc is located on the symmetry line at a distance $\frac{4}{3} \frac{R}{\pi}$ from the base. We can now determine $I_{\bar{x}}$ for the given area from I_x. From Fig. A13.13,

$$I_x = \int y^2 dA = \int_o^\pi R^2 \sin^2\theta \, x dy$$

$$= \int_o^\pi R^2 \sin^2\theta \, R \cos\theta \, R \cos\theta d\theta$$

$$= R^4 \int_o^\pi \sin^2\theta \cos^2\theta \, d\theta = \frac{R^4}{8} \pi$$

Since $I_x = I_{\bar{x}} + \bar{y}^2 A$,

$$I_{\bar{x}} = I_x - \bar{y}^2 A = \frac{\pi R^4}{8} - \left(\frac{4}{3} \frac{R}{\pi}\right)^2 \frac{\pi R^2}{2} = R^4 \left[\frac{\pi}{8} - \frac{8}{9\pi}\right]$$

Tables A13.1 and A13.2 show the center of gravity and the second moment of area of some typical sections.

Table A13.1 Center of gravity and area of some typical sections.

Title	Shape	$\bar{x}$	$\bar{y}$	Area
Quarter circular area		$\dfrac{4R}{3\pi}$	$\dfrac{4R}{3\pi}$	$\dfrac{\pi r^2}{4}$
Semicircular area		0	$\dfrac{4R}{3\pi}$	$\dfrac{\pi r^2}{2}$
Rectangle		$\dfrac{L}{2}$	$\dfrac{B}{2}$	$L.B$
Triangle		$\dfrac{a+b}{3}$	$\dfrac{h}{3}$	$\dfrac{1}{2}bh$

Table A13.2 Second moment of area for some typical sections.

Title	Shape	I_x	I_y
Circular cross-section		$\dfrac{\pi}{4}R^4$	$\dfrac{\pi}{4}R^4$
Semi-circular section		$\dfrac{\pi}{8}R^4$	$\dfrac{\pi}{8}R^4$
Quarter circle		$\dfrac{\pi}{16}R^4$	$\dfrac{\pi}{16}R^4$
Ellipse		$\dfrac{\pi}{4}ab^3$	$\dfrac{\pi}{4}a^3b$
Triangle		$\dfrac{1}{12}bh^3$	

14. INTRODUCTION TO FLUID MECHANICS

14.1 What is a Fluid?

A fluid is that substance which can flow. Water flows down a river, blood flows through arteries and veins, and crude oil flows through pipelines. Thus, liquids are fluids. So also are gases, since inhaled air flows down the bronchial passages to the lungs and back, cooking gas flows through a tube from the cylinder to the stove, as does a cool breeze on a monsoon day.

To understand the property of a fluid that enables it to flow, consider a block of rubber attached to two metal plates, the lower one being fixed as shown in Fig. 14.1. Let us apply a shear force

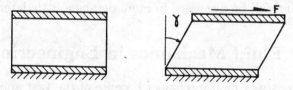

Fig. 14.1 A solid under the action of a shear force.

F to the upper plate. Under the action of this force, the rubber block deforms, and the system comes to rest with a distortion, measured by the shear angle γ. The more the force F, the more is the distortion.

Since the plate is at rest after the deformation γ, a force equilibrium must prevail, with the rubber block providing a shear force opposing the applied force. Thus, a solid, such as this

461

rubber block at rest, opposes applied shear forces by undergoing deformation with the opposing shear force increasing with deformation.

This is valid only for solids. Clearly, a fluid, be it a liquid or a gas, can not restrain the applied shear by deforming to an equilibrium position. We, therefore, say that: *a fluid at rest can not sustain shear.*

This may be used as a definition of a fluid. But fluids can, and do, sustain shear when they are not at rest, but are continually deforming. Consider the flow of a liquid through a pipe under the action of a pressure difference between the two ends (Fig. 14.2).

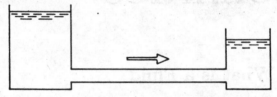

Fig. 14.2 A fluid flowing through a pipe.

If there was absolutely no external resistance to shear, the pressure difference would not have been resisted at all, and the fluid would have continually accelerated in the pipe. But that is not what we find. We see that the pressure difference controls the flow rate. For a constant pressure difference, there is a fixed flow rate for a given geometry, and that flow rate increases with the pressure difference. This suggests that the shear resisting the pressure difference arises from and depends on the rate of flow. This point will be discussed in more details a little later.

14.2   Fluid Mechanics in Engineering

Engineers study fluid mechanics because of the vast applications fluids find in engineering practice. The earliest need to understand fluids arose in connection with the transport of the life-giving water from the rivers and lakes to cities where human civilizations developed. To make fluids flow, we need to apply force or pressure. Thus, pumps are required to make water flow in municipal water mains. An animal heart is also a pump that makes the blood flow through arteries and veins. Fans and blowers are required to make the air flow down air-conditioning and ventilation ducts. An engineer needs to specify and design the fans, blowers and pumps, and therefore, needs to understand the forces required to move fluids.

Similarly, a moving fluid when it comes in contact with a body applies a force on it. We are well familiar with wind blowing leaves, uprooting trees and even demolishing structures like bridges and towers. An aeroplane is able to defy gravity because the air moving over its wings applies on it a force which we call lift.

Similar forces generated by fluid motion are responsible for propulsion and steering of boats and ships, rockets, aeroplanes, kites and gliders. The production of sound in our vocal chords, trumpet and harmonica are all similar applications where moving air applies forces on vocal chords and reeds setting them vibrating and radiating noise or music. An engineer must understand these forces if he has to play a role in the design of such system.

Another application is for transmission of force and motion. Levers, chains, belts and gears are familiar mechanical devices to change the point of application or the direction of a force. The same can be achieved by a fluid much more efficiently and conveniently. Thus, the fluid confined in a hydraulic press (Fig. 14.3) effectively transmits the force applied at piston A to piston B, changing its direction and magnitude in the process. Such a prin-

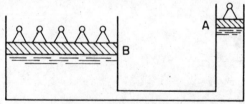

Fig. 14.3 A hydraulic press.

ciple is used in diverse applications such as brakes of automobiles, hydraulic and pneumatic controls of machinery in industry, hydraulic lifts, automatic transmission in automobiles, etc.

Another vast area of application of fluid mechanics depends upon the ability of the fluid flow to modify the rates of heat and mass transfer by the process known as *convection*. We are familiar with how a heated fluid rises up, bringing colder fluid in contact with a hot surface, thereby increasing the rate of heat loss from the hot surface. Similarly, an evening breeze washes the foul, polluting gases away from a city street. It also increases the rate of evaporation of sweat from our bodies, providing it a cool relief. This principle finds use in design of evaporative air coolers, spin dryers of milk and slurries, cooling towers, doping of semi-conductors, and a host of other applications.

Thus, we see that fluids at rest and in motion have a vast scope of applications. In these introductory chapters we would attempt to understand only the most elementary aspects of fluids at rest and in motion.

15. FLUID PROPERTIES

15.1 Introduction

A fluid has many properties whose values are required to specify its state. Some of the properties it has in common with solids are:
- o Density
- o Specific volume
- o Specific gravity
- o Temperature

Some other properties defined for fluids alone are:
- o Pressure
- o Vapour pressure
- o Viscosity
- o Surface tension
- o Compressibility

All of the above properties may be constant throughout the bulk of a fluid body, or may change from point to point. You are quite familiar with the definition of the first group of properties, their units and common methods of their measurements. In this book the standard SI units are used, though a set of conversion factor to the other commonly used units are included in the Appendices.

The SI units of density are kg/m^3. Water has a density of about 1,000 kg/m^3 at 4° C, while density of air is 1.29 kg/m^3 at 0°C and at standard atmospheric pressure.

Specific volume is defined as the volume of a unit mass of the fluid, and is the reciprocal of density. Thus, specific volume v of water is 10^{-3} m^3/kg or 1 litre/kg and that of air is 0.77 m^3 /kg or 770 litre/kg.

Specific gravity is a dimensionless quantity denoting the weight of a volume of fluid relative to the weight of an equal volume of water (at 4°C). Thus, specific gravity of water at 4°C is necessarily 1.0, and that of air at NTP is 1.29 x 10^{-3}.

Temperature of a substance is a measure of the level or intensity of the its heat content. Thus, a body at higher temperature looses heat to a body at lower temperature. The SI unit of temperature is **Kelvin (K)**. The temperature of melting ice on this scale is 273.16 K, and that of boiling water (at one atmospheric pressure) is 373.16 K. A more convenient scale is the **Celsius scale (°C)** on which ice point is 0°C and steam point is 100°C. Note that 1 K temperature *interval* is exactly the same as 1°C temperature *interval* but 1 K temperature is much lower than 1°C.

In the following sections we introduce some properties specific to the fluids.

15.2   Pressure

The fact that a fluid at rest applies a force normal to any surface in contact has been known for a long time. In view of the present knowledge of the incessant motion of molecules in a fluid, this force may be explained as the force due to the constant bombardment of the surface by the fluid molecules.

In a fluid at rest, this force must necessarily be normal to the surface, therefore such a fluid can not sustain a shear (Sec. 14.1). Thus this force will depend on the area, and therefore, we define pressure as the normal force due to fluid on a unit area. Clearly the dimensions of pressure are force over area. In SI units, it is Newton/metre^2, or N/m^2. This unit has been named as **Pascal (Pa)** in, honour of the great seventeenth century French mathematician *Blaise Pascal* who was among the first scientists to study the behaviour of pressure in fluids.

The SI unit of pressure **Pa** is not widely used in engineering practice of english-speaking countries where the old British unit of pound force per square inch or **psi** is still commonly used and understood. The conversion factor between the two is easily obtained as shown below:

$$1\ Pa\ =\ 1\frac{N}{m^2} = 1(\frac{kg.m}{s^2}).\frac{1}{m^2}$$

$$=\ (2.205\frac{lb}{kg}).(kg).(3.281\frac{ft}{m}).(m).(\frac{1}{sec^2}).(0.0254\frac{m}{in})^2.(\frac{1}{m^2})$$

$$=\ 0.00467(\frac{lb\ ft}{sec^2}.\frac{1}{in^2}) = 0.00467 \times (\frac{1}{32.17}\frac{lbf}{lb.ft/sec^2}).\frac{1}{in^2}$$

$$=\ 1.451 \times 10^{-4}\ \frac{lbf}{in^2} = 1.451 \times 10^{-4}\ psi$$

Thus, a Pascal is a very small unit. The standard sea-level atmospheric pressure of 14.7 psi becomes 1.0132×10^5 Pa, usually written as 101.32 kPa. Some other common measures of pressure will be introduced later in Sec. 16.2.5.

15.2.1 Pascal Principle

One of the first definite principles about pressure is credited to *Blaise Pascal* who showed that pressure at a point within a static fluid is the same in all directions. Thus, if we locate a small surface at a point P within a fluid (Fig. 15.1), and are

Fig.15.1 A small surface element at point P with the direction θ.

somehow able to measure the force exerted by the fluid on it, we would find that the measured force,and hence pressure, does not depend on the inclination of the surface.

Pascal established this by conducting an experiment in which the pressure at a point in a given direction was measured by introducing at that point a small bulb covered with a diaphragm. The other end of the air-filled bulb was connected to a pressure measuring device (such as a U−tube manometer). He showed that the changing orientation of the bulb at the given point does not change the measured pressure, establishing that the pressure at a point is independent of direction.

This fact has far reaching consequences. It permits the use of a fluid to change the direction of application of a force very easily. Thus, the downward push on the brake pedals of an automobile is transmitted as the braking force to the four wheels through hydraulic pressure. Various industrial presses, actuators and transmissions exploit this property of pressure.

15.2.2 Absolute and Gauge Pressures

Almost everything that an engineer is concerned with is submerged in the vast atmosphere consisting of many gases. This atmosphere exerts a fairly large pressure on all things. Very often engineers are interested only is pressure applied by fluids *over and above* this atmospheric pressure. Since most mechanical devices used to measure pressure read zero when open to atmosphere and thus indicate pressures above (or below) the atmospheric pressure, it is common to call such pressures as *gauge pressures*. Thus, an absolute pressure of 150 kPa will be indicated as (150 kPa − atmospheric pressure) or (150 − 101.32) kPa = 48.68 kPa on such a gauge.

To clarify that this is a gauge pressure, it will be reported as 48.68 kPa gauge. In common English engineering units of pressure, i.e., **psi** which stands for pound force per square inch, the letter 'g' is appended to indicate a gauge pressure, and a letter 'a' to indicate absolute pressure. Thus,

$$24 \text{ psig} = (24 + 14.7) \text{ psia} = 38.7 \text{ psia}$$

Negative gauge pressures are called vacuum, thus

$$10 \text{ psi vacuum} = -10 \text{ psig} = (-10 + 14.7) \text{ psia} = 4.7 \text{ psia}$$

The absolute, gauge and vacuum pressures are illustrated graphically in Fig. 15.2.

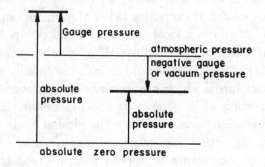

Fig. 15.2 The absolute, gauge and vacuum pressures.

15.3 Vapour Pressure

The molecules of a fluid, whether a liquid or a gas, are moving constantly in random directions. If a small quantity of a liquid is placed in a container, some molecules near the free - surface will have sufficient energy to escape into the space above. This is the phenomena of evaporation. As the liquid evaporates, the concentration of vapour in the space above the liquid increases. The motion of molecules in the vapour phase results in a reverse phenomenon also i.e. in condensation of some vapour. It is conceivable that after some time, the rate of liquid evaporating will be equal to the rate of vapour condensing, and an equilibrium will be achieved. The space above the liquid is then said to be saturated with the vapour. The pressure that the vapour exerts at that time is termed as the *saturated vapour pressure*. The saturated vapour pressure of a liquid is constant at a given temperature, and increases with temperature. Water has a saturated vapour pressure of only 2.34 kPa at 20°C which rises to 7.38 kPa at 40°C. Boiling of a liquid occurs when its saturated vapour pressure equals the atmospheric pressure above it. Thus, at sea level water boils at 100°C because its saturated vapour pressure at that temperature is 101.32 kPa, equal to the standard atmospheric pressure at sea level.

The phenomena of cavitation associated with hydraulic machines such as pumps and turbines is related to vapour pressure.

When a liquid flows over a surface at high speeds, its pressure may decrease, and if it decreases below the corresponding saturated vapour pressure, a local boiling of liquid occurs leading to formation of small vapour bubbles. If at a subsequent location the pressure rises again, these bubbles collapse rapidly giving rise to high impact forces which may damage the surface.

The formation of dew on a damp cold evening is also related to the vapour pressure. When the ground cools rapidly in the evening, the actual vapour pressure in air may end up being higher than the saturated vapour pressure near the ground leading to condensation of atmospheric water vapour as dew on ground.

15.4 Viscosity

As stated earlier, a fluid at rest does not support any shear stress, but a moving fluid can and does sustain shear stresses by deforming continually.

Consider a layer of a fluid confined between two surfaces, the upper one moving and the lower stationary (Fig. 15.3).

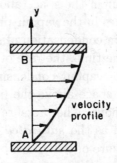

Fig. 15.3 Velocity profile in flow between two surfaces.

Let the fluid have a horizontal velocity which changes from layer to layer. The layer of fluid in immediate contact with plate A is at rest so that its velocity is zero. As we move up to the upper plate B the velocity changes. The local velocity variation with y has been plotted. The dark line shown is termed as the velocity profile and it graphically depicts the velocity V as a function of the vertical coordinate y.

15.4.1 Newton Law of Viscosity

Consider a fluid element as shown Fig. 15.3. The fluid above the upper surface CD is moving faster than this element, and tends to drag this element along, i.e., a shear force is applied on the surface CD in the direction of the flow. Similarly the fluid below EF is moving slower than the element and hence, a shear force acts on the face EF in the reverse direction. Newton examined this phenomenon and stated that the shear stress, i.e., the shear force per unit area in a fluid varies linearly with the local gradient of velocity, i.e., with the local slope of the velocity profile. Thus,

$$\tau = \mu \frac{dV}{dy} \qquad (15.1)$$

where μ is termed as the coefficient of viscosity or simply the viscosity.

This is known as the *Newton law of viscosity* and applies strictly to only a small class of fluid which are called *Newtonian fluids* for this reason. Most common fluids such as air, water, mineral oils are quite Newtonian. But there are some fluids such as paints, slurries and suspension in which τ does not vary linearly with dV/dy and which are therefore termed as non-Newtonian fluids (Fig. 15.4).

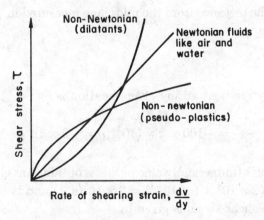

Fig. 15.4 Newtonian and non-Newtonian fluids.

15.4.2   Units of Viscosity

To obtain the units of μ, note that τ is a stress with the units of N/m^2 or Pa, and dV/dy has units of (m/s)/m or s^{-1}. Thus, the units of viscosity are Ns/m^2 or Pa.s. Water at 20°C has a viscosity of 1.002×10^{-3} Pa.s, while air has a much lower viscosity of 1.77×10^{-5} Pa.s at the same temperature. A commonly used unit of viscosity is **centipoise (cp)** which is equal to 10 Pa.s, so that the viscosity of water has a value of about 1 cp and that of air about 0.0177 cp.

Viscosity of most liquids decreases' with increasing temperatures. For example, we all know that engine oil become thicker (i.e., less prone to flow on account of increased viscosity) in winter than it is in summer. But the viscosity of gases behave in the opposite manner. Viscosity of air rises from 1.84×10 Pa.s at 300 K (27°C) to 2.30×10^{-5} Pa.s at 400 K (127°C).

Example 15.1

Hydraulic fluid (μ= 200 cp or 0.2 Pa.s) flows in a channel of width 0.2 cm under pressure. Its velocity at any cross-section varies with y such that it is given as

$$V_x = 100\{1 - (10^3 y)^2\}$$

where y is the distance from the midplane measured in metre and V_x is in m/s.

Solution

The velocity gradient at any given location is

$$\frac{dV_x}{dy} = -100 \times 2 \times (10^3)^2 y = -2 \times 10^8 y$$

The maximum (numerical) value of this is obtained at either plate where it is $(2 \times 10^8 \times 1 \times 10^{-3}) = 2 \times 10^5 (\text{s}^{-1})$ and is zero at the centre. The shear stress is given by

$$\tau = \mu \frac{dV}{dy} = 0.2 \times (-2 \times 10^8) y = -4 \times 10^7 y \quad (\text{Pa})$$

This gives a linear variation of shear stress across the channel shown in Fig. 15.5. Maximum shear stress occurs at the two walls, and is zero at the centre.

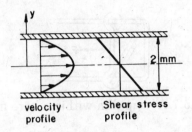

Fig. 15.5 Linear variation of shear stress.

Note that at the lower wall $y = -1 \times 10^{-3}$ m so that τ there is $+4 \times 10^4$ Pa. The positive sign indicates that shear stress on this wall is such that it tries to drag the wall in the flow direction. On the upper wall however, the value of τ is the same but it has a negative sign. This does not imply that the force on the wall is now in the up-stream direction. But its because the value of τ obtained from this relation has the correct sign for a surface with outward normal in the positive coordinate direction, which for this example is a surface facing upwards. The stress on a face with opposite orientation is of course equal and opposite to it by the law of action and reaction. Thus the negative at the upper plate location means a backward force on the fluid surface at this location but with normal facing upwards. On the plate, however, the stress points in the opposite direction, i.e., it tends to drag the plate along the low (Fig. 15.6).

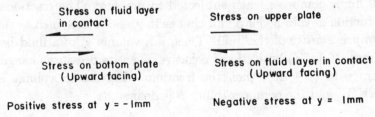

Fig. 15.6

15.5   Surface Tension

Consider a wire frame carrying a soap film as shown in Fig. 15.7. The soap film has a tendency to pull back on the wire and a

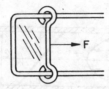

Fig. 15.7 A soap film within a wire-frame.

force F is required to maintain its equilibrium. The internal force that tends to pull back the wire is termed as the force of surface tension, and because of it a free surface of a liquid behaves like a stretched membrane. A mosquito or a small insect sitting on the surface of a water pond is held up by this force. The surface tension is also responsible for giving spherical shape to small drops of liquid. Similarly, the force of surface tension gives rise to the phenomenon of capillarity which makes wax rise up the wick of a candle or sap rise up the stem of a plant.

It depends on the liquid and the gas pair forming the interface. The force expressed on a per unit length basis is termed as *surface tension* and is denoted by σ. Its unit clearly is N/m and its value for pure water in contact with air is 7.56×10^{-2} N/m at $0°$C. The value of surface tension changes drastically with any impurities.

15.6   Compressibility

All fluids compress when subjected to pressure. The fractional reduction in volume for a unit change in pressure is termed as the *compressibility* of the fluid. Thus, if a volume V of a fluid becomes $(V + dV)$ with dV as negative when its pressure changes from p to $(p + dp)$, then the fractional decrease in volume is $-dV/V$, and the compressibility β is defined as

$$\beta \; = \; \frac{\text{Fractional } decrease \text{ in volume}}{\text{Increase in pressure}} = -\frac{1}{V}\frac{dV}{dp}$$

where dV/dp is the derivative of volume V with respect to the pressure p.

The unit of compressibility is the inverse of that of pressure, i.e., Pa^{-1}.

The compressibility of a perfect gas can easily be evaluated. It depends on the nature of the compression process, viz. whether the compression is isothermal, adiabatic or some thing else. We know that for isothermal conditions, i.e., with temperature held constant (usually associated with slow compression),

$$pV = C, \text{ a constant}$$

or

$$V = C/p$$

Therefore,

$$\frac{dV}{dp} = -\frac{C}{p^2}$$

$$\beta = \frac{1}{V}\frac{dV}{dp} = \frac{1}{p} \tag{15.2}$$

Since the commonly encountered pressures are of the order of 10^5 Pa, the isothermal compressibility of gases is of order 10^{-5} (Pa^{-1}). This value decreases as pressure increases.

For adiadatic processes (usually associated with compression so fast that heat cannot be conducted away from the gas under consideration), the relation between p and V is

$$pV^\gamma = \text{Constant} \tag{15.3}$$

where γ is the ratio of specific heats, C_p/C_v, whose value is 1.4 for diatomic gases like air. It can be shown, that for this condition,

$$\beta = -\frac{1}{V}\frac{dV}{dp} = \frac{1}{\gamma p} \tag{15.4}$$

Since $\gamma > 1$, adiabatic compressibility is less than the isothermal compressibility.

Compressibility of liquids is much less than that of gases, i.e., the volume changes produced by a given pressure is much smaller in the case of liquids. Water has a value of β of about

5×10^{-10} (Pa^{-1}) at normal pressures, which is about 1/20,000th of that of air.

Another commonly used term is the bulk modulus of elasticity E, defined as the reciprocal of the compressibility. Thus,

$$E = \frac{1}{\beta} = -V \frac{dp}{dV} \qquad (15.5)$$

Its unit is that of pressure, i.e., Pa and it can be interpreted as the pressure imposed divided by the fractional decrease in volume. It follows that for prefect gases, the isothermal bulk elasticity is simply p, and the adiabatic bulk elasticity is γp.

Example 15.2

How much pressure (in standard atmospheres) must be imposed on water to reduce its volume by 1%?

Solution

We know that compressibility of water is about $5 \times 10^{-10} (Pa^{-1})$. Compressibility given by Eq. (15.2) is

$$\beta = -\frac{1}{V} \frac{dV}{dp}$$

1% reduction in volume translates to $dV/V = -0.01$, so that

$$dp = -\frac{1}{\beta} \frac{dV}{V} = -\frac{1}{5 \times 10^{-10}} \times (-0.01) = 2 \times 10^7 \text{ Pa}$$

Since one standard atmosphere is 1.013×10^5 Pa, the additional pressure imposed is $(2 \times 10^7 / 1.013 \times 10^5) = 1.97 \times 10^2$ or about 197 standard atmospheres.

Problems

15.1 Glycerin has a density of 1.261×10^3 kg/m^3 at 20°C. What is its specific volume, specific weight and specific gravity at that temperature?

15.2 During a test of a steam turbine, the observed vacuum in the condenser was 900 Pa when the actual atmospheric pressure was 102.3 kPa. What was the absolute pressure in the condenser?

15.3 The run pressure of a wind tunnel is 81 psig. What is the absolute pressure in SI units?

15.4 An oil of viscosity 0.96 Pa.s fills the space between two parallel plates 6 mm apart. If the upper plate moves with a velocity of 1.5 m/s while the lower plate is stationary, determine the shear stresses acting on the plate, if the velocity gradient is assumed constant.

15.5 What is the viscosity of the fluid in Problem 15.4 in centipoise?

15.6 The bulk modulus or elasticity of sea-water is given as 2.37×10^9 Pa. Estimate the density of sea water at a depth of 8 km where the pressure is 82.75 MPa.

16. PRESSURE IN A STATIC FLUID

16.1 Introduction

In this chapter we obtain the relations required for evaluation of
the fluid pressure in a stationary or a static fluid. These same
relations are also valid if the fluid is moving at a uniform and
constant velocity. But these are not valid if the fluid is undergoing any acceleration, as is the case when a tank car moves along
a straight railway track at an increasing speed, or along a curved
railway track at constant speed, or when a bucket full of water is
spinning about its own axis. In the last two cases the fluid particles experience centripetal acceleration and hence the relations
for pressure variation derived here are not applicable.

16.2 Variation of Pressure in a Static Fluid

16.2.1 Along a Vertical Direction

Consider a point A in a fluid at rest, and another point B at a
distance δz directly above it, as shown in Fig. 16.1. We want to
evaluate the pressure difference $\delta p = p_B - p_A$.

Since the fluid is at rest, all forces acting on it, or on any part
of it must be in equilibrium. Consider a thin vertical cylinder of
fluid as shown, with one end of area δS around point A and the
other around point B. The length of this vertical cylinder is δz.
Let us consider the vertical forces acting on the cylindrical free-
body. There are pressure forces $p_A \, \delta S$, acting upwards at A, and

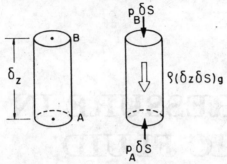

Fig. 16.1 Pressure variation along a vertical direction.

another $p_B\, \delta S$ acting downwards a B. The weight of this column of fluid is $\rho(\delta S\ \delta z)g$ where $\delta S\ \delta z$ is the volume, ρ is the fluid density, and g is the acceleration due to gravity (also the weight per unit mass).

There is no other vertical force on this control volume since the normal to the lateral curved surface, and hence the pressure acting on it, is horizontal everywhere.

Equilibrium of vertical forces on this free-body requires that algebraic sum of the three vertical forces be zero, that is

$$p_A \delta S - p_B \delta S - \rho(\delta S.\delta z)g = 0$$

or

$$\delta p = p_B - p_A = -\rho g \delta z$$

Thus, the pressure decreases as we go up in a stationary fluid at a rate equal to ρg per unit distance moved.

This relation is expressed as a differential equation

$$\frac{\partial p}{\partial z} = -\rho g \qquad (16.1)$$

where $\partial p/\partial z$ represents the partial derivative of p with respect to z, with x and y held constant, and is, therefore, the variation of pressure in the vertical direction. The combination ρg can be interpreted as the weight of fluid per unit volume (= mass per unit volume × weight per unit mass) and is usually given a symbol γ. For water, the value of γ, termed as the specific weight is 9,810 N/m^3.

If the density of the fluid is taken as constant,

$$p_z = p_o - \rho g z \qquad (16.2)$$

where p_o is the pressure at $z = 0$, and p_z is the pressure at an altitude z.

The densities of most liquids can be taken as constant, and therefore, in any liquid, the pressure varies linearly with height. In liquids, we generally specify depth h measured downwards from its free-surface where the pressure is the atmospheric pressure (Fig. 16.2). Clearly $z = -h$, and therefore,

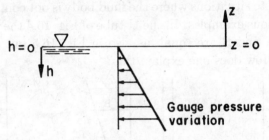

Fig. 16.2 Hydrostatic pressure variation.

$$p = p_o + \rho g h \qquad (16.3)$$

The gauge pressure is the pressure measured above the atmospheric pressure, i.e., $p - p_o$. Therefore,

$$p_{gauge} = \rho g h \qquad (16.4)$$

This is termed as the hydrostatic pressure distribution. Note again that though Eqs. (16.1) and (16.2) in this section have been obtained for a stationary fluid, they are valid for non-accelerating fluid, as in Sec. 16.1.

16.2.2 Along a Horizontal Direction

Consider two points A and B along a horizontal line in the x-direction in a fluid at rest (Fig. 16.3).

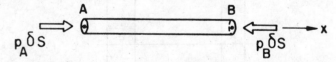

Fig. 16.3 Pressure variation along a horizontal direction.

Consider a cylindrical free–body of fluid as shown. Let us consider the equilibrium of this free–body under forces in the

x–direction. The only x–direction forces that act on this cylinder are the pressure forces acting on the circular faces at A and B, namely $p_A \delta S$ and $p_B \delta S$, acting in opposite directions as shown. Equilibrium therefore requires that $p_A = p_B$, or that the pressure does not vary in a horizontal direction.

Here again, the result apply to any non-accelerating fluid as well. This equal level – equal pressure principle can also be generalized to situations where the fluid body is not continuous as in the previous examples. In the U-tube of Fig. 16.4 the pressures at points A and B are equal, as also in Fig. 16.5 but not so in Fig. 16.6. How does one explain this?

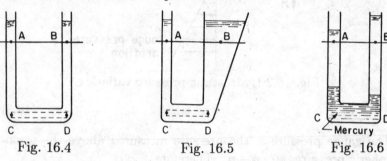

Fig. 16.4   Fig. 16.5   Fig. 16.6

In Figs. 16.4 and 16.5, we can consider a thin cylinderical horizontal liquid free-body from C to D and show that $p_C = p_D$. And since the vertical distance AC and BD are the same and the liquid in the two limbs has the same density, decrease in pressures in going from C to A is the same as in going from D to B, and hence $p_A = p_B$.

But in Fig. 16.6, though $p_C = p_D$, one traverses different fluids through different heights in going from C to A and from D to B, and thus the pressure changes will not be the same, so we should not expect p_A and p_B to be equal.

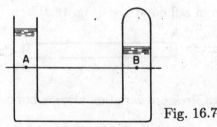

Fig. 16.7

Similarly, in Fig. 16.7 where the right limb is closed, enclos-

equal as long as points A and B are both below the liquid levels, but in Fig. 16.8 where a volume of gas is trapped in the middle columns $p_A \neq p_B$, since $p_C \neq p_D$, there being no fluid body whose equilibrium can be considered to establish that $p_C = p_D$.

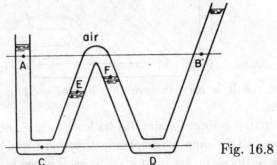

air

Fig. 16.8

Since the density in a gas is very small, the pressure variations within a gas (unless it is spread over distances of hundreds of meters, as in earth's atmosphere) can be neglelcted. Thus, in Fig. 16.8, $p_E = p_F$ both being equal to the gas pressure, even if the levels of E and F are different.

16.2.3   The Free Surface of a Liquid

Consider two points A and B along a horizontal line in a liquid as shown in Fig. 16.9.

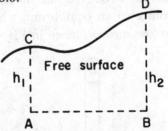

Free surface

Fig. 16.9 Construction to show that the free surface of a liquid at rest must be horizontal.

If point A is at depth h_1 from point C on the free surface of the liquid, and point B at a depth h_2 from point D on the free surface, then the gauge pressures at A and B are $p_{gauge,A} = \rho g h_1$, $p_{gauge,B} = \rho g h_2$ Since the fluid is stationary (or non-accelerating), and points A and B are along a horizontal line, the pressures p_A

and p_B should be equal, and hence $h_1 = h_2$, or the free surface itself must be horizontal. This is summarized by saying that a *liquid maintains its level*, even if the free surface is not continuous as in Fig. 16.10.

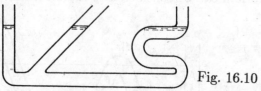

Fig. 16.10

This result is used in levelling devices used in construction projects.

Of course, a liquid maintains its level only if the free surface is exposed to the same atmosphere everywhere, and the points on the free surface can be connected to each other by lines within the same fluid. Thus, levels in various limbs in Fig. 16.7 and Fig. 16.8 are different.

16.2.4   Barometer

A barometer is a very simple device used to measure the atmospheric pressure. In principle it consists of a long tube (typically about a metre long) closed at one end which is filled with mercury and inverted into a mercury cup. As the mercury drains into the cup, a vacuum is built up at the closed end, and the mercury column comes to an equilibrium when it is supported by the atmospheric pressure as shown in Fig. 16.11.

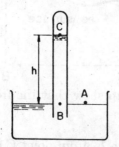

Fig. 16.11 A barometer.

Clearly, the pressure at point B is the same as that at point A where it is atmospheric, because A and B are at the same level and are connected by the same liquid. Now, within the vertical column,

$$p_B = p_{atm} = p_C + \rho g h \qquad (16.5)$$

where ρ is the density of mercury. The absolute pressure at C is the saturated vapour pressure of mercury at the ambient temperature (2.47 Pa at 0°C, 16.0 Pa at 20°C and 81.05 Pa at 40 °C).

If we measure h accurately, we can determine the atmospheric pressure using Eq. (16.5). Commercial barometers are based on this simple principle, but include some features which permit accurate and simple measurement of h.

Example 16.1

On a hot day when the temperature was 40°C, the mercury column in a barometer was 75.89 cm tall. What was the atmospheric pressure at that instant?

Solution

The atmospheric pressure will be given by Eq. (16.5). The vapour pressure of mercury at 40°C is given as 81.05 Pa, so that

$$\begin{aligned} p_{atm} &= 81.05 \text{ Pa} + 13.5 \times 10^3 (\frac{\text{kg}}{\text{m}^3}) \times 9.81 (\frac{\text{m}}{\text{s}^2}) \times 0.7589 \text{ m} \\ &= 100.586 \text{ kPa} \qquad (16.6) \end{aligned}$$

which is less than the standard atmospheric pressure of 101.35 kPa. Note that we have used 13.5×10^3 kg/m^3 for the density of mercury at 40°C.

16.2.5   Pressure Head

Consider a vertical column of liquid with point A on the surface and B at a depth h below the surface (Fig. 16.12). Now

$$p_{B,gauge} = p_B - p_A = \rho g h$$

At a given depth h, pressure $p_{B,gauge}$ is fixed, once the fluid density ρ is fixed. This permits us to introduce another measure of the

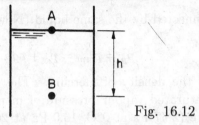

Fig. 16.12

gauge pressure, the *head h* of the given fluid. Thus, a head h of water corresponds to a gauge pressure $\rho_w g h$ or $\gamma_w h$, where ρ_w and γ_w are the mass density and specific weight of water, respectively. Similarly, a head h of mercury corresponds to a gauge pressure $\rho_m g h$ or $\gamma_m h$, where ρ_m and γ_m are the mass density and the specific weight of mercury, respectively.

Example 16.2

Express 101.3 kPa gauge as a head of water and as head of mercury.

Solution

We know that $p_{gauge} = \rho_A g h_A$, where h_A is the head of fluid A and ρ_A is the density of fluid A.

For water $\rho = 10^3$ kg/m^3, and for mercury $\rho = 13.6 \times 10^3$ kg/m^3. Therefore, head of water

$$h_w = \frac{p_{gauge}}{\rho_w \cdot g} = \frac{101.3 \times 10^3 (\text{N/m}^2)}{10^3 (\frac{\text{kg}}{\text{m}^3})} \times 9.8 (\frac{\text{N}}{\text{kg}}) = 10.34 \text{ m of water}$$

and head of mercury

$$h_m = \frac{p_{gauge}}{\rho_m \cdot g} = \frac{101.3 \times 10^3 (\text{N/m}^2)}{13.6 \times 10^3 (\frac{\text{kg}}{\text{m}^3}) \times 9.8 (\frac{\text{N}}{\text{kg}})} = 0.760 \text{ m}$$

or 760 mm of mercury.

Since 101.3 kPa is about one standard atmospheric pressure, this pressure expressed as head is 760 mm of mercury or 10.34 m of water.

A pressure of 100 kPa (cf. standard atmospheric pressure of 101.35 kPa) is called 1 bar and is used routinely in meteorological practice. The preferred unit is a millibar equal to 10^{-3} bar which is 100 Pa.

16.3 Capillarity

Take a very thin glass tube open at both ends and insert it into a beaker containing water. We see that water rises up the tube beyond the level in the beaker (Fig. 16.13).

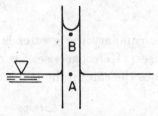

Fig. 16.13 Rise of liquid in a capillary.

This phenomenon is known as capillarity. Clearly, the pressure at point A which is at the same level as the fluid outside must be the atmospheric pressure and, therefore, that at point B must be less than p_{atm}. How is this possible?

It is the force of surface tension that gives rise to this phenomenon. Since water wets the glass capillary, a concave miniscus is formed in the tube as shown in Fig. 16.14. Consider the plug

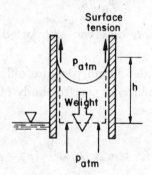

Fig. 16.14 Free-body diagram of the column that rises in a capillary.

of liquid between this miniscus and the beaker level. The vertical forces on this plug are as shown.

The pressure at both sides is atmospheric, as discussed above, so the weight of the plug of height h is balanced by the surface tension forces alone. If σ denotes the surface tension, then the total force is $\sigma \times 2\pi R$, where $2\pi R$ is the circumference of the tube where the surface tension forces acts. The weight of the fluid supported is $\rho g \times \pi R^2 \times h$. Thus,

$$\rho g \pi r^2 h = 2\pi R \sigma \quad \text{or} \quad h = \frac{2\sigma}{\rho g R} \tag{16.7}$$

The thinner the capillary the greater is the height through which the liquid rises. Kerosene rises up a wick by the same principle, and so does the sap up the trunk of a tree.

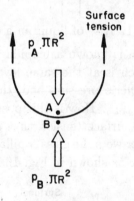

Fig. 16.15 Pressure difference supported by a meniscus.

One could have also found pressure difference supported by the miniscus by considering the free-body of the miniscus alone as in Fig. 16.15. Hence

$$p_A.\pi R^2 = p_B.\pi R^2 + \sigma.2\pi R \quad \text{or} \quad (p_A - p_B) = \frac{2\sigma}{R} \tag{16.8}$$

Example 16.3

How high can water rise in a capillary of diameter 1 mm and in another of diameter 0.1 mm?

Solution

For water, the value of σ may be taken as 71.2×10^{-3} N/m at 30°C.

Using the formulation of Sec. 16.2.5, the capillary rise is given by Eq. (16.6) as

$$h = \frac{2\sigma}{\rho g R}$$

For 1 mm diameter tube, $R = 0.5 \times 10^{-3}$ m, and

$$h = \frac{2 \times 71.2 \times 10^3 (\text{N/m})}{10^3 (\frac{\text{kg}}{\text{m}^3})} \times 9.8 (\frac{\text{N}}{\text{kg}}) \times 0.5 \times 10^{-3} (\text{m})$$

$$= 0.029 \text{ m or } 29 \text{ mm of water}$$

For 0.1 mm tube, we can show that the capillary rise is 10 times as much, or 29 cm.

Example 16.4

A manufacturer making sintered storage vessels for petrol tests them for leakage by filling them with water. One such vessel passed the test with water, but when it was used to store petrol, it leaked. Explain. If a vessel shows leakage of water only under a head of more than 0.50 m, determine the maximum diameter of the microholes.

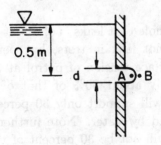

Fig. 16.16

Solution

Let Fig. 16.16 show an exaggerated view of the situation. Let d be the diameter of the microhole. Water before leaking out forms a miniscus as shown, and this miniscus can support a pressure difference. The gauge pressure on the inside point A just before leakage of water is

$$p_{A,gauge} = \rho_{water}gh = 10^3(\frac{kg}{m^3}) \times 9.8(\frac{N}{kg}) \times 0.5(m)$$
$$= 4.9 \times 10^3 N / m^2 \text{ (or Pa)}$$
$$p_{B,gauge} = 0$$

Therefore, the pressure difference supported is

$$p_A - p_B = 4.9 \times 10^3 \text{ Pa}$$

Equation (16.7) relates $(p_A - p_B)$ to σ and R. Thus,

$$(p_A - p_B) = \frac{2\sigma}{R}$$

or

$$R = \frac{2\sigma}{p_A - p_B}$$
$$= \frac{2 \times 71.2 \times 10^{-3}(N/m)}{4.9 \times 10^2(N/m^2)}$$
$$= 29.07 \times 10^{-6} \text{ (m)}$$

Thus the smallest hole that leaks is of about 29 micron size.

If water does not leak in tests, but petrol does, this can be because the surface tension of petrol at 30°C is about 20.8 N/m, which is only about 30% of that of water, and hence a petrol miniscus will support only 30 percent of the pressure difference supported by water. Note further that, the head of petrol supported will not be 30 percent of that of water, since the density of petrol is lower. A pressure difference of 30 percent of 4.9×10^3 Pa is 1.4×10^3 Pa which in terms of head of petrol is 1.4×10^3 Pa/[694 kg/m^3 (the density of petrol) $\times 9.81$ (N/kg)] = 0.20 m, which is 40 percent of the head of water supported.

16.4 Pressure Measurement

Measurement of fluid pressure is the commonest measurement made in practice. Many things such as the flow rate of a fluid, the mass of the gas or liquid stored in a container, the atmospheric conditions, the forces developed on bodies and surfaces as they move through a fluid body or as a fluid moves past them, are all related to either pressures, the changes of pressure or to the rates of change of pressure. This section describes some of the more frequently used pressure measuring systems.

16.4.1 Piezometer

The simplest device for measuring pressures in liquids is a piezometer, consisting of an open-ended tube connected to the liquid at the level where the pressure measurement is desired (Fig. 16.17).

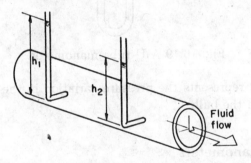

Fig. 16.17 Piezometric tubes along a pipe.

The liquid rises in the piezometer tube to a level h, and becomes stationary. The fluid is stationary within the tube, and is open to atmosphere at the upper end. The gauge pressure at the lower end is γh ($= \rho g h$), where γ is the specific weight of the fluid and h is the height of the liquid in the tube above the measurement point. Note that h itself is a measure of the pressure as the head of the given liquid. In Fig. 16.17, since $h_1 > h_2$, the pressure along the axis of the pipe has decreased.

In Fig. 16.18 note that p_A is γh_1 and p_B is higher at γh_2, even though the liquid rises to the same level in the two piezometers attached to the bulb. The pressure difference $\gamma(h_2 - h_1)$ which

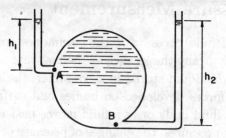

Fig. 16.18 Liquid rises to the same level in two piezometers connected to a cross-section.

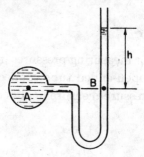

Fig. 16.19 A U-tube manometer.

is $(p_B - p_A)$ represents the pressure variation in the static fluid contained in the bulb.

16.4.2 Manometers

A simple manometer consists of a U-tube as shown in Fig. 16.19. Clearly $p_A = p_B$, since the two points are at the same level and can be connected by a line lying entirely within one fluid. The pressure p_B can be determined from the piezometeric relation $p_B = \gamma h$, where h is the height of the liquid column above point B and γ is the specific weight of the fluid. Thus,

$$p_A = \gamma h = \rho g h \quad (gauge)$$

If the fluid in the bulb is a light fluid (having a small value of γ_1), and p_A is very large, then height h may become inconveniently large. In that case we use a heavy fluid (such as mercury) as the gauge fluid in an arrangement as shown in Fig. 16.20.

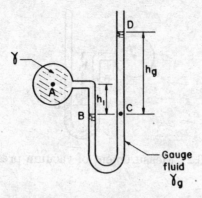

Fig. 16.20 A U-tube manometer using a gauge fluid.

To evaluate $p_{A,gauge}$, we proceed sequentially as shown below:

$$p_A = p_B - \gamma h_1$$
$$p_B = p_C$$
$$p_C = p_D + \gamma_g h_g$$

Combining the three relations,

$$p_A = p_D + \gamma_g h_g - \gamma h_1$$

Since p_D is the atmospheric pressure, therefore,

$$p_{A,gauge} = \gamma_g h_g - \gamma h_1 \qquad (16.9)$$

This same device is also suitable for measurement of gas pressures as well, but in that case, since γ (the specific weight of the gas) will be much smaller than γ_g,

$$p_{A,gauge} = \gamma_g h_g \qquad (16.10)$$

This is equivalent to neglecting the pressure variations within the gas column, and setting $p_A = p_B$.

This arrangement is also suitable for measuring pressures less than the atmospheric, i.e., for measuring vaccum pressures as in Fig. 16.21. In this case, point B is above point D.

Then,

$$p_A = p_D - \gamma h_1, \quad p_B = p_C - \gamma_g h_g, \quad p_C = p_D$$

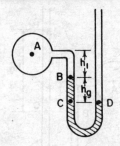

Fig. 16.21 Measurement of vacuum pressures.

So that

$$p_A = p_D - \gamma h_1 - \gamma_g h_g$$

$$p_{A,gauge} = -\gamma h_1 - \gamma_g h_g \qquad (16.11)$$

This has a negative value, signifying that point A has a vacuum pressure.

16.4.3   Differential Manometers

A U-tube manometer can also be used for measuring the pressure difference between two points A and B. Fig. 16.22 shows one such

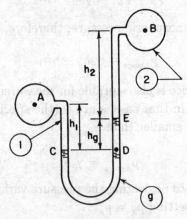

Fig. 16.22 A differential manometer.

application. As before, we proceed sequentially from points A to B, through C, D and E and write

$$p_A = p_C - \gamma_1 h_1, \quad p_C = p_D$$

$$p_D = p_E + \gamma_g h_g, \quad p_E = p_B + \gamma_2 h_2$$

Combining the four relations,

$$p_A - p_B = -\gamma_g h_g + \gamma_2 h_2 - \gamma_1 h_1 \tag{16.12}$$

Figure 16.23 shows another possible arrangement for a differential manometer. It can easily be shown that in this case

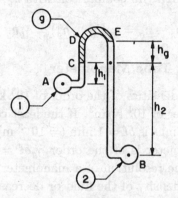

Fig. 16.23 Another differential manometer.

$$p_A - p_B = \gamma_g h_g + \gamma_1 h_1 - \gamma_2 h_2 \tag{16.13}$$

16.4.4 Reservoir Manometer

In a simple U-tube manometer, the levels in two limbs must be read to determine the gauge pressure, or the differential pressure. A reservoir manometer shown in Fig. 16.24 is designed to remove this difficulty. Here, level 0–0 indicates the level of the gauge

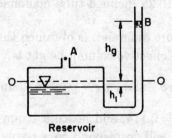

Fig. 16.24 Reservoir manometer.

fluid when $p_A = p_B$. As pressure p_A is increased, the level in the reservoir is pushed down through a distance h_1, while that in the right-hand limb rises through a distance h_g. The volume of the gauge fluid displaced from the reservoir must be equal to the volume of the additional fluid pushed into the limb. Therefore, if the cross-sectional area of the reservoir is much larger than that of the limb, the length h_1 will be much smaller than the length h_g. Thus, neglecting h_1 in comparison with h_g,

$$p_A - p_B = \gamma_g(h_1 + h_g) = \gamma_g h_g \qquad (16.14)$$

16.4.5   Inclined Tube Manometer

Most liquids have densities of the order of 10^3 kg/m^3, or specific weights of the order of 10^4 N/m^3. If the least count of measurement of liquid columns is $\delta\ell = 1$ mm $(= 10^{-3}$ m), then resolution of a simple manometer is of the order $\gamma_g\delta\ell = 10 \times 10^{-3} = 10$ Pa. To improve the resolution of a manometer then, we can either decrease the density of the fluid or decrease the least count of length measurement, both rather difficult if the complexity of the equipment involved is not to be unreasonably increased. One simple method consists of inclining the manometric tube through an angle θ as in Fig. 16.25.

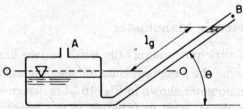

Fig. 16.25 Inclined-tube manometer.

Since the pressure difference is obtained through the vertical column height, the effective column height is

$$h_g = l_g \sin\theta \quad \text{and} \quad p_A - p_B = \gamma_g l_g \sin\theta \qquad (16.15)$$

For $\theta = 10°$, $\sin\theta = 0.174$, and hence a 1 mm resolution on the measurement of l_g will correspond to a resolution of 0.174 mm on h_g, which corresponds to a pressure resolution of 10^4(N/m^3)

×0.174 × 10⁻³ (m) or 1.74 Pa, an improvement in resolution by a factor better than 5.

16.4.6 Reducing Errors in Manometry

The measurement of pressure using manometers is so direct that there are few sources of error associated with it. Two main causes of errors are the changes in the density of the gauge fluid because of changes in the ambient temperature, and the effects of capillarity.

Water and mercury are the most commonly used gauge fluids. Density of water changes from 999.7 to 992.2 kg/m³ when the temperature increases from 10°C to 40°C, the typical low and high temperatures in a laboratory. This represents about 0.75 percent variation. The density of mercury over the same temperature range changes from 13.57 × 10³ to 13.50 × 10³ kg/m³, which is even less of a percentage change.

The capillarity, arises because of surface tension and 'pulls up' the water column in a thin tube forming a concave miniscus. It was shown in Sec. 16.3 that the rise is given by Eq. (16.6) as

$$h = \frac{2\sigma}{\rho g R} = \frac{2 \times 71.2 \times 10^{-3}(\text{N/m})}{10^3(\text{kg/m}^3) \times 9.81(\text{N/kg}) \times D/2}$$

$$= \frac{2.9 \times 10^{-5}(\text{m}^2)}{D}$$

where D is the diameter of the tube.

To restrict error of measurement to less than 10 Pa, the capillary rise should be less than 1 mm (or 10^{-3} m) which gives a value of D of at least 2.9 cm. This is too large for practice.

One factor that retrieves the situation is that in most cases we either use a U-tube with the two limbs of approximately the same bore, or we read the changes in level within one tube. In either case, the errors due to capillarity cancel out. The capillarity effects either in the two limbs of the U-tube, or at two locations in the same tube can be assumed to be the same, if the bore of the tubes are uniform, and if the degree of cleanliness is the same. To minimize the effect of normal manufacturing variations, a bore of atleast 10 mm is recommended. Presence of grease affects surface tension and capillarity drastically, and therefore, the tubes must be de-greased frequently. Alcohol de-greases the tube by itself, and is sometimes preferred over water as the gauge fluid. But relatively higher volatality of alcohol and its larger variation of density with temperature (3.2 percent compared to 0.7 percent for water over 10-40°C) work against it.

Another precaution in the use of manometers for pressure measurement is to remove all air bubbles carefully from the liquid columns and to tap the manometer before reading it.

Problems

16.1 A large closed tank is partially filled with water at 80°C and evacuated so that there is no air left in the tank. If the depth of water in the tank is 0.2 m, what is the absolute pressure at the bottom of the tank at 20°C and at 83°C, given that the vapour pressure of water at the two temperatures is 2.33 kPa and 51.0 kPa, respectively?

16.2 Calculate the pressure at the bottom of the closed tank shown in Fig. 16.26.

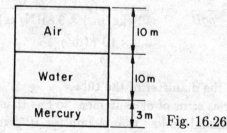

Fig. 16.26

16.3 A mercury manometer indicates 720 mm when the temperature is 27°C. What is the atmospheric pressure, in bar, if the density of mercury is 13.53×10^3 kg/m^3?

16.4 Convert the following pressures:

(a) 10 m head of water to head of mercury

(b) 2.8 mm head of water to Pascal, gauge

(c) 720 mm of mercury to Pascal, vacuum

(d) 28 kPa,g to head of water

(e) 230 kPa,g to absolute pressure in Pascal.

16.5 Manometer practice recommends a minimum bore 10 mm for mercury manometers to avoid capillary errors. Estimate the depression of mercury miniscus at 20°C with this bore. Given the value of surface tension of mercury at 20°C as 0.475 N/m.

16.6 Tanks A and B of Fig. 16.27 are filled with air and sealed. Pressure gauge shown reads 200 kPa,g and the mercury column measures 1.8 m. What is the value of mercury column h_B?

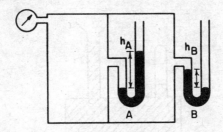

Fig. 16.27

16.7 In Fig. 16.28, a closed U-tube is filled with water and mercury, as shown. What is the height h? Given that the vapour pressure of water at the appropriate temperature is 20 kPa and that the barometric pressure is 750 mm of mercury.

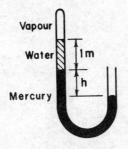

Fig. 16.28

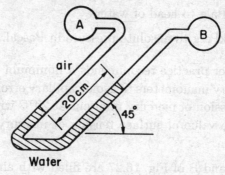

Fig. 16.29

16.8 Determine $(p_A - p_B)$ as shown by the inclined tube manometer of Fig. 16.29.

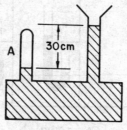

Fig. 16.30

16.9 Determine the pressure of air inside the tube A of Fig. 16.30.

17. PRESSURE FORCES ON SUBMERGED SURFACES

17.1 Introduction

Dams, sluice gates, submarines and ships are among the many structures subjected to pressure forces due to static fluids, and an engineer needs to calculate such forces for successful design.

We studied that the gauge pressure at a point depends upon the depth below the free surface of a static fluid, and that the pressure force acts normal to a surface.

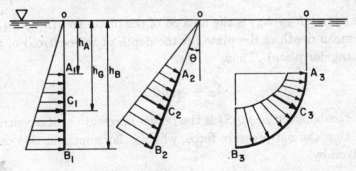

Fig. 17.1 Three surfaces immersed in a fluid.

If we take elemental areas dA at points C_1, C_2 and C_3 on the three surfaces shown in Fig. 17.1, the pressure forces on the three elements are identical in magnitude, but act in different directions, as shown. The force distributions on the three plates are entirely different, even though the pressures at identical depths are identical.

On flat plates 1 and 2 the pressure forces at different locations are parallel. Hence the total force can easily be evaluated by simple integration. But on plate 3 the forces are in different directions. Therefore direct integration will not work.

17.2   Pressure Forces on Flat Surfaces

To evaluate the total force on plate 1, divide the plate into small elemental areas dh, calculate the force on each of them as a function of h, and sum (or integrate) over the entire range of h from h_A to h_B. This gives,

$$F_p = \int_{h_A}^{h_B} (p_{gauge}.dS) = \int_{h_A}^{h_B} (\rho_f gh)bdh$$

where b is the width of the plate.
Thus,

$$F_p = \rho_f gb \int_{h_A}^{h_B} hdh = \rho_f gb \frac{(h_B^2 - h_A^2)}{2}$$

$$= \rho_f gb \frac{(h_B - h_A).(h_B + h_A)}{2} \tag{17.1}$$

The term $b(h_B - h_A)$ is the area S_1 of the plate, and $(h_B + h_A)/2$ the mean depth of the plate (or the depth of the centroid of the rectangular plate). Thus,

$$F_p = \rho_f gh_G S_1$$

The combination $\rho_f gh_G S_1$ is the (gauge) pressure at the centroid and thus the net pressure force, which is horizontal in this case, is given by

$$Fp = \text{(gauge pressure at centroid)} \times \text{(area of the plate)} \tag{17.2}$$

This is the pressure force due to the liquid alone, and is in addition to any force due to the atmosphere.

The pressure force on plate 2 is found in the same manner, except that the area of the small element of 'height' h is now $b(dh/\cos\theta)$, where θ is the angle that the plate makes with the vertical. The integration then gives

$$F_p = \rho_f g b \frac{h_B^2 - h_A^2}{2\cos\theta} = \rho_f g (\frac{h_B + h_A}{2}) b (\frac{h_B - h_A}{\cos\theta}) \qquad (17.3)$$

Here, $\rho_f g(h_B + h_A)/2$ is the depth of the centroid while $b(h_B - h_A)/\cos\theta$ is now the area of the plate S_2, and therefore the pressure force due to the liquid is again given by Eq. 17.2. Note also that the horizontal component of this force is

$$F_{p,x} = F_p \cos\theta = \rho_f g (\frac{h_B + h_A}{2}) b (h_B - h_A)$$

exactly the same value in magnitude as that on plate 1.

From this, we conclude that the *horizontal force on an inclined flat plate is equal to that on the "vertical projection"* of that plate.

17.3 Pressure Forces an Curved Surfaces

If the plate is curved, as plate 3 in Fig. 17.1, the force cannot be determined by direct integration, since the different elemental forces are not in the same direction (Sec. 17.1). We can, however, take the horizontal and vertical components of the elemental forces and integrate them individually to obtain the horizontal and vertical components of the net pressure force.

Here adopt a slightly different approach to bring out the physical aspects of the problem a little more clearly.

Consider the free-body of the fluid element A_3B_3D bounded by the surface A_3B_3 on one side, a vertical surface DB_3, and a horizontal surface A_3D on the other sides as shown in Fig. 17.2(a). Since the fluid is at rest, this element must be in equilibrium under the action of all the external forces acting on it. Fig. 17.2(b) shows the free-body diagram of this element.

Here F'_{AB} is the force on the fluid surface in contact with the plate, which by Newton principle of action and reaction is equal

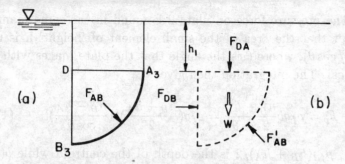

Fig. 17.2 (a) A curved surface immersed in a fluid and (b) the free-body diagram of water element A_3B_3D.

and opposite to the fluid force F_{AB} on the curved plate $A_3 B_3$, the force that we want to determine. The other forces on this fluid element are the horizontal force on the vertical side DB_3, the vertical force on the horizontal side DA_3, and the vertically downwards weight force.

The equilibrium of the horizontal forces acting on the element gives that the horizontal component of force F'_{AB} is equal in magnitude (and opposite in direction) to the horizontal force on the vertical surface DB_3. But vertical surface DB_3 is the vertical projection of the curved surface A_3B_3. Thus, the horizontal component of the pressure forces on a curved surface in a static fluid is equal to the horizontal force (which is the only force) on the vertical projection of that surface. This is easily evaluated by Eq. (17.2) where the 'area of the plate' is replaced by the 'vertically projected area of the curved plate'.

To find the vertical component of pressure force F_{AB}, note that it is equal to the force on the horizontal surface DA_3 *plus* the weight of the fluid element DA_3B_3. The pressure on DA_3 is constant, equal to $\rho_f g h_1$, and therefore the pressure force is equal to $\rho_f g h_1 \times$ (Area DA_3). The product $h_1 \times$ Area DA_3 is interpreted as the volume of the liquid column standing on area DA_3, and therefore the force $\rho g(h_1 \times$ Area $DA_3)$ can be seen as the weight of the liquid column above the surface DA_3. It then is a simple matter to see that the net vertical component of force on $A_3 B_3$ is the sum of the weight of the fluid element $DB_3 A_3$ and that of the liquid column above DA_3, or that *the vertical force component on a surface is equal to the weight of the liquid column standing on that surface (upto the free-surface of the liquid).*

This result has wide applications, if care is exercised in identifying the liquid column standing on the surface. Following are certain rules as 'facts' whose 'proofs' are left to the reader:

1. On a thin plate, the force on the upper surface is the same (though in the opposite direction) as that on the lower surface. Note that pressures on points P and P′ on either sides of the plate are equal (Fig. 17.3).

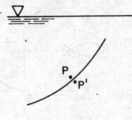

Fig. 17.3

2. The vertically upward force on the bottom surface AB (or any part of it) of a solid body is the same as on the bottom of a plate of the same geometry at the same location, which in turn is equal in magnitude to the weight of the liquid supported by it, but acting upwards (Fig. 17.4).

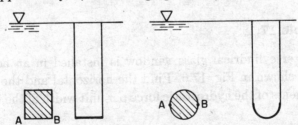

Fig. 17.4 Pressure on the bottom surfaces AB of solid bodies is equal (but opposite) to the weight of the fluid mass that can be supported on the surfaces AB.

3. The upward vertical force on wall AB of a container filled with a liquid (Fig. 17.5) upto the level shown is equal in magnitude to the weight of the liquid that would have filled

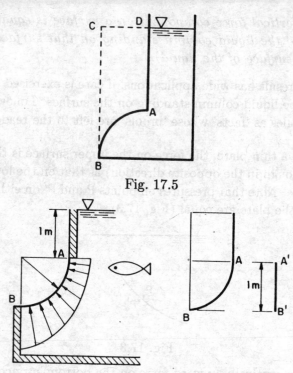

Fig. 17.5

Fig. 17.6

the volume ABCD to the same level, even though there is no fluid there.

Example 17.1

A quarter cylindrical glass window is installed in an aquarium tank as shown in Fig. 17.6. Find the horizontal and the vertical components of the hydrostatic force per unit width of the window.

Solution

The horizontal force on the window is the same as that on the vertically projected area A'B', which by Eq. (17.2) is

$$F_H = \rho_g h_G S_v$$

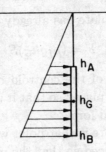

Fig. 17.7 Pressure distribtuion on a vertical plate.

$$= 10^3 \left(\frac{kg}{m^3}\right) \times 9.81 \left(\frac{N}{kg}\right) \times 1.5(m) \times 1(m)$$

per m width of window

$$= 1.45 \times 10^4 \text{ N per m width}$$

This force is to the left.

The vertical force on the window is equal to the weight of water that could stand on surface AB, if possible. Thus,

$$F_v = \rho g(\text{volume of water column that could stand on AB})$$
$$= 10^3 \left(\frac{kg}{m^3}\right) \times 9.81 \left(\frac{N}{kg}\right) \times (1(m) \times 1(m) + \frac{\pi}{4}(1 \text{ m}^2))$$

per m width

$$= 7.7 \times 10^3 \text{ N per metre width}$$

This acts downwards.

The resultant force has a magnitude of

$$\sqrt{(1.45 \times 10^4)^2 + (7.7 \times 10^3)^2} = 1.64 \times 10^4 \text{ N/m width of window}$$

17.4 Centre of Pressure on a Rectangular Surface

Consider a rectangular vertical surface (of width b) immersed in a fluid of density ρ_f (Fig. 17.7). The hydrostatic pressure distribution on the surface is as shown, increasing with depth h.

The total force on the plate has already been evaluated as

$$F = (\rho_f g h_G) S$$

where h_G is the depth of the centroid of the area. But where should this resultant be placed so that it produces the same effect as the actual distributed force?

The centre of pressure is that point where the resultant pressure force should be placed so that the moment produced about any point is the same as the moment of the actual distributed pressure forces.

Let us take the moment of all the forces about point O, where the extended surface meets the free-surface of the liquid. If we take an element dS at depth h, the pressure force is $\rho_f g h \, dS$, and the moment contribution about point O is $(\rho_f g h).h.dS$, clockwise. The net moment about O is obtained by integrating it as

$$\int_{Area} \rho_f g h^2 dS = \rho_f g \int_{Area} h^2 dS$$

If the centre of pressure, where the force $\rho_f g h_G S$ may be placed, is at the depth h_{cp}, then its moment about O is $(\rho_f g h_G S) h_{cp}$, and the equivalence requirement becomes

$$(\rho_f g h_G S) h_{cp} = \rho_f g \int_{Area} h^2 ds$$

or

$$h_{cp} = \frac{\int_{Area} h^2 dS}{h_G S} \qquad (17.4)$$

The integral in the numerator is the second moment of area about the free-surface of the liquid.

17.4.1 Parallel Axis Theorem

This theorem permits us to relate the second moment of an area about the free surface of the liquid to the second moment of the area about its own centroid, which is a property of the area and is usually tabulated. Table 17.1 provides the value of I_G, the moment of area about centroid, for some common shapes. The theorem states that

$$I_O = I_G + S h_G^2 \qquad (17.5)$$

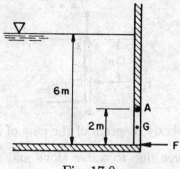

Fig. 17.8

where I_G is the second moment of area S about the centroidal axis, and I_O is the moment about an axis parallel to the centroidal axis, but a distance h_G away.

Introducing Eq. 17.5 in Eq. 17.4,

$$h_{cp} = \frac{I_G + Sh_G^2}{h_G S} = h_G + \frac{I_G}{h_G S} \qquad (17.6)$$

where h_{cp} and h_G are the depths of the centre of pressure and the centroid from the free-surface of the liquid.

This equation indicates that the centre of pressure is always below the centroid, but the separation between the two reduces as h_G increases.

Example 17.2

A sluice gate AB of width 3 m is hinged at A, as shown in Fig. 17.8, and holds water behind it upto a depth of 6 m. The pressure of water on the gate tends to swing it open, and a force F is applied near point B to hold it in place. Estimate the value of the force F required.

Solution

The total hydrostatic force on the gate is given by Eq. (17.2) as

$$F_p = (\rho g h_G)S = [10^3(\frac{\text{kg}}{\text{m}^3}) \times 9.81(\frac{\text{N}}{\text{kg}}) \times 5(\text{m})] \times [2(\text{m}) \times 3(\text{m})]$$

$$= 2.94 \times 10^5 \text{ N}$$

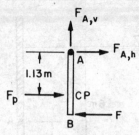

Fig. 17.9 Free-body diagram of the gate of Example 17.2.

This is the force due to water alone and acts to the right. The force due to the atmospheric pressure is in addition to it, but we do not need to calculate it since the atmospheric pressure acts on both sides of the plate, and thus cancels out. This is typical of most cases, but there may be a few situations where the atmospheric force may also have to be accounted for.

The point of application of this force can be determined from Eq. (17.5) as

$$h_{cp} = h_G + \frac{I_G}{h_G S}$$

I_G of a rectangular plate of width $b = 3$ m and height $h = 2$ m, is $bh^3/12 = 3 \times 2^3/12 = 2$ m^4 from Table 17.1. Therefore,

$$h_{cp} = 5 + 2/(6 \times 2) = 5.13 \text{ m}$$

only 13 cm below the centroid.

[Note that if the top edge of the plate was touching the free-surface, h_G would have been 1 m, and $h_{cp} = 1 + 2/(6 \times 1) = 1.27$ m or 27 cm below the centroid, clearly showing that separation between CP and CG decreases as the plate descends further. Can you explain it with reference to the pressure prism on the plate shown in Fig. 17.7?]

The free-body diagram of the gate is shown in Fig. 17.9. Force F_p acts on CP which is 5.13 m from the free-surface, or 1.13 m from hinge A. $F_{A,h}$ and $F_{A,v}$ are the horizontal and vertical components of the hinge force at A.

The externally applied force F acts at B which is at 2 m from A. If we consider the moment equilibrium about the hinge point A, we can directly determine force F. Thus,

$$F \times 2(\text{m}) - F_p \times 1.13(\text{m}) = 0$$

or $\quad F = \dfrac{F_p \times 1.13}{2} = \dfrac{2.94 \times 10^5 (\text{N}) \times 1.13}{2} = 1.55 \times 10^5 \text{ N}$

17.4.2   Centre of Pressure on an Inclined Plate

Consider an inclined plate submerged in a liquid as shown in Fig. 17.10. As in Sec. 17.2, the net pressure force on the plate is given by the product of the area of the plate and the pressure at the centroid. Thus,

$$F_p = (\rho_f g h_G) S_{plate}$$

This force acts normal to the plate as shown.

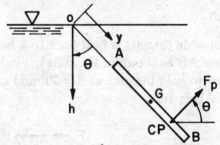

Fig. 17.10 Pressure forces on an inclined plate submerged in a liquid.

We next find the location CP of the centre of pressure. To do this, it is most convenient to work in terms of y, the distance along the plate length measured from point O on the free-surface as shown. Clearly, the pressure at any y is $\rho_f g y \cos \theta$, so that the moment contribution of an elemental area $dS (= b \, dy)$ along the plate is $(y \rho_f g y \cos \theta) dS$.

The total moment is found by integrating this over the whole area as

$$\int_{Area} (y \rho_f g y \cos \theta) dS = \rho_f g \cos \theta \int_{Area} y^2 dA = \rho_f g \cos \theta I_o$$

where I_o is the second moment of inertia of the area about a normal axis through O. Equating this to the moment of the resultant force acting at the centre of pressure gives

$$y_{cp}(\rho_f g y_G \cos \theta S) = \rho_f g \cos \theta I_o$$

or

$$y_{cp} = \frac{I_o}{y_G S} \qquad (17.7)$$

where S is the total area of the plate. Writing I_o in terms of I_G through the parallel axis theorem $(I_o = I_G + y_G^2 S)$ gives

$$y_{cp} = y_G + \frac{I_o}{y_G S} \qquad (17.8)$$

This is similar to Eq. (17.6) except that y's replace the vertical depths h's. In fact, Eq. (17.6) can be seen as a special case of Eq. (17.8) since y's are equal to h's when a plate is vertical.

Example 17.3

Water in a 2 m wide channel is held back to a height $h = 1.2$ m by a hinged gate AB as shown in Fig. 17.11. Find the horizontal force F required to hold the gate at $\theta = 30°$ and at $\theta = 16°$ if the length of gate AB is 3m.

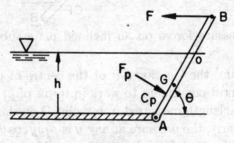

Fig. 17.11

Solution

The length AO of gate in contact with water is $h/\cos\theta$. The pressure at the centroid of the wetted portion of the gate is $\rho_f g h_G$ where h_G is the depth of the centroid which is $h/2$. Thus, the pressure at the centroid is $\rho_f g h/2$. The area of the plate is $bh/\sin\theta$ so that the total pressure force is $Fp = \rho_f g b h^2 /2\sin\theta$. The location of the centre of pressure is given by

$$y_{cp} = y_G + \frac{I_G}{y_G S}$$

where y's are measured along the plate from O. Clearly $y_G = h/2\sin\theta$, $S = bh/\sin\theta$ and $I_G = b(h/\sin\theta)^3/12$. Thus,

$$y_{cp} = \frac{h}{2\sin\theta} + \left(\frac{bh^3}{12\sin^3\theta}\right)\left(\frac{2\sin\theta}{h}\right)\frac{\sin\theta}{bh}$$

$$= \frac{h}{2\sin\theta} + \frac{h}{6\sin\theta} = \frac{2h}{3\sin\theta}$$

Moment due to force F_p about O, then is

$$F_p\left[\frac{h}{\sin\theta} - \frac{2h}{3\sin\theta}\right] = \frac{\rho_f gbh^2}{2\sin\theta}\frac{h}{3\sin\theta} = \frac{\rho_f gbh^3}{6\sin^2\theta}$$

Force F must provide the same moment about O. Thus,

$$Fl\sin\theta = \frac{\rho_f gbh^3}{6\sin^2\theta}$$

where l is the length $AB = 2$ m, and $F = \frac{\rho_f gbh^3}{6l\sin^3\theta}$
For given values of $\rho_f = 10^3$kg/m^3, $b = 2$ m, $h = 1.2$ m, $l = 3$ m, and $\theta = 30°$ we get

$$F = \frac{10^3\left(\frac{kg}{m^3}\right) \times 9.81\left(\frac{N}{kg}\right) \times 2(m) \times (1.2)^3 \times (m^3)}{6 \times 3(m)\sin^3 30°} = 15.1 \text{ kN}$$

and for $\theta = 60°$, this becomes $F = 2.9$ kN.
Thus, less force is required as the plate becomes more upright.

Example 17.4

Water is flowing upto 0.5 m deep over the flash boards on a dam as shown in Fig. 17.12. Find the moment produced at the base A of the flash board.

Solution

This is not a problem in fluid statics since the water is flowing. If the water was flowing exactly horizontally, as it would far upstream of the flash boards, we could still apply hydrostatic equations in the vertical direction because the fluid will not be

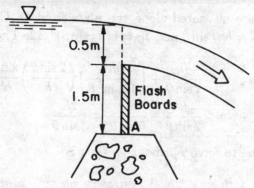

Fig. 17.12 Flow over the flash boards on a dam.

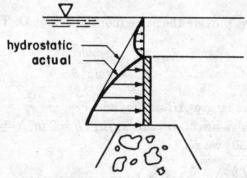

Fig. 17.13 The hydrostatic and the actual pressure distribution over the flash boards on a dam.

accelerating in that direction and the equilibrium equations in the vertical directions will give the same results as if the fluid was static. But even this simplification is not available here because the fluid is definitely accelerating near the top of the dam. Yet all along the lower length of the boards the fluid is essentially at rest, and the actual pressure differs from the hydrostatic distribution only at a few locations as shown in Fig. 17.13.

It is therefore possible to obtain a reasonable estimate of the forces by approximating pressure distribution by the ideal hydrostatic pressure variation along the flash board.

The total hydrostatic force then is

$$(\rho_f g h_G)S = 10^3 \left(\frac{\text{kg}}{\text{m}^3}\right) \times 9.81 \left(\frac{\text{N}}{\text{kg}}\right) \times 1(\text{m})$$
$$\times (1.5 \text{ m}^2/\text{m width})$$
$$= 1.47 \times 10^4 \text{ N/m width}$$

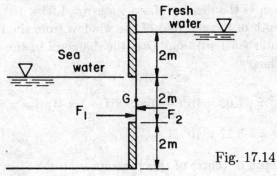

Fig. 17.14

The location of the centre of pressure from the free surface is obtained from Eq. (17.6) as

$$h_{cp} = h_G + \frac{I_G}{h_G S}$$

For a rectangular plate of 1.5 m height, $I_G = (1.5)^3/12$ m^4 per m width $= 0.24$ m^4 per m width, and thus,

$$h_{cp} = (0.50 + 0.75) + \frac{0.24}{(0.50 + 0.75) \times 1.5} = 1.38 \text{ m}$$

Thus, the centre of pressure thus is $(2.0 - 1.38) = 0.62$ m from the base, and the moment produced by the hydrostatic force about the base, is 1.47×10^4 (N/m width) $\times 0.62$ (m) $= 9.1 \times 10$ Nm/m width of the flash boards.

Example 17.5

A circular window 2 m in diameter has sea water (specific gravity 1.03) on one side and fresh water on the other as shown in Fig. 17.14. Find the resultant thrust on the window, and its point of application.

Solution

The force on the left-hand side of the window is given by

$$F_1 = (\rho_1 g h_{G1}) \times \left(\frac{\pi D^2}{4}\right)$$

where ρ_1 is the density of sea water $(= 1.03 \times 10^3 \text{ kg/m}^3)$, h_{G1} the depth of the centroid of the window from the free surface of sea water $(= 1 \text{ m})$, and D is the diameter of the window $(= 2 \text{ m})$. Thus,

$$F_1 = 1.03 \times 10^3 (\frac{\text{kg}}{\text{m}^3}) \times 9.81 (\frac{\text{N}}{\text{kg}}) \times 1(\text{m}) \times \pi \times \frac{(2)^2(\text{m})^2}{4}$$
$$= 3.17 \times 10^4 \text{ N}$$

The depth of centre of this pressure is

$$h_1 = h_{G1} + \frac{I_G}{h_{G1}S}$$

where I_G is the second moment of area of a circular window about a horizontal diameter, which from Table 17.1 is

$$I_G = \frac{\pi D^4}{64} = \pi(2^4)(\text{m}^4)/64 = 0.79 \text{ m}^4$$

Then,

$$h_1 = 1(\text{m}) + \frac{0.79(\text{m}^4)}{1(\text{m}) \times \pi(\text{m}^2)} = 1.25 \text{ m}$$

which is 0.25 m below the centre of the window. The force on the right-hand side is

$$F_2 = (\rho_2 g h_{G2}) \times (\pi D^2/4)$$

where $\rho_2 = 1 \times 10^3 \text{kg/m}^3$, h_{G2} is the depth of centroid from the water level on this side $(= 3 \text{ m})$. Thus,

$$F_2 = 10^3 \left(\frac{\text{kg}}{\text{m}^3}\right) \times 9.81 \left(\frac{\text{N}}{\text{kg}}\right) \times 3(\text{m}) \times \pi \times \frac{(2)^2(\text{m})^2}{4}$$
$$= 9.25 \times 10^4 \text{ N}$$

and its point of application is at depth

$$h_2 = h_{G2} + \frac{I_G}{h_{G2}S} = 3(\text{m}) + \frac{0.79(\text{m}^4)}{3(\text{m})) \times \pi(\text{m}^2)} = 3.08 \text{ m}$$

from the fresh water free surface, which is 0.08 m below the centre of the window.

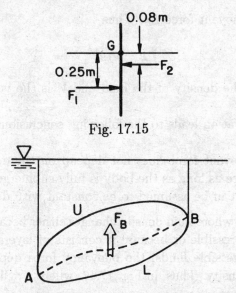

Fig. 17.15

Fig. 17.16 A body fully-submerged in a liquid.

The resultant pressure force magnitude is

$F_2 - F_1 = 9.25 \times 10^4$ (N) $- 3.17 \times 10^4$ (N) $= 6.08 \times 10^4$ N

to the left. Its point of application can be found by taking the moments about the centroid G. From Fig. 17.15,

$$h = \frac{F_2 \times 0.08(\text{m}) - F_1 \times 0.25(\text{m})}{F_2 - F_1}$$

$$= \frac{9.25 \times 10^4 \times 0.08 - 3.17 \times 10^4 \times 0.25}{9.25 \times 10^4 - 3.17 \times 10^4} = -0.009 \text{ m}$$

or 9 mm *above* the centroid.

17.5   Buoyancy

Consider a fully-submerged body as shown in Fig. 17.16. Imagine the body surface to be divided into an upper and a lower surface by a meridianol line AB.

The downwards force on upper surface AUB is equal to the weight of the liquid standing on this surface. Similarly the upwards force on lower surface ALB is equal to the weight of the fluid that can stand on ALB. Thus, net upward force on the body is equal to the weight of the fluid of volume equal to the difference of the two volumes, which is the volume of the body itself.

This is the buoyant force F_B. Thus,

$$F_B = \rho_f g V$$

where ρ_f is the density of the fluid and V is the volume of the body.

This derivation leads to the following conclusions:

1. The buoyant force does not depend on the depth of submergence as long as the body is fully submerged, and fluid density can be assumed to be constant with depth.

2. In cases where fluid density changes either because the fluid is compressible or because it consists of layers of different incompressible fluids, the buoyancy force depends on the local density. Thus, in Fig. 17.17, where a cylinder of base

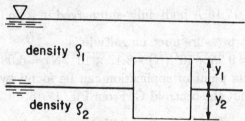

density ρ_1

density ρ_2

y_1

y_2

Fig. 17.17 A vertical cylinder submerged at the interface of two liquids.

area S is held at the interface of two fluids of density ρ_1 and ρ_2, the buoyancy force F_B is given by

$$\begin{aligned} F_B &= g[\rho_1 V_1 + \rho_2 V_2] \\ &= g[\rho_1 y_1 S + \rho_2 y_2 S] \end{aligned}$$

3. A body floats when the buoyancy force equals its weight. Since buoyancy force depends upon ρ_f, and the weight on the average body-density ρ_b, a body whose density ρ_b is less than fluid density ρ_f floats with only a part of its volume submerged. A body floats fully submerged when and only when its average density is equal to the fluid density ρ_f, and then it can be at equilibrium at any depth.

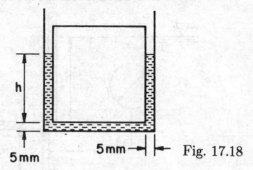

Fig. 17.18

Example 17.6

A steel cube of sides 1 m is to be floated in a container as shown in Fig. 17.18, with a clearance of 5 mm on each side and at bottom. Find the weight of the mercury required.

Solution

When the steel cube is floating, the weight of mercury 'displaced' is equal to cube's total weight. If height of the cube submerged is h, the volume of mercury displaced is $h \times 1(\text{m}) \times 1(\text{m})$, its weight, and hence the buoyancy force is

$$\rho_m g V = 13.6 \times 10^3 \left(\frac{\text{kg}}{\text{m}^3} \right) \times 9.81 \left(\frac{\text{N}}{\text{kg}} \right) \times h(\text{m}^2) = 1.33 \times 10^5 h \ (\text{N/m})$$

This must be equal to the weight of the cube which is 7.6×10^3 $(\text{kg/m}^3) \times 9.81 \ (\text{N/kg}) \times 1(\text{m}^3) = 7.46 \times 10^4$ N. The equivalence of the two forces gives

$$h = 7.46 \times 10^4 (\text{N}) / 1.33 \times 10^5 (\text{N/m}) = 0.56 \text{ m}$$

Thus, mercury upto a depth of $(0.56 + 0.0005) = 0.0565$ m is required. The cross-sectional dimensions of the container are $(1.0 + 2 \times 0.005)$ m $= 1.01$ m square, and therefore, the volume of mercury required is

$$0.565(\text{m}) \times 1.01(\text{m}) \times 1.01(\text{m}) - 0.56(\text{m}) \times 1(\text{m}) \times 1(\text{m}) = 0.0163 \text{m}^3,$$

the total weight of which is 13.6×10^3 (kg/m) $\times 9.81$ (N/kg) $\times 0.0163(\text{m}^3) = 2.17$ kN. Thus, a 2.17 kN (or 221.9 kg) weight of mercury can float a steel cube of weight 74.6 kN (or 7600 kg) in this geometry.

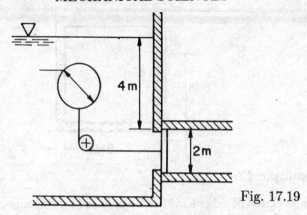

Fig. 17.19

Example 17.7

A rectangular gate (2 m × 1 m) can slide without friction as shown in Fig. 17.19. It is held in place by a thin cable which pulls it to the left due to buoyancy on a balloon of diameter D (and of negligible weight). The gate just slides open with the level of water in the tank as shown. What is the value of the balloon diameter D? Where should the cable be attached to the gate, if it is to slide smoothly?

Solution

The net force on the gate due to hydrostatic pressure must balance the force due to buoyancy on the baloon. The force of pressure on the gate is

$$\rho_w g h_G S = 10^3 \left(\frac{kg}{m^3}\right) \times 9.81 \left(\frac{N}{kg}\right) \times (4+1)(m) \times [2(m) \times 1(m)]$$

$$= 9.81 \times 10^4 \ N$$

Buoyancy force on a balloon of diameter D is

$$\rho_w g [4/3 \pi (D/2)^3] = 10^3 \left(\frac{kg}{m^3}\right) \times 9.81 \left(\frac{N}{kg}\right) \times 4/3 \times \pi \times D^3/8$$

$$= 5.14 \times 10^3 D^3 \ (N/m^3)$$

Equating the two gives

$$D = \left[\frac{9.81 \times 10^4 (N)}{5.14 \times 10^3 (N/m^3)}\right]^{1/3} = 2.67 \ m$$

If the gate is not to foul in sliding, the net moment applied on it must be zero. This requires that the cable should be attached to the centre of pressure. The depth h_{cp} of the centre of pressure is given by

$$h_{cp} = h_G + \frac{I_G}{h_G S}$$

where I_G is the second moment of area of the rectangular plate (of $b = 1$ m and $h = 2$ m) about its centroidal axis. This is obtained, with reference to Table 17.1 as $bh^3/12$ or $1(\text{m}) \times [2(\text{m})]^3/12 = 0.67$ m^4. The depth of centre of pressure from the free-surface is then,

$$h_{cp} = 5 + \frac{0.67}{5 \times 2} = 5.067 \text{ m}$$

or 6.7 cm below the centre of the plate. This is where the cable should be attached.

Table 17.1   Centroidal location and the second moment of area about centroidal axis for some common shapes.

Geometry	Area, S	Centroid location, y_G	I_G
y_G [rectangle, base b, height h]	hb	$\dfrac{h}{2}$	$\dfrac{bh^3}{12}$
y_G [triangle, base b, height h]	$\dfrac{hb}{2}$	$\dfrac{h}{3}$	$\dfrac{bh^3}{36}$
y_G [circle, diameter d]	$\dfrac{\pi d^2}{4}$	$\dfrac{d}{2}$	$\dfrac{\pi d^4}{64}$

Problems

17.1 Determine the total force and location of the centre of pressure for a circular plate of 2 m diameter immersed vertically in water with its top edge flush with the free surface of water.

17.2 Repeat problem 17.1, except that the top edge is now 1 m below the water surface.

17.3 Again, solve problem 17.1 except that the top edge is now 20 m below the water surface. Draw conclusions from the results of these three problems.

17.4 Determine the total force and position of the centre of pressure on an equilateral triangular plate immersed with its one side lying on the surface of water and the plane of the plate making an angle of 60° to the vertical.

17.5 A concrete wall 7 m high and 3 m wide retains water on one side to a depth of 5 m. What is the toppling moment exerted by water on this wall per metre length of the wall?

17.6 Compute the force F required to hold the 1 m wide gate shown in Fig. 17.20.

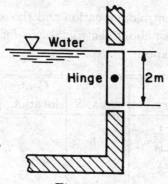

Fig. 17.20

17.7 Find the total vertical and horizontal hydrostatic forces on the cylindrical gate of length 1 m shown in Fig. 17.21.

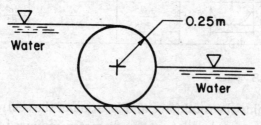

Fig. 17.21

17.8 What will be the force required to hold the semi-cylindrical

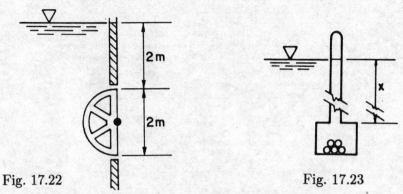

Fig. 17.22   Fig. 17.23

gate of Fig. 17.22 in place? What will be the vertical hydrostatic force acting on this gate?

17.9 The cross-section of a barge can be approximated as a rectangle 8 m × 12 m. What will be the draft when it carries 500 tonne of load? Its own weight is 200 tonne.

17.10 An aluminium cube of 1 m on side requires a force of 17 kN to lift it, when it is submerged in water. What is its specific gravity?

17.11 A hydrometer can be modelled as a cylinder 13 mm in diameter and 50 mm long with a 250 mm long stem of 3 mm diameter shown in Fig. 17.23. The total mass of the hydrometer is 90 g. To what depth x will this float in glycerine of density 1.25×10^3 kg/m^3.

18. FLUIDS IN MOTION

18.1 Field Description

The fluid velocity may vary from particle to particle in a moving fluids. Since a very large number of particles is involved, it usually is quite inconvenient to keep track of the velocity of each fluid particle. Instead, we can specify the velocity as a function of the spatial location x (and of time t) and not worry about what particle is at that location at time t. Thus, velocities $V(x_o, t_1)$ and $V(x_o, t_2)$ at the same location x_o at two different times t_1 and t_2 may refer to the velocities of two different particles that happen to occupy the given location x_o at the respective times.

Consider a hypothetical flow produced by a sphere of mean radius R_o whose surface pulsates at a frequency f with velocity $U_o \sin 2\pi ft$. Then, fluid velocity at radius r and time t is given by

$$V(r,t) = \frac{R_o^2 U_o^2}{r^2} \sin 2\pi ft \qquad (18.1)$$

Hence at a given radius r, the velocity is a sinusoidal function of time t, but at two different times, the velocity specified is not the velocity of the same particle, but of two different particles occupying the given location at the respective times.

This method of description wherein a property is specified as a function of the space location, and is not tied to specific particles, is termed as the field description. Thus Eq. (18.1) specifies a *velocity field*.

18.2   Time Rate of Change

The method of describing velocities as fields should be handled with extreme care when accelerations are to be determined. Since the velocities specified for a given value of x at two different times are not the velocities of the same particle, the rate of change with time of the velocity calculated at a specified location cannot at all be interpreted as the acceleration.

Further, consider an incompressible fluid flowing through a channel (Fig. 18.1) whose width changes as $b(x) = b_o(1 - kx)$. Clearly, as the fluid flows down the channel, its velocity would

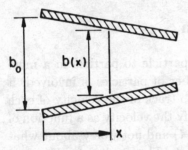

Fig. 18.1 A converging channel.

increase. If Q' be the volume flow per unit depth of the channel, the velocity $v(x)$ at location x is given by

$$v(x) = \frac{Q'}{b(x)} = \frac{Q'}{b_o(1 - kx)}$$

which changes from one x–location to another.

Clearly, v is not a function of time t. A 'velocity' probe at a fixed location x will record unchanging velocity with time. Such a field is termed as a steady field. But the unchanging velocity at a fixed x does not indicate that the fluid particles have constant velocities and are not accelerating. In fact, a particle that leaves one x–location moves on to a downstream location in time δt, where velocity is higher and thus undergoes an acceleration. We see that in this steady case, the δV experienced by a particle in time δt is the change in velocity across the distance $V\delta t$ that the particle covers in time δt, and if $\partial V/\partial x$ represents the gradient of velocity along x, the velocity $V\delta t$ away is $V + (\partial V/\partial x)(V\delta t)$. Thus, this particle experiences a change in velocity of magnitude

$\delta V = (\partial V/\partial x)(V\delta t)$ in time δt, thus representing an acceleration of $\delta V/\delta t = (\partial V/\partial x)V$ or $V.\partial V/\partial x$. This acceleration can be attributed to motion of the particle in the presence of a spatial gradient of velocity and is termed as *convective* acceleration. If, in addition, the local velocity is also changing with time, at a rate $\partial V/\partial t$, the net acceleration experienced by the particle can be obtained as

$$\text{Acceleration} = \frac{\partial V}{\partial t} + V\frac{\partial V}{\partial x} \qquad (18.2)$$

This is known as Euler acceleration formula. Here $\partial V/\partial t$ is the partial derivative of velocity with respect to time with the position x held constant (called the *local* acceleration), while $\partial V/\partial x$ is the spatial gradient with time held constant (and gives rise to the *convective* acceleration).

18.3 Pathlines, Streamlines and Streaklines

It is often convenient to represent velocity fields graphically. A number of concepts are used for the purpose, the most straightforward being the concept of *pathlines*. Its a line in the flow field representing the trajectory of a specified fluid particle. Experimentally, these lines may be obtained by injecting some tracers (such as a small puff of smoke in air, or a blob of dye in a liquid) and then following their progress with time, say by long - exposure photography.

Another concept used is that of a *streakline*. If a tracer is continuously injected at a fixed point in the field, the line formed by the tracers at a later time is the streakline. For example, visible smoke plume from a chimney is a streakline.

However, the most commonly used graphical concept however is that of a *streamline*. It is a line in the flow field drawn such that the trangent to it at every point is in the direction of the local, instantaneous velocity vector. If we mix small, slender and neutrally buoyant needles in the fluid, each one will align with the direction of the local velocity, and the overall picture will give the pattern of streamlines. To clarify these three concepts and to bring out their differences clearly, consider the flow of air from west to east. Obviously streamlines in this flow will run from west to east as shown in Fig. 18.2

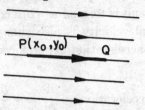

Fig. 18.2 Streamlines in a uniform steady flow.

If a tracer is injected at $P(x_o, y_o)$, then in a time t_o it will come to a point Q, a distance $V_o t_o$ away, where V_o is the velocity of air. Clearly the pathline, streakline and the streamline, as defined, above should coincide with the line segment PQ.

Now let the wind direction change at time t_o to become northerly, i.e., from north to south. Now for any time $t > t_o$ to streamlines run from north to south as shown in Fig. 18.3. But the path of the particle which was at $P(x_o, y_o)$ at $t = 0$

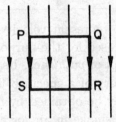

Fig. 18.3 Streamlines, streaklines and pathlines in a flow that is not steady.

is now represented by PQR, since the particle changes direction only after reaching point Q. The streakline through point P is distinct from both streamline and pathline through that point. Obviously, the line of particles that passed through P between time $t = 0$ and t_o, and which occupied the line PQ at $t = t_o$ in Fig. 18.3, will move down southwards parallel to itself, and at $t = 2t_o$ occupy the line SR. The particles which passed point P in time $t > t_o$ will move directly southwards and occupy the line segment PS. Thus, PSR is the streakline at $t = 2t_o$.

We thus see that in an unsteady flow, streamlines, streaklines and pathlines can all be distinct, whereas in a steady flow they are necessarily identical.

18.4   Mass Balance

Consider a steady flow of an incompressible fluid in a duct of varying section as shown in Fig. 18.4. Since the flow is steady,

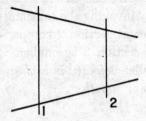

Fig. 18.4 A duct of varying section.

there will be no build up of mass contained within the duct between any two sections 1 and 2, and therefore, the rate of mass flow in across section 1 must exactly equal the rate of mass flow out across section 2. If V_1, A_1 and V_2, A_2 represent the velocities and areas at sections 1 and 2, respectively, then

$$\rho V_1 A_1 = \rho V_2 A_2 \qquad (18.3)$$

where ρ is the constant density of the fluid. Also,

$$V_1 A_1 = V_2 A_2 \qquad (18.4)$$

This statement for the conservation of mass is known as the continuity equation for incompressible flows.

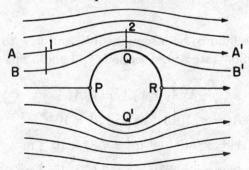

Fig. 18.5 Streamlines in a steady flow past a cylinder.

Figure 18.5 shows the flow pattern of a steady incompressible non-viscous fluid past a circular cylinder. Since flow is steady, the lines can be taken to represent streaklines or pathlines or streamlines. If we consider lines as streamlines, the local velocity vector is always tangential to the streamlines, and therefore, no fluid crosses these lines. Thus, any fluid that enters the 'channel' formed by streamlines AA' and BB' must remain within it. It can then easily be seen by invoking the continuity Eq. (18.5) that the fluid velocity is larger at section 2 compared to that at section 1, since the area at section 2 is smaller. In fact, this can be stated as a general rule: *velocity is higher in the region where streamlines are closer together.*

Example 18.1

Figure 18.6 shows a device in which the piston is retracting at the rate of 0.1 m/s in a cylinder of diameter 10 cm. Find the velocity at minimum section 2 where the diameter is 1 cm and at section 3 where the diameter is 3 cm.

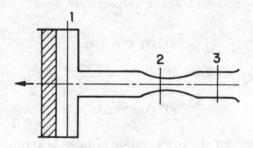

Fig. 18.6

Solution

If we take the fluid to be incompressible, and flow to be steady, then volume flow rate across sections 1, 2 and 3 must be equal. Therefore,

$$V_1 A_1 = V_2 A_2 = V_3 A_3$$
$$A_1 = \pi/4(0.1)^2 \text{ m}^2 = 7.85 \times 10^{-3} \text{ m}^2$$
$$A_2 = \pi/4(0.01)^2 \text{ m}^2 = 7.85 \times 10^{-5} \text{ m}^2$$
$$A_3 = \pi/4(0.03)^2 \text{ m}^2 = 7.07 \times 10^{-4} \text{ m}^2$$

The fluid velocity V_1 must be equal to the piston velocity 0.1 m/s. Therefore,

$$V_2 = V_1 A_1 / A_2 = 10 \text{ m/s},$$

$$V_3 = V_1 A_1 / A_3 = 1.11 \text{ m/s},$$

Thus, the velocity is maximum at the minimum section 2.

18.5 Bernoulli Equation

Perhaps the most commonly used equation in fluid mechanics is the Bernoulli equation which is a statement of conservation of energy in a non-viscous, steady and incompressible flow.

Consider two neighboring streamlines in a steady flow (Fig. 18.7). Since flow is steady, the streamline pattern will not change

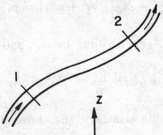

Fig. 18.7 Flow between two streamlines close together.

with time, and as was argued above, any fluid flowing between the two streamlines will always remain between them.

The statement of energy principle would require that rate of energy flowing out at the open section 2 will be equal to rate of energy flowing in across the open section 1 *plus* the net rate of work done while flowing from Section 1 to 2 in the absence of any heat transfer.

The energy associated with the fluid is its kinetic energy and potential energy. Here the internal energy is neglected because in the absence of compressibility and viscosity (it has the same kind of action as friction) mechanical energy will not change to internal energy or vice versa, and thus, we need to consider only the mechanical energy balance.

For a volume flow rate of Q, the rate of flow of energy is $\rho Q[\frac{1}{2}V^2 + gh]$, where $1/2V^2$ and gh represent the kinetic and potential energy per unit mass, respectively.

Work is done on the fluid by surface forces. Since the flow is assumed to be non-viscous, the only surface forces that are present are pressure forces. On lateral surfaces (i.e. along the streamlines) the flow velocity is tangential to the surface, while pressure is normal. The rate of doing work is, therefore, zero on these surfaces. Only at the cross-sections 1 and 2 are the pressure forces in line with the velocity vector, and the next rate of doing work on the fluid then is $(p_1 A_1 V_1 - p_2 A_2 V_2)$. The second contribution is negative since pressure and fluid velocity on surface 2 are opposed to each other.

The energy balance then requires,

$$\rho Q \left[\frac{1}{2}V_2^2 + gh_2\right] - \rho Q \left[\frac{1}{2}V_1^2 + gh_1\right] = p_1 A_1 V_1 - p_2 A_2 V_2$$

Since $Q = A_1 V_1 = A_2 V_2$, dividing by ρQ, and rearranging, gives

$$\frac{1}{2}V_2^2 + gh_2 + \frac{p_2}{\rho} = \frac{1}{2}V_1^2 + gh_1 + \frac{p_1}{\rho}$$

Sections 1 and 2 being arbitrary, the above can be written as

$$\frac{V^2}{2} + gh + \frac{p}{\rho} = \text{constant} \tag{18.5}$$

along a streamline in a steady, incompressible flow of a non-viscous fluids.

This is known as *Bernoulli equation* in honour of the swiss scientist who first obtained it.

Example 18.2

Consider the flow of an incompressible fluid past a circular cylinder whose streamlines were shown in Fig. 18.5. It was argued there that the changing spacing of the streamline along the circumstance of the cylinder denotes that flow velocity is changing. At the farthest upstream point P on the cylinder where the fluid comes to rest is termed as the upstream stagnation point. As the flow moves downstream, its velocity increases till it is maximum at shoulder points Q and Q', and then decreases again to the stagnation point at R. By an analysis, beyond the scope of this text, it can be shown [See, of example, Gupta, V., and Gupta, S.K., *Fluid Mechanics and Its Application*, Wiley Eastern, p.365] that the surface velocity at any angular location θ is given by

$$V = 2V_o \sin \theta \qquad (18.6)$$

where V_o is the far upstream velocity and θ is the clockwise angle measured from the upstream stagnation point P.

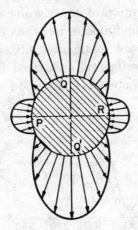

Fig. 18.8 Pressure distribution on a circular cylinder.

We can use Bernoulli equation to determine the distribution of pressure along the surface of the cylinder.

Consider the central streamline. Far upstream, flow velocity is V_o and pressure is p_o, the ambient pressure. Assuming the flow to be steady, incompressible and non-viscous (the very assumptions for which Eq. (18.6) is obtained), then neglecting the effects of gravity, $V^2/2 + p/\rho$ should be constant along any streamline. This gives

$$\frac{V^2}{2} + \frac{p}{\rho} = \frac{V_o^2}{2} + \frac{p_o}{\rho}$$

or

$$(p - p_o) = \frac{1}{2}\rho[V_o^2 - V^2] = \frac{1}{2}\rho V_o^2[1 - 4\sin^2\theta] \tag{18.7}$$

This distribution is plotted in Fig. 18.8.

Note that (i) the pressure difference is positive near the nose and tail of the body, denoting that the pressures there are larger than the ambient, while $(p - p_o)$ is negative near the shoulders, denoting that the pressures there are less than the ambient; (ii) the pressure distribution is completely symmetric about both horizontal and vertical axes. This gives zero net force on the cylinder, a result in sharp contract to the physically observed large values of the drag force on such a cylinder held in a fluid stream. This departure, both in theory and experimental observation arises because of the assumption of non-viscous flows, which is not a good assumption for bluff bodies like the cylinder.

In fact, if we compare the theoretically obtained pressure distribution with the experimentally observed distribution (Fig. 18.9) we find that the agreement is quite good on the 'front' of the

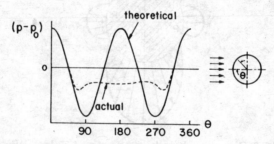

Fig. 18.9 Theoretical and actual pressure distributions on a circular cylinder.

cylinder, but breaks down in the 'rear'. If the body was not bluff but streamlined, that is, slowly tapering towards the tail, the agreement in pressures is remarkably improved over nearly the whole of the body (Fig. 18.10). Thus, on streamlined bodies the assumptions of non-viscous flows is quite reasonable and such bodies experience very low drags while moving in fluids. A bomb is given such a shape so that it doesn't slow down too much in flight and a dolphin has a similar shape so that it expends very little energy in swimming.

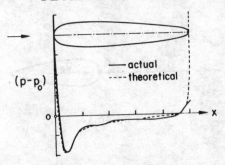

Fig. 18.10 Theoretical and actual pressure distributions on a streamlined body.

Gauge pressures are positive in regions where the surface velocity is larger than the free-stream value V_o, and is negative where surface velocity is smaller than V_o. Therefore, if we change the shape of the cylinder such that it is asymmetrical resulting in larger velocities on the top than at the bottom, an asymmetrical pressure distribution would result with lower pressures on top than on bottom. This could be exploited to generate a sideways force, normal to the drag direction. An aircraft wing is so shaped that this net sideways force is directed upwards and it helps balance the weight of the aircraft and makes flight possible. For this reason, this sideways force is termed as lift (Fig. 18.11).

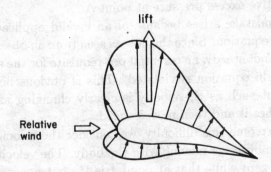

Fig. 18.11 Generation of lift on an aerofoil.

Example 18.3

Consider a bomb–shaped body travelling in air at a speed of 100 m/s. Find the excess pressure at the nose A of the body shown in Fig. 18.12.

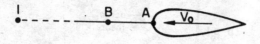

Fig. 18.12

Solution

The atmospheric air is still, and we can take the flow to be in-compressible and non–viscous. On a streamline connecting a far away point 1 to point A, we note that the velocity far away is zero and, the pressure there is the ambient pressure p_{atm}. On the nose of the aircraft, air particles travel with the body at velocity V_o.

If we are not careful and apply the Bernoulli equation between the far point 1 and point A using the above data, excess pressure on point A is obtained as

$$p_A - p_{atm} = -\frac{1}{2}Vo^2$$

This negative excess pressure is an obvious violation of the experience of any motorbike rider, which suggests that there should be a positive excess pressure at point A.

This mistake arises because of an invalid application of the Bernoulli equation. Since the flow as seen from an observer on the ground is not steady, an essential pre-requisite for the application of Bernoulli equation is violated. This is obvious because at a fixed point such as B, velocity is clearly changing as the body appproaches it and then moves ahead.

To overcome this difficulty, we consider the velocity fields as seen by an observer moving with the body. The velocity at point A is then zero while that at point 1 is V_o, but now to the right. The field as seen by this observer is unchanging with time, i.e., it is steady and Bernoulli equation can safely be applied. Then in this frame of reference,

$$p_A + \frac{1}{2}\rho V_A^2 = p_{atm} + \frac{1}{2}\rho V_1^2$$

Here $V_A = 0$ and $V_1 = V_o$, so that

$$p_A - p_{atm} = \frac{1}{2}\rho V_o^2 = \frac{1}{2} \times 1.2\left(\frac{\text{kg}}{\text{m}^3}\right) \times 100^2 \left(\frac{\text{m}}{\text{s}}\right)^2 = 6 \times 10^2 \text{ Pa}$$

which is positive, as it should be.

Example 18.4 (Torricelli Theorem)

Water fills a large reservoir as shown in Fig. 18.13. At a depth h from the free surface is a well-rounded orifice of diameter d. What is the flow velocity out of the orifice?

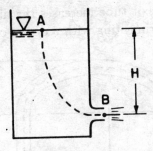

Fig. 18.13

Solution

Consider a streamline which runs from the free surface to the orifice. Its exact shape within the reservoir in material. If the reservoir is large (as compared to the orifice diameter d) the velocity of water at A is negligible as compared to the velocity at B. Using free surface of water as the datum for h, an application of Bernoulli equation between points A and B gives.

$$\frac{V_B^2}{2} + \frac{p_B}{\rho} + gh_B = \frac{V_A^2}{2} + \frac{p_A}{\rho} + gh_A$$

Here $V_A = 0$, $p_A = p_B = p_{atm}$, $h_A = 0$ and $h_B = -H$. Therefore,

$$V_B^2 = 2gH \quad \text{or} \quad V_B = \sqrt{2gH} \tag{18.8}$$

This is the famous *Torricelli theorem*.

The volume flow rate through the orifice will be $(\pi d^2/4)\sqrt{2gH}$.

The jet of water after issuing from the orifice take a parabolic trajectory (same as of a projectile), and when it reaches the ground, H_G meter below the free surface, it can be shown that velocity there will be $\sqrt{2gH_G}$, but discharge rate will be the same $(\pi d^2/4)\sqrt{2gH}$. This implies, that the jet diameter contracts as it falls.

If instead of falling as a free jet, the issuing water is confined into a flexible tube, of diameter d, connected to the orifice and running to the ground, a simple application of Bernoulli equation will still give the velocity at the ground (now the tube outlet) as $\sqrt{2gH_G}$, but now the volume flow rate will be larger at $(\pi d^2/4)\sqrt{2gH_G}$, instead of the earlier $(\pi d^2/4)\sqrt{2gH}$. Thus, attachment of the exit tube increases the discharge rate, as can be verified by experimenting with a desert-cooler tank.

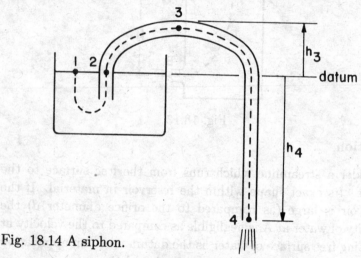

Fig. 18.14 A siphon.

Similar results apply on siphons (Fig. 18.14) as well. Applying Bernoulli equation between points 1 and 4, gives

$$V_4 = \sqrt{2gH_4}$$

Here, the pressure at point 2 at the water level, but inside the siphon tube, will be lower than p_{atm} and

$$\frac{p_2}{\rho} + \frac{V_2^2}{2} = \frac{p_1}{\rho} + \frac{V_1^2}{2}$$

Since V_1 is very small, and $p_1 = p_{atm}$,

$$(p_2 - p_{atm}) = p_{2g} = -\rho V_2^2/2$$

where $V_2 = V_4$, by continuity.

The lowest pressure will be reached at point 3 (the highest point) where

$$p_{3g} = p_3 - p_{atm} = -\frac{\rho V_3^2}{2} - \rho g h_3$$

Using the value of $V_3 = V_4 = \sqrt{2gh_4}$, $p_{3g} = -\rho g(h_3 + h_4)$ where $(h_3 + h_4)$ is the length of the longer limb of the siphon. This result can be obtained directly as well by applying Bernoulli equation between points 3 and 4.

The siphon will work as long as the pressure at point 3, the lowest pressure in the system, is above the vapour pressure of water at the ambient temperature. At 20°C, the vapour pressure of water is 2.34 kPa, so that the maximum length of the long limb of a siphon is

$$h_3 + h_4 = \frac{p_{atm} - p_3}{\rho g} = \frac{(101.31 - 2.34) \times 10^3 (\text{Pa})}{10^3 (\text{kg/m}^3) \times 9.81 (\text{m/s}^2)} = 10.09 \text{ m}$$

This is the condition to be must be satisfied for the siphon to flow full. But if we allow water to separate from the tube walls inside the longer limb, we can make only h_3 as large as 10.09 m. In this case, the delivery rate of water will be determined by the shorter limb, and water jet will contract in some portion of the longer limb.

Example 18.5

Three short pipes are draining from a water reservoir with a 'head' of 5 m as shown in Fig. 18.15. Find the discharge rate and the velocity of water at the exit in each of the three cases. Pipe diameters at section B are 50 mm each, while that at C is 40 mm.

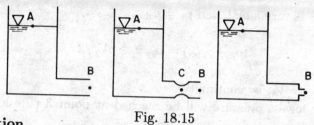

Solution   Fig. 18.15

Since pipe sections are specified as short, we neglect the viscous
losses and apply Bernoulli equation along a streamline from a
point A on the reservoir surface to a point B at the exit. Clearly,
$V_A = 0$, $p_A = p_B = p_{atm}$, $z_A = 5$ m and $z_B = 0$. Application of
Bernoulli equation then gives

$$V_B^2/(2 \times 9.81) = 5$$

or

$$V_B = \sqrt{2 \times 9.81 \times 5} = 9.90 \text{ m/s}$$

Exactly the same calculations apply in each of the three cases.
Hence, exit velocity is the same in each, including in the last case
where the pipe diameter is reduced drastically.

This is contradictory to the common observation that when
you pinch the free end of a garden hose the water jet comes out
at much increased velocity. This is sometimes (but erroneously)
explained by invoking the continuity equation that since the flow
area is now reduced, the velocity must increase ! The correct
explanation depends upon the viscous losses in the garden hose,
which definitely cannot be assumed to be 'short'. This will be
explained in the next chapter.

The pressure at point C in the reduced 40 mm section of
the second case can be obtained by applying Bernoulli equation
between points C and B. The velocity at point C is obtained
in terms of the velocity at point B by the mass conservation
equation

$$V_C S_C = V_B S_B$$

or

$$V_C = V_B . \frac{S_B}{S_C} = 9.90 \text{ (m/s)} \times \frac{\pi/4 \times (0.050)^2}{\pi/4 \times (0.040)^2} = 15.47 \text{ (m/s)}$$

Bernoulli equation now gives

$$\frac{(15.47)^2}{2 \times 9.8} + \frac{p_c}{1000 \times 9.8} = \frac{(9.90)^2}{2 \times 9.8} + \frac{p_{atm}}{1000 \times 9.8}$$

or

$$p_{c,g} = p_c - p_{atm} = \frac{1}{2} \times 1000[(9.90)^2 - (15.47)^2] = -70.7 \text{ kPa}$$

which is quite a severe vacuum.

In absolute terms, this corresponds to $(101.35 - 70.7)$ kPa or 30.65 kPa absolute. Vapour pressure of water at 70°C is 31.2 kPa, and hence a vapour bubble forms at this location if the temperature of water is around 70°C or more.

If asked to determine the minimum area of constriction to avoid a vapour bubble at 20°C where vapour pressure of water is only 2.34 kPa, we could work backwards from this pressure and calculate V_c using Bernoulli equation between point C and B. Therefore,

$$\frac{V_C^2}{2g} + \frac{p_C}{\rho} = \frac{V_B^2}{2g} + \frac{p_B}{\rho}$$

or

$$V_C^2 = V_B^2 + \frac{2}{\rho}(p_B - p_C)$$

Substituting the values gives

$$V_C^2 = 9.90^2 + \frac{2}{1000} \times (101.35 \times 10^3 - 2.34 \times 10^3) = 296.01 \text{ (m/s)}^2$$

This implies a velocity of 17.20 m/s at C. Using the continuity equation between B and C gives

$$V_C \frac{\pi}{4} d_C^2 = V_B \frac{\pi}{4} d_B^2$$

and

$$d_C = d_B \sqrt{\frac{V_B}{V_C}} = 0.050 \times \sqrt{\frac{9.90}{17.20}} = 0.038 \text{ m}$$

Thus, if the diameter at C is reduced only slightly more (to 38 mm from the present 40 mm) vapour bubbles would appear at 20°C (instead of 70°C earlier). But pinching the end at B in the last case does not result in this problem at all.

18.6   Engineering Energy Equation

The essential requirement of Bernoulli equation that the flow
be nonviscous restricts its usefulness, even though the class of
problems where flow can be assumed to be non-viscous is fairly
large. The presence of viscosity results in shear stresses on the
walls of the streamtube, shown in Fig. 18.7 (Sec. 18.5), which
extract energy from the flow such that the energy per unit mass
flow $(V^2/2 + p/\rho + gz)$ does not remain constant but decreases
in the downstream direction.

But first, lets cast the energy terms on the basis of per unit
weight of the fluid by dividing each term by g. This gives
$(V^2/2g + p/\gamma + z)$ as the energy per unit weight where γ is the
specific weight of the fluid. For water at 4°C, $\gamma = 9.8 \times 10^3 \text{ N/m}^2$,
and that for air at NTP is 12.6 N/m^2. Note that each term in
the above expression has a unit of length, the same as that of the
pressure head.

If we denote the energy lost per unit weight in traversing from
point 1 to point 2 by h_L (termed as the head loss), then

$$\left(\frac{V^2}{2g} + \frac{p}{\gamma} + z\right)_2 = \left(\frac{V^2}{2g} + \frac{p}{\gamma} + z\right)_1 - h_L \qquad (18.9)$$

The head loss h_L, represents the mechanical energy lost in
viscous action on a per unit weight throughout basis and is valid
only along a streamline and for steady and incompressible flows
only, but accounts for viscous losses.

The evaluation of h_L is not easy for it depends, in a very complicated manner, on the geometry of the flow arrangement, the
flow velocity and the properties of the fluid. Some approximate
methods of estimating head losses will be provided later, but
let us introduce first one more generalization of the engineering
energy equation.

We know that pumps, blowers and compressors increase the
energy of fluids, while a turbine extracts energy from the flowing fluid. If we express the work done by a pump or blower on
the fluid per unit weight throughput by h_s, Eq. (18.9) may be

modified as

$$\left(\frac{V^2}{2g} + \frac{p}{\gamma} + z\right)_2 = \left(\frac{V^2}{2g} + \frac{p}{\gamma} + z\right)_1 + h_s - h_L \qquad (18.10)$$

where h_s is the energy supplied - in by the pumps per unit weight throughput. It is termed as the head created by the pump, and has units of length, the same as that of any other pressure head. Here subscript 2 refers to the downstream or outlet conditions, and 1 to the upstream or inlet conditions.

A turbine extracts energy and, therefore, Eq. (18.10) should work in such case too with h_s taken as negative and representing the energy removed per unit weight of the fluid throughput. The exit energy in that case is the less than the input energy.

Example 18.6

A blower in a duct is used to create a jet of air of velocity 20 m/s. If the duct area is 0.5 m^2, determine the power of the blower employed (Fig. 18.16).

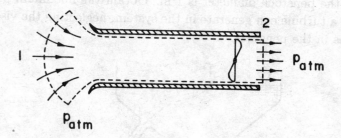

Fig. 18.16 A blower in a duct.

Solution

The blower sucks in air from stagnant ambient air and produces a jet emerging in the stagnant air. We take point 1 for upstream and point 2 on the same streamline just past the duct exit, and apply the engineering energy equation Eq. (18.10) between the two points to get

$$\frac{V_2^2}{2g} + \frac{p_2}{\gamma} + z_2 = \frac{V_1^2}{2g} + \frac{p_1}{\gamma} + z_1 + h_s - h_L \qquad (18.11)$$

Here, $V_1 = 0$, $z_1 = z_2$ and $p_1 = p_2 = p_{atm}$ as the jet is emerging in the ambient air whose pressure will be imposed on it. V_2 is given as 20 m/s, and γ is the specific weight of air $= 1.22(kg/m^3) \times 9.81(N/kg) = 11.96$ N/m^3 at 20°C. Neglecting viscous losses

$$\frac{(20)^2}{2 \times 9.8} = h_s = 20.4 \text{ (J/N)}$$

The blower therefore needs to input 20.4 J/N weight of air flowing through the duct. The rate of flow of air through the duct in weight terms is (specific weight) × (velocity) × (area of duct), i.e., 11.96 (N/m^3)× 20 (m/s)× 0.5 (m^2)=119.6 N/s. The blower power required is 20.4 (J/N) × 11.6 (N/S) = 2.44 kW, which is about 3.25 horsepower.

Example 18.7

Water from behind a dam passes through a tube (called a *penstock*) and runs a turbine. The difference in elevation between the reservoir level and the discharge point is 50 m (Fig. 18.17) and the penstock diameter is 1 m. Obtain the maximum power that a turbine can generate in the system, neglecting the viscous losses in the penstock.

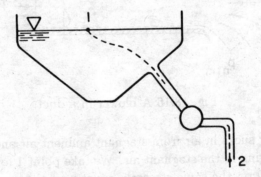

Fig. 18.17

Solution

Let us select a streamline with point 1 on the free surface of the reservoir and point 2 at the exit from the penstock. The various terms in the engineering equation Eq. (18.10) are $V_1 = 0$, $p_1 = p_2 = p_{atm}$, $z_1 - z_2 = 50$ m, and $h_L = 0$, so that

$$\frac{V_2^2}{2g} + \frac{p_2}{\gamma} + z_2 = \frac{V_1^2}{2g} + \frac{p_1}{\gamma} + z_1 + h_s - h_L$$

becomes

$$\frac{V_2^2}{2g} + z_2 = (z_2 + 50) - h_T$$

where h_s has been replaced by the negative of h_T, the head extracted by the turbine.

Thus,

$$h_T = 50 - \frac{V_2^2}{2g} \tag{18.12}$$

Clearly, as V_2 increases, h_T the head extracted by the turbine per unit weight of fluid decreases because more of the energy is appearing at the exit of the turbine as kinetic energy of the fluid. But increasing V_2 increases the volume throughout and may result in an increase in the power produced which is the rate of energy extraction per unit time. The power generated is therefore given by

$$
\begin{aligned}
W_T &= h_T \times \text{(weight throughput per second)} \\
&= h_T \times \text{(area) (velocity) (specific weight)} \\
&= h_T \times (\pi/4) \times (1)^2 \ (\text{m})^2 \times V_2 \times 9.81 \times 10^3 \ (\text{N/m}^3) \\
&= 7.7 \times 10^3 h_T V_2 \ (\text{N/m})
\end{aligned}
$$

where h_T is measured in m and V_2 in m/s. This gives the units of W_T as (N/m).m.(m/s)=(N.m/s)=J/s=Watt.

Use of the expression for h_T in terms of V_2 from Eq. (18.12) gives

$$W_T = 3.85 \times 10^6 V_2 - 3.92 \times 10^6 V_2^3 \tag{18.13}$$

The result is plotted as Fig. 18.18.

The power output of the turbine is zero when $V_2 = 0$, i.e., no flow takes place through the turbine. A maximum value of velocity $V_2 = 31.22$ m/s is obtained when extraction of head in the turbine is zero, all the potential energy in the reservoir being converted to kinetic energy at the exit of the tube. The

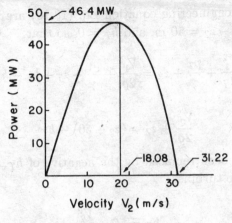

Fig. 18.18

same velocity is also obtained from Toricelli theorem wherein the losses and extraction are all assumed to be zero. The power then is obviously zero. Between these two limits of velocity V_2, the power extraction first increases and then decreases as shown. A value of 18.08 m/s for V_2 represents the velocity for the maximum rate of power output, which in this case is 46.6 MW. Ofcourse, the actual power output would be less than this because of the viscous and other losses within the penstock and the turbine which have all been neglected in this analysis.

The volume flow rate required for this output will be 18.02 (m/s) $\times (\pi/4) \times (1)^2$ (m)^2 = 14.2 m^3/s or 5.11×10^7 kg/hr.

Example 18.8

A household pump sucks water from a tubewell with water table 5 m deep and pumps it to an overhead tank 7 m above ground through a 25 mm tube. Find the power of the pump required for a modest delivery of 100 litre/minute. Neglect all losses.

Solution

Taking point 1 on the free surface of the well water, and point 2 at the tube exit in the overhead tank (Fig. 18.19), and datum at ground level, the value of the various terms in the engineering energy equation are $V_1 = 0$, $p_1 = p_{atm}$, $z_1 = -5$ m; $p_2 = p_{atm}$

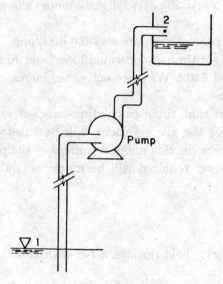

Fig. 18.19

and $z_2 = +7$ m.

The exit velocity V_2 is obtained from the flow rate as

$$V_2 = \text{(volume flow rate) / (pipe area)}$$
$$= \frac{100 \times 10^{-3}}{60} (\text{m}^3/2)/\frac{\pi}{4}(0.025))^2(\text{m}^2) = 3.40 \text{ m/s}$$

Using these value in the energy equation gives

$$\frac{(3.40)^2 \text{ (m/s)}^2}{2 \times 9.81 \text{ (m/s}^2)} + 7 \text{ (m)} = -5 \text{ (m)} + h_s$$

or

$$h_s = 0.59 + 7 + 5 = 12.59 \text{ m}$$

This should be the head produced by the pump. Note that this is dominated by the elevation head (12 m), and is little affected by the volume flow rate. A pumping rate of 250 litre/minute will increase it to 15.67 (m), only a 24% increase for 150% increase in flow rate. To calculate the power of the pump required (in absence of losses), multiply the head h_s by the rate of weight throughput, which is $(100 \times 10^{-3}/60)(\text{m}^3/\text{s}) \times 9.81 \times 10^3(\text{N/m}^3) = 16.35$ N/s. Thus, the power required for 100 litre/minute is 12.50

(m) × 16.35 (N/s)=205.5 W, slightly more than a quarter horse power.

But if the volume flow rate was 250 litre/min., (an increase of 150%) the weight throughout would have been 40.87 N/s and the power required 640.5 W, representing an increase of more than 200%.

Note again that these calculations neglect all losses in the pipe (which for the given velocity will be substantial) and the pump, and hence provide under-estimates of the power required. The actual power required will be much more than the values obtained here.

Problems

18.1 The velocity field through a 1-d device can be expressed as

$$V = 0.1 \sin(\pi/2t)(1 + 0.2x^2)$$

where t is measured in second, x in meter and the velocity is given in m/s. Is the flow steady? What is its velocity at $t = 1$ and $x = 2$ m? What is its acceleration at the same time and location ?

18.2 The velocity field is given in m/s as

$$V(x) = 0.32(1 + 0.01x^2)$$

where x is in metres. Is the fluid accelerating ? If yes, what is its acceleration at $x = 10$ m?

18.3 Water flows in a 5 cm diameter tube with a mass flow rate of 2 kg/s. Compute the average velocity.

18.4 Water flows between two parallel flat plates 1 cm apart. The velocity profile is given as

$$V(x) = 5[1 - (y/0.005)^2]$$

where V is in m/s and y is in m. Find the average velocity and the mass flow rate per metre width of the channel so formed.

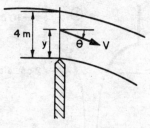

Fig. 18.20

18.5 The velocity V of water measured at various vertical positions over a 2-D obstacle shown in Fig. 18.20 is given below. Compute the flow rate per unit width.

y (m)	V (m/s)	θ (degree)
0.5	9	4
1.0	7.95	6.5
1.5	7.0	9.0
2.0	6.4	12
2.5	5.9	16
3.0	5.4	18
3.5	5.0	20

18.6 Water flows in a duct of reducing cross-section. If at a point where the velocity is 10 m/s, the pressure is 80 kPa, absolute, find the velocity to which it can be accelerated before cavitation starts, given that the vapour pressure at the given temperature is 8.2 kPa.

18.7 A jet emerges from a nozzle as shown in Fig. 18.21. If the pressure at section 1 is 100 kPa gauge, find the velocity and diameter of the jet 10 cm from the exit.

18.8 A tank is filled up to a depth of 3 m (Fig. 18.22). Two holes are punched, 1 m and 2 m from the free-surface. Find the distances x_1 and x_2 where the two jets would hit the ground.

18.9 A pump is used to pump water from a reservoir at ground to a height of 10 m. If the flow rate is 100 litre per minute

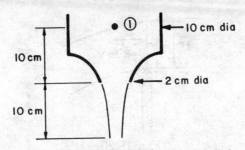

Fig. 18.21

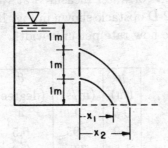

Fig. 18.22

and the discharge tube has a diameter of 5 cm, calculate
the minimum pump power required.

18.10 A water turbine draws from a reservoir and discharges 100
m below through a tube of diameter 2 m. If the flow rate is
25 m³/s, find the power developed by the turbine ? What
percentage of the potential energy of the stored water does
it represent?

19. SOME PRACTICAL APPLICATIONS

19.1 Measurement of Flow Rates

The straight forward method of finding the rate of flow of a liquid through a duct or a channel is to direct the liquid into a container for a measured interval of time and to determine the amount of liquid collected by weighing the container with and without the liquid. But this method is not convenient in most cases, and is not applicable to the gases. Therefore, other methods have been devised for finding the flow rates.

19.1.1 Venturimeter

It consists of a converging-diverging length of a pipe with two taps, A at the constant area inlet and B at the throat (Fig. 19.1). The outlet diverging portion is much longer than the inlet converging portion to reduce the pressure losses associated with expansions. The taps are connected to the two limbs of a U-tube manometer containing a gauge fluid (usually mercury) with density ρ_g.

Since the section at 1 (where tap A is located) has the same diameter as the original pipe, the velocity therein is the same, equal to Q/S_1, where Q is the volume flow rate. Velocity at section 2 (where tap B is located) has an increased value equal

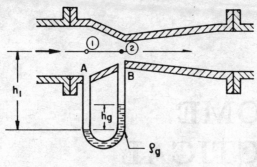

Fig. 19.1

to Q/S_2. Bernoulli equation between sections 1 and 2, gives

$$\frac{V_1^2}{2g} + \frac{p_1}{g} + z_1 = \frac{V_2^2}{2g} + \frac{p_2}{g} + z_2$$

Here $z_1 = z_2$, so that

$$\frac{Q^2}{2gS_1^2} + \frac{p_1}{g} = \frac{Q^2}{2gS_2^2} + \frac{p_2}{g}$$

or

$$Q = S_2\sqrt{\frac{2(p_1 - p_2)/\rho}{(1 - S_2^2/S_1^2)}} \tag{19.1}$$

The pressure difference $(p_1 - p_2)$ can be evaluated using the manometric formulation (Sec. 16.4). Thus,

$$p_2 = p_1 + \rho g h_1 + \rho_g g h_g - \rho g(h_1 - h_g)$$

or

$$(p_1 - p_2) = (\rho_g - \rho)g h_g \tag{19.2}$$

Using Eq. (19.2) in Eq. (19.1) gives

$$Q = \left[S_2\sqrt{\frac{2g(p_g/\rho - 1)}{1 - S_2^2/S_1^2}} \right]\sqrt{h_g} \tag{19.3}$$

Though this result has been obtained for the case of horizontal venturi, it can easily be verified that the same result holds even when it is inclined to the horizontal. Note that in Eq. (19.3), the expression within the square brackets is a constant for a given fluid and a venturi, so that the flow rate is proportional to the square root of the gauge height h_g.

This equation has been obtained for an idealized frictionless fluid. In any actual venturimeter, there would be some losses. This is accounted for by inserting an experimentally determined coefficient C_v in the equation such that

$$Q = C_v Q_{ideal}$$

$$Q = \left[C_v S_2 \sqrt{\frac{2g(p_g/\rho - 1)}{1 - S_2^2/S_1^2}} \right] \sqrt{h_g} \qquad (19.4)$$

Figure 19.2 shows a typical plot of C_v as a function of flow velocity expressed as Reynolds number $\rho V_2 d_2/\mu$, where subscript 2 refers to the throat conditions. For most venturimeters the value of C_v for practical speeds is larger than 0.95.

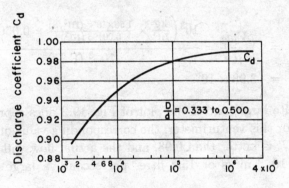

Fig. 19.2

Example 19.1

A venturimeter consists of a 125 mm diameter inlet reducing to a 50 mm throat. The gauge fluid used is mercury (sp.gr. 13.6), and the gauge height is 250 mm. Determine the flow rate of water through the venturimeter.

Solution

Substitute the following values in Eq. (19.3):

$$S_2 = (\pi/4).(0.050)^2 = 0.00196 \text{ m}^2$$

$$S_2/S_1 = (0.050/0.125)^2 = 0.16 \quad \rho_g/\rho = 13.6$$

$$h_g = 0.250 \text{ m.}$$

This gives

$$Q = (0.00196) \text{ m}^2 \sqrt{\frac{2 \times 9.81(\text{m/s}^2)(13.6 - 1)}{(1 - 0.16^2)}} \sqrt{0.250 \text{ (m)}}$$

$$= 1.56 \times 10^{-2} \text{ m}^3/\text{s}$$

This has to be corrected by the appropriate coefficient C_v. Let us calculate, just for the sake of curiosity, the non-dimensional throat velocity as Reynolds number.

$$\text{Re}_2 = \frac{\rho V_2 d_2}{\mu} = \frac{10^3 \left(\frac{\text{kg}}{\text{m}^3} \times \frac{1.56 \times 10^{-2} \text{ (m}^3/\text{s})}{0.00196 \text{ (m}^2)} \times 0.050 \text{ (m)}\right)}{10^{-3} \text{ (Pa.s)}}$$

$$= 3.98 \times 10^5$$

It is quite large, and if the plot of Fig. 19.2 does represent the values for this venturimeter, the corresponding value of the coefficient C_v is better than 0.98, and the actual flow will be about $1.52 \times 10^{-2} \text{ m}^3/\text{s}$ or 15.2 litres per second, a large flow rate indeed.

19.1.2   Orifice Plate Meter

This consists of a plate with a sharp-edged orrifice inserted in the pipe. The flow has to accelerate to pass through the reduced area orifice. The streamlines for this flow are as shown in Fig. 19.3.

After emerging from the orifice, the streamlines are still converging and it is only some distance later that the minimum flow area is reached. This area is called *vena contracta*. If the orifice area is S_o, the area A_2 at the vena contracta can be written as $C_c S_o$, where C_c is termed as the contraction coefficient. If

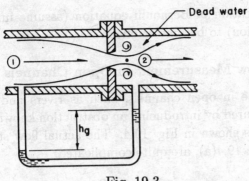

Fig. 19.3

we apply Bernoulli equation between the point 1 upstream of the orifice and the point 2 at the vena contracta, and write the pressure difference $(p_1 - p_2)$ in terms of the gauge height h_g on the manometer, we obtain exactly the same equation as for a venturimeter. Thus,

$$Q = \left[C_v S_2 \sqrt{\frac{2g(\rho_g/\rho - 1)}{1 - S_2^2/S_1^2}}\right]\sqrt{h_g}$$

$$= \left[C_v C_c S_o \sqrt{\frac{2g(\rho_g/\rho - 1)}{1 - (C_c S_o/S_1)^2}}\right]\sqrt{h_g}$$

In a typical orifice meter, $S_o < S_1$ so that $(C_c S_o/S_v)^2 << 1$, and

$$Q = C_v C_c S_o \sqrt{2g(\rho_g/\rho - 1)} . \sqrt{h_g} \qquad (19.5)$$

The product of velocity coefficient C_v and contraction coefficient C_c is written as the discharge coefficient C_d. For a given orificemeter it is plotted as the function of the Reynolds number $\rho V_1 d/\mu$, a dimensionless velocity. The typical range of C_d is between 0.6 to 0.8, suggesting that an orifice plate meter introduces much larger losses than a venturimeter. But since an orifice plate is much cheaper to manufacture and install, it is widely used in situations where the variations in flow rates are relatively small.

Note that an orifice plate or a venturi should be installed in a pipe after a fairly long (say about 20 diameters) section of straight pipe. Valves, bends, expansion and reducers, all introduce eddies, swirls and other irregularities in the flow causing the simple application of Bernoulli equation (assume uniform flow across a section) to become erroneous.

19.1.3   Flow Measurement in Open Channels - Weirs

The flow rate in open channels, such as rivers and canals, is usually measured by introducing an obstruction known as a *weir* in the flow as shown in Fig. 19.4. The actual flow streamlines, shown in Fig. 19.4(a), are quite complicated.

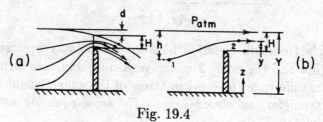

Fig. 19.4

It is possible to idealize this flow picture by the one shown in Fig. 19.4 (b). The velocity at a height y above the crest of the weir can then be found by applying the Bernoulli equation on the streamline from point 1 to 2. But first, it should be noted that the vertical velocities and accelerations far upstream of the weir are small and therefore the pressure variations there can be taken as essentially hydrostatic, that is,

$$p_1 = p_{atm} + \rho g h$$

Bernoulli equation between points 1 and 2 now gives

$$\frac{V_2^2}{2g} + \frac{p_2}{\rho g} + z_2 = \frac{V_1^2}{2g} + \frac{p_1}{\rho g} + z_1$$

or

$$\frac{V_2^2}{2g} + \frac{p_{atm}}{\rho g} + (Y - H + y) = \frac{p_{atm} + \rho g h}{\rho g} + (Y - h) \quad (19.6)$$

since pressure in the free jet at Section 2 will be atmospheric all across, and the velocity V_1 is negligible small compared to V_2. This gives

$$\frac{V_2^2}{2g} = H - y$$

or

$$V_2 = \sqrt{2g(H - y)}$$

Thus, the velocity at point 2 varies with the location of the point in the vertical plane at the weir location. The total discharge rate can then be readily found by integration. Thus,

$$Q = \int_o^H V_2(y)b\,dy$$

where b is the width of the weir.

Introducing $V_2(y) = \sqrt{2g(H - y)}$ in the above equation, the indicated integration gives

$$Q = \frac{2b}{3}\sqrt{2gH^3} \qquad (19.7)$$

Thus, by simply measuring H, the depth of flow over the weir (a simple length measurement), one can determine the flow rate in the channel.

To account for the various assumptions involved, we multiply the ideal flow rate given by Eq. 19.7 by C_d, a discharge coefficient, to obtain the actual flow rate. The value of C_d is determined experimentally and is typically about 0.6.

19.2 Measurement of Flow Velocity at a Point

The discharge rate obtained through methods such as those described in Sec. 19.1 give only the gross information from which the average velocity across the channel cross-section can be obtained. For determining the details of the velocity variations across the section, we need some other device. Various such devices or techniques are available, some more sophisticated than others. The simplest and perhaps the most widely used device consists of a bent tube inserted head-on into the flow as shown in Fig. 19.5. Its open end is carefully aligned normal to the flow and the other end of the tube is connected to a pressure measuring device such as a manometer. This is called a *Pitot tube*.

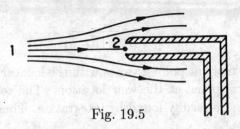

Fig. 19.5

Consider the streamline at the centre of the tube. Let the velocity at point 1 be the velocity V, and the local pressure be p. Since the other end of the tube is blocked by some pressure measuring device, the flow cannot proceed down the tube indefinitely, and it comes to rest such that $V_2 = 0$. The pressure at V_2 is read off from the manometer and is termed as the stagnation pressure p_s, since fluid at point 2 is at rest, i.e., stagnant. Clearly then,

$$\frac{p}{\rho g} + \frac{V^2}{2g} + z_1 = \frac{p_s}{\rho g} + 0 + z_2$$

Since $z_1 = z_2$

$$V = \sqrt{2\rho(p_s - p)} \tag{19.8}$$

Thus, measurement of the pressure difference $(p_s - p)$ gives the local velocity V.

A combination probe, known as the Pitot-static tube gives the pressure difference $(p_s - p)$ directly when connected to a differential manometer. It consists of two concentric tubes, the central tube opening to the flow and acting as the Pitot tube, while the outer tube has orifices some distance down-stream from the front end (Fig. 19.6). The pressure sensed by this tube is the actual pressure in the flow. This pressure is termed as the static pressure (hence the name, static tube). From the gauge height h_g on the differential manometer, the pressure difference $(p_A - p_B)$ can be calculated as

$$p_A - p_B = (\rho_g/\rho - 1)\rho g h_g$$

where ρ_g and ρ are the densities of the gauge fluid and the flowing fluid, respectively. Then the velocity of the flow is given by Eq. (19.8) as

$$V = \sqrt{2\rho g(\rho_g/\rho - 1)h_g} \tag{19.9}$$

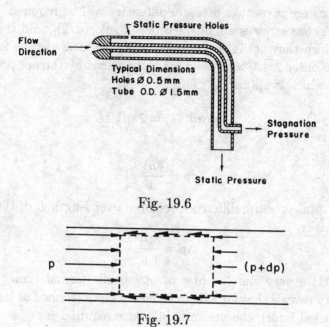

Fig. 19.6

Fig. 19.7

19.3 Flow through Pipes

Fluids are carried from one location to another through pipes. If the elevation of the fluid rises in the process, one would obviously need to apply a higher pressure at the upstream end than at the downstream end. The required pressure difference can be obtained by Bernoulli equation due to the fact that the flow velocity through the pipe will be constant if the cross-section of the pipe does not change. Thus, neglecting viscous losses and taking points 1 and 2 at the entrance and the exit of the pipe, respectively,

$$p_1 - p_2 = \rho g(z_2 - z_1)$$

If $z_2 > z_1$, a pump will be required to create the positive pressure difference $(p_1 - p_2)$.

The neglect of viscous losses is a very bad assumption in pipes. Consider a horizontal length of pipe of diameter d carrying a flow with an average velocity V as in Fig. 19.7. Take a slug of fluid of length $\delta \ell$ inside the pipe. If viscous stresses are neglected, no pressure difference is required to maintain a flow. But if viscous stresses are present, a pressure difference will be required. Since the viscous stresses are of the order of $\mu[V/(d/2)]$, and the area on which they act is of the order $\pi d\delta\ell$, the total viscous force is of the order $2\pi\mu V\delta\ell$. This must be balanced by the net pressure force which is $\delta p(\pi d^2/4)$. Thus,

$$\delta p(\pi d^2/4) = 2\pi\mu V\delta\ell$$

or

$$\delta p = \frac{8\mu V}{d^2}\delta l$$

The total pressure difference required over length L of the pipe then is

$$\Delta p = \frac{8\mu V L}{d^2}$$

This is a very simple order of magnitude formulation. Exact theory reveals that for a certain class of flows (termed as laminar, discussed later), the pressure difference required is

$$\Delta p = \frac{32\mu V L}{d^2} \qquad (19.10)$$

This represents a decrease in pressure along the flow direction due to the viscous action as a fluid flows through a length L of a pipe. It is most convenient to express this equation as

$$\frac{\Delta p}{\frac{1}{2}\rho V^2} = 64(\frac{\mu}{\rho V d})(\frac{L}{d}) \qquad (19.11)$$

If we recognize $\rho V d/\mu$ as the Reynolds number Re, the non-dimensional velocity measure introduced in the last section, Eq. (19.10) becomes

$$\Delta p = \frac{64}{\text{Re}}\cdot\frac{1}{2}\rho V^2\cdot\frac{L}{d} \qquad (19.12)$$

This result is known as Hagen-Poiseuille law and is valid for low values of Reynolds numbers (less than about 2200) and for sufficiently long tubes (L/d greater than about 100). At higher values of Reynolds numbers, the pressure drop is much larger than predicted by this equation.

Example 19.2

A fine capillary of bore 0.5 mm is used as a flow measuring device
in the arrangement as shown in Fig. 19.8. The head difference

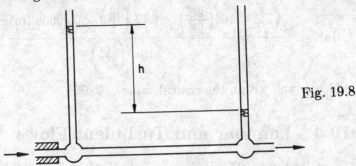

Fig. 19.8

as measured by piezometric tubes across a 10 cm length of this
capillary is 15 cm of water. What is the flow rate through the
tube?

Solution

Since the tube diamter D is quite small, $L/d = 200$, and the
Reynolds number is likely to be small such that Eq. (19.12) may
apply. A head difference of Δh corresponds to a pressure difference $\Delta p = \rho g \Delta h$, where ρ is the density of the fluid. Using
Eq. (19.12) or its equivalent from Eq. (19.10), the pressure drop
is obtained as

$$\Delta p = \frac{32 \mu V L}{d^2}$$

or

$$V = \frac{\rho g \Delta h d^2}{32 \mu L}$$

and the volume flow rate

$$Q = \frac{\pi d^2}{4} . V = \frac{\pi \rho g \Delta h d^4}{128 \mu L}$$

This shows that volume flow rate is linear in Δh, the head drop
across the capillary. For the data given, the flow rate is

$$Q = \frac{\pi \times 10^3 (\frac{kg}{m^3}) \times 9.81 (\frac{N}{kg}) \times 0.15 \, (m) \times (0.0005)^4 \, (m^4)}{128 \times 10^{-3} (\frac{Ns}{m^2}) \times 0.10 \, (m)}$$

$$= 2.26 \times 10^{-8} \text{ m}^3/\text{s} = 2.26 \times 10^{-2} \text{ ml/s} = 1.35 \text{ ml/min}$$

which is a very low rate indeed. The corresponding flow velocity is 1.72 m/s, and the Reynolds number is

$$\text{Re} = \frac{\rho V D}{\mu} = \frac{10^3 \left(\frac{\text{kg}}{\text{m}^3} \right) \times 1.72 \left(\frac{\text{m}}{\text{s}} \right) \times 0.00055 \text{ (m)}}{10^{-3} \left(\frac{\text{Ns}}{\text{m}^2} \right)} = 860$$

which is well within the critical value of 2200.

19.4   Laminar and Turbulent Flows

To understand the difference in behaviour of fluids at low and high speeds, consider the following experiment first performed by the famous British scientist Osborne Reynolds in the year 1883.

Water is made to flow through a uniform-bore long glass tube with the pressure difference across the two ends maintained constant (Fig. 19.9). The entrance to the tube is well-rounded. To

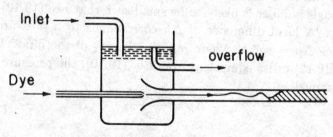

Fig. 19.9

visualize the nature of the flow, a thin filament of dye is introduced at the centre-line of the tube. When the velocity of the flow is small, the dye filament appears as a straight unbroken line spanning the entire length of the tube. But as the velocity is increased (by increasing the level of water in the head tank), the filament first becomes a bit wavy and then breaks down completely. The dye then fills the entire flow area.

The flow is said to have undergone a transition from the orderly 'laminar' regime, where it moves smoothly, to the 'turbulent' regime, wherein there is considerable amount of the random fluctuations. The laminar motion at low velocities keeps the dye filament intact, whereas the random motion of the turbulent flow forces it to break down. Careful experiments with different tubes and different fluids have established that though the transition occurs at different values of velocities, but when the values are non-dimensionlized, as Reynolds number $\rho V d/\mu$, the transition is seen to occur at almost a constant value of the Reynolds number, irrespective of the actual diameter of the tube, or the values of the fluid properties ρ and μ.

Reynolds number Re= $\rho V d/\mu$, plays a significant role in deciding whether the flow is laminar or turbulent. For flow through pipes, the transition value of the Reynolds number $\simeq 2300$, signifying that for Re < 2300 the flow is laminar while above this value it is generally turbulent.

19.5   Pipe Losses in Turbulent Flows

The Hagen Poiseuille relation Eq. (19.12) is valid only when the flow through a pipe is laminar, i.e., Re= $\rho V/d\mu$ is less than about 2300. For larger values of Re, the flow is generally turbulent and this equation no longer applies. Careful experiments have established that in such cases, the pressure drop Δp across a pipe of length L can be written as

$$\Delta p = f \frac{1}{2}\rho V^2 \frac{L}{d}$$ (19.13)

where f is constant for very large values of Re, and depends only on the roughness of the pipe walls. The coefficient f is termed as the friction factor.

Note that Eq. (19.12) for laminar flow has been casted in the same form, and Eq. (19.13) can be applied to laminar flow with

$$f = 64/\text{Re}$$ (19.14)

Thus, the value of friction factor f is constant for very large values of Re, and is 64/Re for low values of Re. For intermediate velocities, the value of f can be read off from experimentally determined plots such as the Moody chart shown in Fig. 19.10.

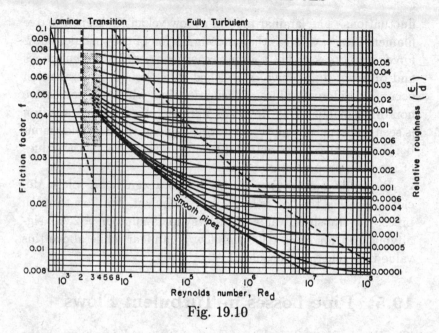

Fig. 19.10

Here, the friction factor is read off against relative roughness ϵ/d of the walls, where ϵ is the mean roughness height on the tube walls. The example below illustrates the use of Moody chart in calculation of the pressure losses.

Example 19.3

Water flows in a garden hose of internal diamter 12 mm and length 30 m. If the flow rate is 0.2 litre/s, determine the pressure in the mains. Take ϵ of the hose walls as 1.5×10^{-4} m (0.15 mm).

Solution

To find the pressure difference by Eq. (19.13) we need the value of V and of friction coefficient f. The value f can be read off the Moody chart of Fig. 19.10 if we know the value of Reynolds number based on diameter Re_d and of the relative roughness ϵ/d.

The value of velocity V is found from the flow rate Q and the pipe diameter d as

$$V = \frac{Q}{\pi d^2/4} = \frac{0.2 \times 10^{-3}(\frac{\text{m}^3}{\text{s}})}{\pi \times (0.012)^2(\text{m}^2)/4} = 1.77 \text{ m/s}$$

The value of the Reynolds number based on diameter then is

$$\text{Re}_d = \frac{\rho V D}{\mu} = \frac{10^3(\frac{kg}{m^3}) \times 1.77 \left(\frac{m}{s}\right) \times 0.012 \, (m)}{10^{-3} \, (Pa.s)} = 2.12 \times 10^4$$

Value of relative roughness coefficient ϵ/d is $1.5 \times 10^{-4}(m)/0.012$ (m) = 0.0125. From Fig. 19.10, the value of f corresponding to $\epsilon/d = 0.0125$ and Re= 2.12×10^4 0.043. Using this value in Eq. (19.15), the pressure drop is obtained as

$$\Delta p = f \frac{1}{2} \rho V^2 \frac{L}{d}$$

$$= 0.043 \times \frac{1}{2} \times 10^3(\frac{kg}{m^3}) \times (1.77)^2(\frac{m}{s})^2 \times \frac{30(m)}{0.012(m)}$$

$$= 1.7 \times 10^5 \, Pa$$

Converting it into the head of water,

$$h_L = \frac{\Delta p}{\rho g} = \frac{1.7 \times 10^5(Pa)}{10^3(\frac{kg}{m^3}) \times 9.81(\frac{N}{kg})} = 17.4 \text{ m of water head}$$

Problems

19.1 A venturimeter with a throat diamter of 0.1 m is fitted with a 0.2 m dia. horizontal pipe carrying water at 0.04 m^3/s. If the pressure in the pipe is 2.5 bar *gauge*, determine the pressure at the throat.

19.2 What is the rate of flow of petrol through a venturimeter placed at 30° to the horizontal if the level difference in a U-tube mercury manometer is observed to be 10 cm? The pipe dia. is 5 cm and the throat dia. is 3 cm. Assume a discharge coefficient of 0.94.

19.3 The mean velocity of water in a 15 cm dia. horizontal tube is 5 m/s. What pressure difference would be read in an orifice meter with a 10 cm orifice? Assume $C_d = 0.61$ and $C_c = 0.70$.

19.4 The discharge over a weir is 0.14 m^3/s when water level is 23 cm above the sill. Assuming a C_d of 0.6, calculate the width of the sill required.

19.5 A pitot-static tube is used to measure the velocity of water in a pipe. The stagnation pressure head recorded is 3 m when the static pressure head is 2 m. What velocity is being indicated?

19.6 A pitot survey over a radius of 20 cm pipe yields the following data.

r/R	:0	0.2	0.4	0.6	0.8	0.9	1.0
(m)	:5.0	4.75	4.35	3.50	2.50	1.50	0

Estimate the discharge through the pipe. [Hint: Flow rate

$$Q = 2\pi \int_o^R Vr\,dr = \pi \int_o^R v\,d(r^2)$$

Plot V against r^2 and find the area under the curve to obtain $\int_o^R V d(r^2)$].

19.7 A 20 cm discharge pipe 30 km long transports oil from a tanker to the shore at $0.0.1 m^3/s$. Determine if the flow is laminar or turbulent. Take $\mu = 0.1$ Pa.s and $\rho = 900$ Kg/m^3 for oil.

19.8 In the pipe flow of problem 19.7 above, find the pumping power required.

19.9 Water is discharged from a tank maintained at a constant head of 5 m above the exit of a straight pipe 100 m long, 15 cm diameter. Estimate the flow rate if the friction factor is given as 0.01.

19.10 A fireman with an 8 cm dia. hose needs a water jet with a speed of 16 m/s. If the length of the hose is 30 m and the nozzle dia. is 4 cm, estimate the pumping head required, if $\epsilon = 0.008$ m.

19.11 In the above example if the nozzle diameter is increased to 6 cm, with f and the head remaining constant, what will be the velocity at the jet exit?

20. INTRODUCTION TO THERMODYNAMICS

20.1 Thermodynamics

Thermodynamics is a science that deals with energy, matter, and with the interactions between these two. Historically, the science of thermodynamics developed to provide a better understanding of the working of the machines, such as steam engines and automobile petrol engines, which converted the energy content of fuels to mechanical energy which could do useful work. The science has come a long way since those early days, and is now recognized as a branch of fundamental knowledge which finds applications in as diverse fields as mechanical engineering, electronics, chemical engineering, materials science, physics, chemistry, and even life sciences. It is now seen more as a science which deals with conditions of equilibrium of material systems and the processes that take place when this equilibrium is disturbed.

Some of the great scientists and engineers who have contributed to its development from an engineering of limited applications to a fundamental science of general application are Sadi Carnot, James Joule, Lord Kelvin, Rudolf Clausius, Max Planck, J W Gibbs, Clapeyron, etc.

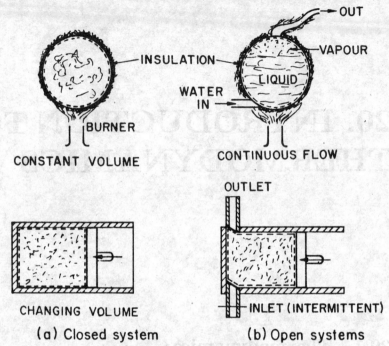

Fig. 20.1 Some examples of closed and open systems.

20.2   Thermodynamic Systems

Whenever we study the equilibrium of a body in mechanics, we draw a "free body", and identify all the forces that act on it. We then apply the basic physical laws to obtain conclusions regarding its equilibrium or otherwise.

In the study of thermodynamics, which involves interactions of matter and energy, we resort to a similar technique. We define a system as a region in space confined within a boundary. Anything within the boundary is part of the *system* and anything outside it is termed as the *surroundings*. The boundary may be a real one, as the walls in a rocket motor, or it may be imaginary, as the one enclosing all the gas that comes out of a balloon when it bursts.

We classify the systems depending upon the nature of the wall and the interactions that it permits between the matter within the system and that without, i.e., the surroundings.

A system may be a *closed* or an *open* system. If the boundary of the system is impervious and does not allow any matter to cross it, it is termed as a closed system. Here, the matter within the system is always the same and, therefore, the simplest conclusion is that its mass remains constant.

Figure 20.1(a) shows two closed systems on the left. Note that in one of them the boundary of the system is itself moving - but since the identity of the gas contained in it is fixed, and there is no exchange of mass with the surroundings, the system is a closed one.

An *open system*, on the other hand, has boundaries that permit exchange of mass, as is shown in Fig. 20.1(b). In the figure at top, the inside of the flask is defined as a system (of fixed volume). On heating, vapour exits through the opening, and hence the system is an open one. In the lower figure on the right, the volume of the system also changes as the piston moves, but the system is an open one because the matter enters or leaves this system of variable volume.

In addition to the matter, energy can also be transferred across the boundaries of a system. The simplest example is the conduction of heat across metal walls of the two flasks shown in Fig. 20.1. Energy can also be transferred across the boundaries into or out of a system as work, discussed later.

A closed system is termed as an *isolated system* if there is no energy exchange between the surroundings and the matter contained within the system. Thus, for all practical purposes, the existence of surroundings is immaterial for an isolated system, since it neither exchanges energy nor matter with it. Any interactions in the case of an isolated system are restricted to those between parts or sub-systems of the isolated system itself.

20.3 Thermodynamic Properties and State of a System

All distinguishing characteristics of a system are called its *properties*. They are quantities that must be specified to define the state of system. The properties which are relevant to describe the thermodynamic state of a system are termed as *thermodynamic properties.* Some of the more important thermodynamic properties are mass, density, pressure, volume, temperature, energy and entropy. We will consider some of these in detail.

The properties of a system may change with time, and as they change, we say that the state of the system is changing. The thermodynamic state of a system is then characterized by a set of unique value of its properties. We will shortly learn that the values of only a few properties need to be specified for the state of the system to be specified completely. It follows, then, that the values of the other properties can be uniquely determined from these specified properties. Take for example a simple compressible substance such as a perfect gas. We need to specify only two of its properties to fix it state. Thus, if the temperature and the density are known, we can find all its other properties such as temperature, internal energy, entropy etc..

The properties of a system may be classified as *intensive* or *extensive*. Those properties which are independent of the mass contained in the system are the *intensive* (from intensity) properties. Such as temperature, density and specific volume (or for that matter all *specific* properties). All other properties are *extensive properties*. Thus, the total volume V of a gas is its extensive property. Given pressure and temperature, we can find the density and, thus, the specific volume without the knowledge of the mass of the gas. But the total volume V depends also on how much matter is contained in the system. As a convention, we use lower-case letters for specific or intensive properties, and capital or the upper-case letters for the corresponding extensive properties. For example, if V is the total volume of a gas, then v is specific volume, i.e., volume per unit mass.

20.4 Thermodynamic Equilibrium

As we know, the thermodynamic state of system changes continually. Under certain circumstances, these changes are negligibly small and the properties of the system are appreciably unchanged with time, with the system exchanging neither mass nor energy with its surrounding (even though the system boundaries may permit it), or between parts of the system itself. The system is, then, said to be in a state of *thermodynamic equilibrium*, characterized by uniformity in the values of the intensive properties throughout the system. Thus, a system in thermodynamic equilibrium will have throughout it the same pressure (precluding mechanical changes), same temperature (precluding heat transfer by conduction or convection), and same chemical composition (precluding intra-species diffusion).

The concept of equilibrium is an important one and is only an abstraction. An isolated system left to itself, i.e., without any external interference, will ultimately reach an equilibrium state when all internal adjustments cease, and the system no longer has a tendency to change its state or the states of any of its component parts.

A system without any unbalanced forces, i.e., with constant pressure throughout, is said to be in *mechanical equilibrium*. A system without any differences in temperatures over it does not exchange heat and is said to be in *thermal equilibrium*. Similarly, a system with uniform chemical composition throughout is said to be in *chemical equilibrium*.

20.5   Thermodynamic Processes

When a system changes from one equilibrium state to another, it is said to have undergone a *process*. A study of processes plays a very important role in engineering, since what is engineering if not the art of changing the state of the system in a desired and controlled manner!

Many different kinds of processes are recognized based on the kinds of interactions that are taking place. For example, an *isothermal* process is one in which the temperature of the system does not change as the process occurs. An *adiabatic* process, on the other hand, is one in which no heat is exchanged between the system and its surroundings as a result of which the temperature of the system generally undergoes a change.

If a process is carried out in such a manner that at every instant of time the system departs only infinitesimally from its equilibrium, the process is termed as *quasi-static*. This can be visualized as an infinite sequence of infinitesimal changes from the initial state to the final state. This process can then be plotted as a sequence of states on a property value chart (state diagram) as in Fig. 20.2. The curve connecting the various quasi-static states,

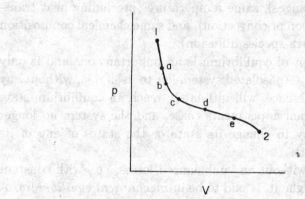

Fig. 20.2 A thermodynamic process.

as the system proceeds from state 1 to the final state 2, is termed as the *path* of the process. Each point on the path represents the intermediate states that the system passes through.

Note that a path can be specified only if each point along it is in equilibrium state. But for a change in state to take place, the system must be displaced from its equilibrium, and therefore, it cannot be in equilibrium all along the path. In any real process the intermediate states would not be equilibrium states, but if the process is considered to take place infinitely slowly, the process may be assumed to occur through a succession of quasi-static states. In any fast process, such as throttling, the departure from equilibrium is significant, and hence the path of the process cannot be specified. Only the end states can be noted in such a case.

20.5.1 Reversible Process

A process is said to be *reversible* if, after it has been carried out, it is possible to undo it completely such that both the system and the surroundings return to their original states.

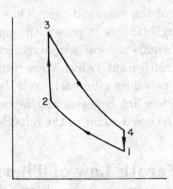

Fig. 20.3 A thermodynamic cycle.

Reversibility of a process requires that it proceeds through a succession of quasi-static states. Any deviation from equilibrium leads to irreversibility of the process.

Note that the above requirement of reversibility implies that the process takes place infinitely slowly. For example, if transfer of heat occurs in a process, the condition of quasi-equilibrium will demand that heat be transferred across an infinitesimal temperature difference which would render the rate of heat transfer infinitesimally small, requiring very large times for the transfer of a finite quantity of heat to take place.

All practical processes are, strictly speaking, *irreversible*. Reversibility is only a thermodynamic idealization, and is used to define the most efficient processes possible. The presence of friction, viscosity, electrical resistance, all contribute to making a process irreversible.

20.5.2 Cyclic Process or Cycle

A *cycle* is simply a sequence of processes that results in the final state of the system to be identical to its initial state, i.e., the net change in *any property* of the system over the whole cycle is exactly zero. Figure 20.3 shows a cycle consisting of four distinct processes.

A process reversed is a special type of cycle in which the end-states of both the systems and the surroundings are the same. In general, a completed cyclic process may result in net changes in the properties of the surroundings. Thus, the refrigerant in an air-conditioning plant as it passes through the compressor, condenser, throttle and evaporator undergoes a cycle in which the properties of the refrigerant (which constitutes the system) are restored to the same values after each cycle, but the surrounding are continually undergoing changes, with heat being continuously pumped out of the cooled room to the outside.

20.6 The Zeroth Law of Thermodynamics and the Concept of Temperature

Like most of the basic laws of science, the zeroth[1] law of thermodynamics is based on generalization and codification of common human experience. The law states that *when two systems A and B are in thermal equilibrium with a third system C* (i.e., no heat exchange takes place when they are brought in thermal contact with system *C*), *they are also in thermal equilibrium with each other.*

This law permits us to introduce the concept of temperature as a measure of the thermal state of the system, such that the systems at same temperature are in thermal equilibrium. A temperature scale is so introduced that a system at higher temperature loses heat to a system at lower temperature.

The Celsius scale of temperature has its zero defined as the temperature of melting ice. The temperature of boiling water (at standard atmospheric pressure) is assigned a temperature of 100°C.

A more fundamental scale of temperature is the Kelvin scale where the zero is fixed at the absolute and the theoretically minimum possible temperature of a substance where all molecular activity ceases. This is determined to be approximately -273 °C. Thus, 0 K temperature on the absolute scale corresponds to -273 °C. One unit temperature difference on the two scales are iden-

---

[1]This strange nomenclature of the law is simply the result of the fact that, historically, first and the second laws of thermodynamics had established their nomenclature extensively before the need was felt to introduce this as a law, more fundamental in nature than the first and the second laws.

tical. Thus, equivalent temperature on the Kelvin scale can be obtained by adding 273 to the temperature value on the Celsius scale. To summarize, 0 K = -273 °C 273 K = 0 °C 373 K = 100 °C

100 K temperature difference = 100°C temperature difference.

20.7 Basic Principles of Thermodynamics

The structure of the science of thermodynamics rests on only a few basic principles. Whenever an engineering system is analyzed from the point of view of thermodynamics accounting for the mass, energy, and entropy in the system is essential.

One of the basic axioms of thermodynamics is the principle of conservation of mass. Mass of a closed system, as defined earlier, remains constant. The mass of an open system (in which matter can cross the boundaries of the system) can of course change, but the *net* change must be equal to the net *influx* of mass into the system, thereby satisfying the principle of conservation of mass.

The accounting for energy is a bit more complex. The first law of thermodynamics, discussed in detail in the next chapter, permits development of strategies through which the net energy of the universe can be treated as constant, so that the change in the energy content of a system is equal to the net flux of energy in its various forms across the system boundaries.

Entropy is a new property introduced in thermodynamics, and its consideration helps in determining whether a thermodynamic process is reversible or irreversible, spontaneous or otherwise. Also, the considerations of entropy shed a new light on the efficiencies of the energy conversion processes. The second law of thermodynamics, another axiomatic law, governs the entropy transactions and plays a very important role in all branches of engineering.

The development of thermodynamics consists, in fact, of casting of the above three principles into a multiplicity of equivalent forms, each convenient for a specific application.

While applying the three principles, we need to determine the values of the properties of the various substances involved at their different states. One of the important simplifying feature is known as the *state postulate* which restricts the maximum number of properties that need to be specified to fix the state of a substance completely. Any other property can then be determined by using what are called the *equations of state*. For example, a gas needs only two properties to be specified independently, say, its pressure and temperature. The density of the gas can then be obtained from the values of pressure and temperature using an appropriate *equation of state*. For a perfect gas we known the equation of state as

$$p = \rho RT \qquad (20.1)$$

where R is the gas constant for the particular gas.

This is a very simple equation of state, and can predict the density approximately only under the restrictive conditions of high temperatures and low pressures. Various other equations of states for gases are available, each only an approximation. One such is the Van der Wall equation of state

$$\left(p + \frac{a}{v}\right)(v - b) = RT \qquad (20.2)$$

where v is the specific volume and a and b are constants that account for the intermolecular attractions, and the minimum volume occupied by a unit mass of the gas in its dense (or liquid) state. This equation has a wider range of validity than the perfect gas equation.

Note that any equation which relates properties of a substance at equilibrium is termed as an equation of state. There is a wide variety of such equations available. The skill consists in selecting the equation appropriate for the problem at hand.

Problems

20.1 A control volume is defined as a region in space. Under what conditions will it be a system? An isolated system?

20.2 Using the equation of stae of a perfect gas, find the specific volume and density of air at 100 kPa and 300 K.

20.3 "Head and energy bear the same relation to a system as rain and water do to a reservoir" Explain the analogy.

20.4 Can the thermodynamic state of a uniform substance be changed without any energy transfer?

20.5 A simple compressible substance is made to undergo first an adiabatic expansion, then a constant pressure compression, and finally a constant volume pressurization which returns the substance to its original state. What is the total change in internal energy for this cycle? Is there a net transfer of energy as work or as heat?

20.6 The normal temperature of the human body is 98.6 °F. What is it in °C and in K?

20.7 Consider a heavy insulated container containing two gases separated by a movable adiabatic partition. Does this constitute a system? (An isolated system?) What kinds of interactions does the partition allow between the two parts?

20.8 Give two examples of reversible (or nearly so) processes from around you.

20.9 Give two examples of patently irreversible processes.

20.10 Give an example of a cyclic thermodynamic process.

21. ENERGY AND THE FIRST LAW

21.1 Principle of Energy Conservation

Historical development of the concept of energy conservation presents an interesting story of how, starting from very primitive observations, physical laws with breath-taking generalities emerge.

The origin of the concept of conservation of energy lies in the common observation that as a force is applied to a moving body, its speed changes, and the change in the kinetic energy of the body is *equal* to the amount of work done by the applied force. In other words, the amount of work done on a body results in an increase in the energy content of the body, and that the energy content of body consists of its kinetic energy (**KE**).

21.1.1 Energy Content

The next development consisted of realizing that the best way to handle the work done by some common force fields (such as gravity of the earth) is to postulate a potential energy (**PE**), the change in which is nothing but the work done by the force field.

The motion of bodies near the earth (like projectiles) were then handled by a simple energy principle which stated that the *total* energy content of a body, which now consists of its kinetic and potential energies, remains constant if there are no external forces (other than those accounted for in the calculation of potential energy) acting on the body. In the presence of other external forces, this statement should read as:

(*Changes in KE + PE = Work done by external forces*)

Remember, that the work done by forces, such as gravity, which contribute to the potential energy is *not* to be included on the right-hand-side of this equation.

Soon limitations of even this equation were apparent. A ball of putty dropped from a height onto a floor comes to rest. What happens to the potential energy it has before it is released? This cannot be explained away by losses, for what kind of conservation is it that allows for losses?

The difficulty is resolved by recognizing that there is one more way of storing energy in a body: the incessant motion of molecules constituting the mass of body hold an enormous amount of kinetic energy not included in the KE of the body calculated on the basis of its *bulk* or macroscopic motion.

It was realized that this energy hidden inside the body can be sensed as the temperature of the body. The molecules move (motion includes linear, spin and vibrational motions) more vigorously at higher temperatures. Thus, the concept of internal energy was introduced to account for these microscopic motions, and the principle of energy conservation was modified to read as: (*changes in kinetic energy, KE + potential energy, PE + internal energy, U = the net energy received by the body from the surroundings.*)

21.1.2   Energy Interactions

Until this development, only one mode of transferring energy to the system was considered, that of doing work by an external force. Consider, for example, a gas contained in a cylinder with a movable piston of area A (Fig. 21.1). If the pressure of the gas at any instant is p, then the external force on the piston should be pA for equilibrium. Considering a quasi-static process, the external force during compression must be only infinitesimally greater than pA. The work done in quasi-static compression, is obtained as

$$\int pA dx$$

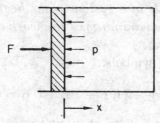

Fig. 21.1 A piston moving in a cylinder against the pressure of the gas.

where x is the motion of the piston. $A dx$ represents the change in volume $-dV$, negative because it is compression. Therefore, the work done in a quasi-static process of compression, (being reversible as explained Chapter 20), is given by

$$- \int p dV$$

This is the work done by external forces in a reversible compression, and clearly it must increase the energy content of the gas. Since the PE and KE of the gas in this operation is not changing, it must be reflected in an increase in the internal energy U, which appears to the senses as an increase in the temperature of the gas.

Since we allow for an increase in temperature, and the temperature of a system also increases by transfer of heat by conduction, it stands to reason that we must also include heat transfer as a mode of transfer of energy to the system from the surroundings. In a very famous series of experiments, Henry Joule established the equivalence of heat and work as energy. Since the beginnings of thermodynamics were with the analysis of heat engines which used heat as input to produce work, it is considered convenient to treat the energy transferred to a system, as heat, as a positive quantity. On the other hand, the work done by a system on the surroundings is treated as a positive quantity.

21.1.3 First Law of Thermodynamics

With the above convention for the signs of the work done and heat transfer terms, the principle of energy now reads as: (*change in the energy contents of a system = the heat transferred to the system from the surroundings - the work done by the system on the surroundings.*)

Using standard symbols,

$$(KE + PE + U)_2 - (KE + PE + U)_1 = Q_{12} - W_{12} \qquad (21.1)$$

where subscript 1 stands for the initial state of the system, while subscript 2 denotes the final state of the system, Q_{12} is the heat transferred to the system in the process from 1 to 2, and W_{12} is the work done by the system on the surroundings in the process. Referring to the convention of signs for the Q and W terms. As before, we use E as a symbol for the total energy content of the system. In terms of E, Eq. (21.1) reduces to

$$E_2 - E_1 = Q_{12} - W_{12} \qquad (21.2)$$

This is the most general formulation of the first law (of thermodynamics) applicable to any closed system, and is valid for any process, quasi-static or not.

21.1.4 First Law Applied to a Cyclic Process

A cyclic process has been defined (Sec. 20.5.2) as one in which the end state of the system is exactly the same as its initial state. (Fig. 21.2).

For the four processes involved, Eq. (21.2) gives

$$E_2 - E_1 = Q_{12} - W_{12}; \quad E_3 - E_2 = Q_{23} - W_{23}$$

$$E_4 - E_3 = Q_{34} - W_{34}; \quad E_1 - E_4 = Q_{41} - W_{14}$$

For the whole cycle,

$$Q_{12} + Q_{23} + Q_{34} + Q_{41} = W_{12} + W_{23} + W_{34} + W_{12}$$

i.e., the *net* heat transferred *to a* system in a complete cycle is equal to the *net* work done *by* the system on the surroundings. This was to be expected from the principle of energy that the *net* energy transferred to a closed system must be equal to zero if there is no net change in the state of the system.

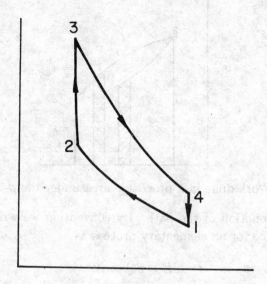

Fig. 21.2

21.1.5 Nature of Work, Heat and Internal Energy

Unlike internal energy, work and heat are defined in the context of a process. Let us consider a system consisting of a substance on which work is done by compression and expansion alone. As before, the work done by a system in a process from state 1 to 2 is

$$W_{12} = \int p dV$$

if the process is quasi-static, i.e., it proceeds so slowly that it deviates from equilibrium only in infinitesimal amounts. Clearly, work done W_{12} is the area under the curve representing the path of the process on a p–V diagram (Fig. 21.3).

It should also be clear that W_{12} depends on the path of the process and not just on the end states 1 and 2 of the system. The work done (by the system) in the process A is larger than in the process B shown.

Thus, W_{12} is different for processes taking path A or B.

We, therefore, cannot define work as a property of state such that we could write W_{12} as $W_2 - W_1$. Similarly, the heat transferred Q_{12} in a process cannot be written as $Q_2 - Q_1$, and is

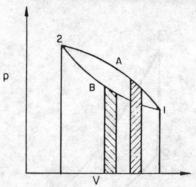

Fig. 21.3 Work done in a process is area under the p-V curve.

clearly a function of the path. By convention we write the energy balance for an elementary process as

$$dE = \bar{d}Q - \bar{d}W \qquad (21.3)$$

where bars over d, with the heat and work terms, denote that W and Q are not functions of state and, therefore, we cannot write exact differentials, or hope to integrate the RHS of Eq. (21.3) directly.

Note that energy E as well as its component KE, PE and internal energy U are all functions of state.

Consequently, the changes in internal energy U, as well as in E in a *cyclic process* are exactly zero since we return to the same state. But the work done is not necessarily zero. If we make a cyclic process out of the two paths A and B shown in Fig. 21.3, taking the system from state 1 to 2 along path A, and back to state 1 along path B, the work done is equal to

$$W_{1A2} + W_{2B1} = W_{1A2} - W_{1B2}$$

This is equal to area enclosed by the cyclic path on the p-V diagram.

Eq. (21.3) which refers to an elementary step process can be integrated over a cycle to give

$$\oint dE = \oint \bar{d}W - \oint \bar{d}Q$$

The LHS of this is zero, since E is a function of state, and hence the energy equation for a cycle is

$$\oint \bar{d}W = \oint \bar{d}Q \qquad (21.4)$$

or, the net work done by a system as it undergoes a cycle is exactly equal to the net heat transferred to it.

Example 21.1

Figure 21.4 shows a piston-cylinder assembly with a spring pressing on the piston. To start with, the piston is at $x = 0.1$ m, and the spring is just touching it. A total of 24 kJ are added to the gas increasing its temperature such that the final pressure of the gas is 6×10^5 Pa (or 6 bar). The piston is now at $x = 0.4$ m. Determine the work done by the gas on the surroundings, and the increase in thermal energy of the gas, if the spring constant k is 60 kN/m.

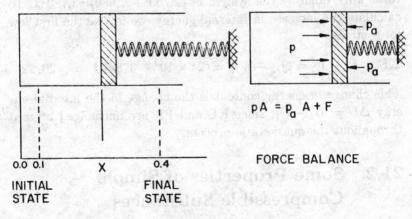

Fig. 21.4

Solution

Work is done by the system against the compression force $F(x)$ of the spring and the atmospheric pressure p_a that acts on the outside surface.

Now, the workdone against the external force F is:

$$\int_{0.01}^{0.4} F\,dx = \int_{0.1}^{0.4} k(x - 0.1)\,dx$$

$$= 60 \times 10^3 \left[\frac{x^2}{2} - 0.1x\right]_{0.1}^{0.4} = 2,700 \text{ J}$$

The work done against the atmospheric pressure is determined from the integral of p_a with volume. But first we need to calculate the initial and the final volumes of the gas. We calculate the area A of the piston using the force balance at the final position:

$$p_f A = p_a A + k(0.4 - 0.1)$$
$$6 \times 10^5 A = 1 \times 10^5 A + 60 \times 10^3 (0.4 - 0.1)$$

or $A = 60 \times 10^3 (0.4 - 0.1)/(6 \times 10^5 - 1 \times 10^5) = 0.036 \text{ m}^2$

This implies that the initial volume is 0.036 $\text{m}^2 \times$ 0.1 m=0.0036 m^3, and the final volume is four times this, i.e., 0.0144 m^3.

The work done against the constant pressure of 1×10^5 Pa, then, is 1×10^5 Pa $\times$ (0.0144-0.0036) m^3 or 1,080 J. Thus, the total work done by the system is (2,700 + 1,080)=3,780 J. To calculate the increase in internal energy, we invoke the first law for systems,

$$\Delta E = E_2 - E_1 = Q_{12} - W_{12} = (24 \times 10^3 - 3780) \text{ J} = 20,220 \text{ J}$$

This change in energy content is the change in the internal energy $\Delta U = U_2 - U_1$, since KE and PE are unchanged at zero throughout the quasi-static process.

21.2   Some Properties of Simple Compressible Substances

Before we proceed with the application of the first law to some specific systems, let us understand the various properties of simple substances required for the application of the law.

One of the ways of classifying substances is on the basis of the modes in which work can be done on substance. Thus, we can do work on a piece of rubber by sketching it, or by deforming its shape. On a perfect gas without any viscosity, work can be done only by expansion or contraction. A simple compressible substance is one on which work can be done only through a change in volume, such as a gas and a non-viscous liquid. In such substances, the state postulate requires the specification of only two intensive properties independently, and then, any other property can be expressed as a function of these properties.

Suppose we take a fixed mass of a simple compressible pure substance, like frozen alcohol, and heat it such that its temperature rises. Further, we allow its volume to change in such a manner that the pressure remains constant. Now plot the temperature T vs. specific volume v curve for a given pressure. We might get a curve like the two shown in Fig. 21.5. As we heat the

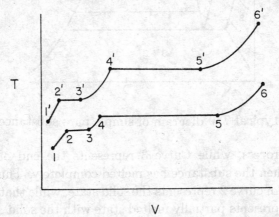

Fig. 21.5 Significant states in the temperature-specific volume curve of a substance as it is heated.

substance, its temperature and volume increase from state 1 to state 2, after which we see an increase of volume while temperature remains constant. This is the melting of the solid substance. At point 3, all of the solid has melted to a liquid, and temperature increases again till point 4 is reached. After this there is a large increase in volume, again at constant temperature. This represents boiling of liquid and its conversion to vapour. The substance is totally at the vapour state at point 5, after which the temperature of the vapour increases again. Fig. 21.5 shows the different curves representing heating at different but constant pressures.

Figure 21.6 shows the locus of the various points 2, 3, 4 and 5 at various pressures. Curve 2 represents the beginning of the

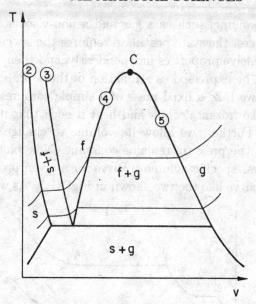

Fig. 21.6 Typical T-v diagram of simple pure substances.

melting process, while Curve 3 represents the end of melting process when the substance has melted completely. Thus, region to the left of curve 2 represents the solid state, while that between 2 and 3 represents partially melted state with the solid s and the liquid state f co-existing. Curve 4 represents the beginning of the evaporation process, and the substance on this curve is termed as saturated liquid. Curve 5 on the other hand represents a saturated vapour state with the substance completely evaporated but still at the boiling temperature $T_{fg} (= T_4 = T_5)$. The area between the curves 4 and 5 represents partially evaporated state at constant temperature T_{fg}. This area is called the vapour dome and is shown separately in Fig. 21.7.

We introduce here the property x, termed as the *quality*, which represents the mass fraction of the substance existing as vapour in the region between the curves 4 and 5. The mass fraction of liquid, then, is $(1 - x)$. Thus, points on curve 4 have $x = 0$ showing pure liquid, while those on curve 5 have $x = 1$, representing pure saturated vapour. The substance in region beyond 5 is termed as *superheated vapour*, with the temperature difference $(T - T_{fg})$ termed as the *degree of superheat*.

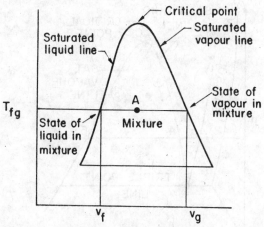

Fig. 21.7 Vapour dome.

The saturated liquid line '4' and the saturated vapour line '5' meet at point C, where the liquid and vapour phases become indistinguishable. This is termed as the *critical point* and the corresponding temperature and pressure are termed as the *critical temperature* and *pressure*, respectively.

Notice from Fig. 21.6 that there is only one temperature at which the three states, solid s, liquid f, and gaseous g co-exist. This is termed as the *triple point temperature*, and corresponding pressure as the *triple point pressure*.

With reference to Fig. 21.7, it is clear that the properties of a mixture of liquid and vapour (also known as *wet vapour*) at a point A can be written in terms of the properties of the corresponding saturated liquid and vapour states. Thus,

$$v_A = (1 - x)v_f + xv_g$$

where x is the quality of vapour, i.e., the mass fraction of vapour in the mixture. Similarly, if we know the internal energy of the liquid and gas phases as u_f and u_g, then internal energy of the mixture is given as

$$u_A = (1 - x)u_f + xu_g$$

Similarly draw p-V and p-T diagram for the various substances.

Figure 21.8 shows the p-V diagram and Fig. 21.9 shows a p-T diagram for a typical substance which expands when it melts.

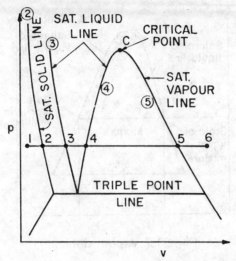

Fig. 21.8 Typical *p-v* diagram for a substance that expands on melting.

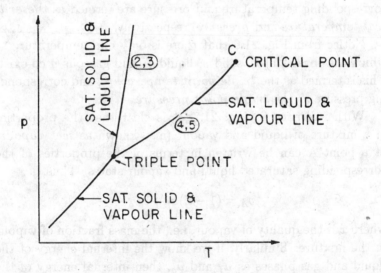

Fig. 21.9 Typical *p-T* diagram for a substance that expands on melting.

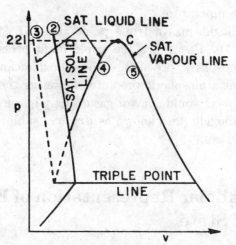

Fig. 21.10 *p-v* diagram for water, a substance that contracts on melting.

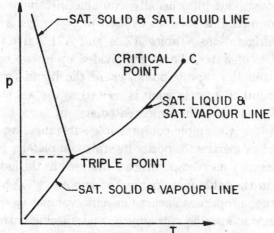

Fig. 21.11 *p-T* diagram for water

Note that triple point appears as a line in the *p-V*, and *p-V* diagrammes, but as a point on the *p-T* diagramme.

Figure 21.10 shows the *p-V* diagramme for water which differs in important details from Fig. 21.8. Water has a peculiar property that its volume as solid is *more* than its volume as liquid. Therefore (Fig. 21.10), the saturated solid line 2 lies to the right of the saturated liquid line 3. Figure 21.11 is the *p-T* diagram for water.

The triple point of water is $0.01°C$ or 273.16 K and 6.11×10^{-3} Pa. At the standard atmospheric pressure of 1.0132×10^5 Pa the melting point of ice is $0°C$ or 273.15 K, lower than the triple point temperature.

Carbon-dioxide has a triple point at -56.6 °C (216.56 K) and 5.147×10^5 Pa. Since the triple point pressure is higher than the atmospheric pressure, carbon-dioxide cannot be in the liquid state at atmospheric pressure. Lowering the temperature to below -56.6 °C would convert gas directly into a solid, and the solid carbondioxide (also known as *dry ice*) sublimates when its temperature rises.

21.3   Tabular Representation of Equations of State

Thermodynamic calculations repeatedly require the properties of substances. These are usually available in tabular forms for common compressible substances used in power generating system and refrigerations. Tables A21.1 and A21.2 give the tables for properties of water. These are divided up in two tables, one for use within the vapour dome where the liquid (water) and vapour (steam) co-exist in what is termed as the wet steam, and the other for use with the superheated steam.

Since water is a simple compressible substance, we need two independent properties to specify its state completely. For convenience, pressure p and temperature T are used as the independent properties in the table for superheated vapours (Table A21.1), and the other properties such as specific volume v, specific internal energy u, specific enthalpy h, and specific entropy s are listed against p and T. The tables may be entered through p and T, or with one of the properties p and T, and another property v, u, h or T. Linear interpolation is resorted to for entering the non-listed values.

Given below are few examples to clarify the use of these table.

Example 21.2

Find the specific volume of super-heated steam at a pressure of 2 bar and a temperature of 200°C.

Solution

A pressure of 2 bar equals 2×10^5 Pa or 200 kPa. The boiling point or saturation temperature of the steam is read as 120.2°C (from the first column of Table A21.2 where it is given in parenthesis under the pressure of 200 kPa). Since the given temperature of 200°C is above this saturation temperature, the substance is superheated, and this is indeed the table to use. We first enter the row corresponding to 200 kPa and go to the right to read the property values in the column marked 200°C at the top. The value of v is read as 1.0804 m³/kg.

Example 21.3

Repeat Example 21.2, except that now the desired temperature is 220°C.

Solution

Table A21.2 show that for pressure of 2000 kPa the properties are listed under temperatures of 200°C and 250°C, but not 220°. Thus, to find the value at 220°C, we resort to linear interpolation.

v at 200°C = 1.0804 m³/kg
v at 250°C = 1.1989 m³/kg
Increase in v for 50°C=1.1989−1.0804=0.1185 m³/kg
Increase in v for 20°C=0.1185× (20/50)=0.0474 m³/kg
Therefore, the value of v at 220°C (and 200 kPa) is

1.0804+0.0474=1.1278 m³/kg.

Example 21.4

Repeat Example 21.3 except that the pressure is 2.2 bar while the temperature is 220°C.

Solution

This is a problem in double linear interpolation, since besides the temperature of 220°C being not given in the tables the tabulated values are only for pressures of 200 kPa (2 bar) and 300 kPa. We already have from Example 21.3 the value of v at 220°C and 200 kPa as 1.1278 m³/kg. We first interpolate for v at 220°C and 300 kPa, and then from the two values of v at 220°C and pressures of 200 kPa and 300 kPa, we interpolate for a pressure of 220 kPa (2.2 bar). Proceeding in a similar fashion as in Example 21.3,

v at 300 kPa and 200°C = 0.7164 m³/kg

v at 300 kPa and 250°C = 0.7964 m³/kg

Therefore,

$$v \text{ at } 300 \text{ kPa and } 220°C = 0.7164 + \frac{2}{5}(0.7964 - 0.7164)$$

$$= 0.7484 \text{ m}^3/\text{kg}$$

Then, interpolating between 1.1278 m³/kg at 200 kPa and 0.7484 at 300 kPa,

$$v \text{ at } 220 \text{ kPa} = 1.1278 + \frac{2}{10}(0.7484 - 1.1278) = 1.0619 \text{ m}^3/\text{kg}$$

The property tables for saturated water and steam (under the vapour dome) pose a problem. Since the temperature and pressure at saturation are not independent, they constitute only one independent variable and, therefore, are sufficient for specifying the state completely.

This problem is solved by listing in Table A21.1 the properties of saturated liquid as well as saturated vapour. For example, at a temperature of 120°C the pressure is fixed at 198.54 kPa. The specific volume of the saturated liquid is given as 1.060×10^{-3} m³/kg while that of the saturated steam is given as 0.8915 m³/kg. We introduce x, the quality of steam as the mass fraction of vapour present in the wet steam, as the other independent variable. Then the specific volume of the wet steam of quality can be found from

$$v = (1 - x)v_f + xv_g$$

where subscript f stands for the saturated liquid (coloquially, fluid) and subscript g for the saturated vapour (coloquially, gas). The quality x, as before, is defined as $m_g/(m_f + m_g)$, where m_g is the mass in gas phase and $(m_f + m_g)$ is the total mass in gas and liquid phases. The other properties of the wet vapour are obtained in a similar fashion. For example, the internal energy u of the vapour of quality x is

$$u = (1 - x)u_f + xu_g$$

We may then find the properties of wet steam (under the vapour dome) by using one of the two properties, temperature and pressure, and either the quality x or v or u of the mixture. The following examples will clarify.

Example 21.5

A sample of wet steam consists of 5 kg of liquid and 20 kg of vapour at a temperature of 40°C. Find the total volume of the sample.

Solution

The quality x of steam is: $x = 20/(5 + 20) = 0.8$ Using Table A21.1, for $T=40°C$, $v_f = 0.0010078$ and $v_g = 19.54$. The specific volume of the wet steam of quality 0.8 then is

$$v = (1 - 0.08) \times 0.0010078 + 0.8 \times 19.546 = 15.6370 \text{ m}^3/\text{kg}$$

Note that the liquid volume can safely be neglected under these conditions. The total volume of 25 kg of wet steam, then, is 390.92 m³.

Example 21.6

What is the quality of wet steam at 120°C if its specific volume is 0.6 m³/kg?

Solution

At 120°C, the pressure of saturated steam is 198.54 kPa (slightly less than 2 atm). The specific volumes of the liquid and gas phases are: $v_f=0.0010606$ m³/kg and $v_g=0.8915$ m³/kg (Table A21.1).

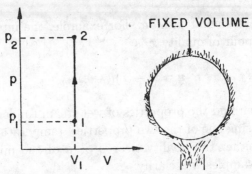

Fig. 21.12 Constant volume heating.

Then, using the relation for the mixture specific volume in terms of the quality x,

$$
\begin{aligned}
v &= (1-x)v_f + xv_g \\
0.6 &= (1-x)1.0606 \times 10^{-3} + x \times 0.8915 \\
x &= (0.6 - 1.0606 \times 10^{-3})/(0.8915 - 1.0606 \times 10^{-3}) = 0.673
\end{aligned}
$$

This is the quality of the wet steam. Note that in this case too we could neglect v_f in comparison to v_g.

21.4 First Law Analysis of Some Important Thermodynamic Processes

21.4.1 Constant Volume Heating

When we heat a substance keeping its volume constant, (Fig. 21.12) no work is done since the boundaries of the system do not move.

In such a situation, all the heat transferred to the system goes to raise its internal energy U. Thus,

$$\Delta U = U_2 - U_1 = Q_{12} \qquad (21.5)$$

Specific heat of a substance is defined as *the quantity of heat required to raise the temperature of a unit mass of the substance through a unit temperature.* Thus,

$$Q_{12} = mc_v \Delta T \qquad (21.6)$$

where m is the mass of the substance and ΔT is the rise in the temperature of the system. Subscript v in the heat capacity c_v signifies that the process is taking place at constant volume.

Combining Eqs. (21.5) and (21.6),

$$\Delta U = m\Delta u = mc_v\Delta T \qquad (21.7)$$

where U, the internal energy of the system, has been expressed in terms of specific internal energy u. From this equation then,

$$C_v = \frac{\Delta u}{\Delta T} = \frac{\partial u}{\partial T}\bigg|_v \qquad (21.8)$$

Thus, the specific heat of a substance is the rate of change of internal energy at constant volume. In general, it varies with the temperature of a substance. But for perfect gases which obey the relation $pv = RT$, the value of C_v is constant $(=R/(\gamma - 1))$ and then

$$u = C_vT = \frac{RT}{\gamma - 1} \qquad (21.9)$$

Example 21.7

Consider a wet steam at a temperature of 100°C with specific volume equal to 1 m³/kg. It is heated to a temperature of 200°C at constant volume. Find the final pressure and the amount of heat that is supplied per unit mass of water.

Solution

In this heating at constant volume, the heat supplied equals the change in specific internal energy. Initially, when temperature is 100°C, the steam is wet and the pressure is 101.33 kPa. At this temperature $v_f=0.0010437$ m³/kg and $v_g=1.6730$ m³/kg (Table A21.1). Since v of the wet steam is 1.0 m³/kg, its quality is obtained from

$$v = (1 - x)v_f + xv_g$$

or $\qquad 1.0 = (1 - x) \times 0.0010437 + x \times 1.6730$

which gives $x = 0.6$

The specific internal energy at the initial stage is found from

$$u = (1 - x)u_f + xu_g$$

But the values of u_f and u_g are not given in Table A21.1. Instead the values of the specific enthalpy h are listed there. The *specific enthalpy h* is defined as

$$h = u + pv$$

so that

$$u = h - pv$$

Using this,

$$u_f = h_f - pv_f = 419.06 \times 10^3 - 101.33 \times 10^3 \times 0.0010437$$
$$= 418.95 \text{ kJ/kg}$$

and

$$u_g = h_g - pv_g = 2676 \times 10^3 - 101.33 \times 10^3 \times 1.673$$
$$= 2506.47 \text{ kJ/kg}$$

The value of u for the wet steam then is

$$u = 0.4 \times 418.95 + 0.6 \times 2506.47 = 1671.46 \text{ kJ/kg}$$

At 200°C, the value of v_g=0.12716 m³/kg, less than 1.0 m³/kg, and hence the steam is superheated at the final state. From Table A21.2 we note at 200°C, v=1.0804 at p=200 kPa and 0.7164 at 300 kPa. By linear interpolation we find that the volume v=1.0 m³/kg will be obtained at a pressure of 200 + 100 (1.0 − 1.0804)/(0.7164 − 10804), or 220.9. At this final pressure the value of u can be found by interpolation from values at 200 kPa and 300 kPa.

At 200 kPa : $u = h - pv$=2870.5−200 × 1.0804=2654.4 kJ/kg
At 300 kPa : $u = h - pv$=2865.5−300 × 07164=2650.6 kJ/kg
Therefore at 220 kPa,

$$u = 2654.4 + \frac{220.9 - 200}{300 - 200}(2650.6 - 2654.4) = 2653.24 \text{ kJ/kg}$$

The heat supplied per unit mass is equal to the change in specific internal energy, which is 2653.24−1671.46=981.8 kJ/kg.

21.4.2   Constant Pressure Process

Consider a process in which 20 kg of air is compressed quasi-statically under a constant pressure of 200 kPa from an initial volume of 15 m^3 to a final volume of 13 m^3 Fig. (21.13). In this process, the work is done on the system by the pressure force. Clearly,

$$W_{12} = -p(V_1 - V_2) = -mp(v_1 - v_2)$$

where 1 and 2 represent the initial and the final states, respectively, and m is the mass of the gas. The sign is negative because this is the work done on the system. Using the first law

$$U_2 - U_1 = Q_{12} - W_{12} = Q_{12} + mp(v_1 - v_2)$$

$$\begin{aligned} Q_{12} &= m(u_2 - u_1) - mp(v_1 - v_2) \\ &= m\left[(u_2 + p_2 v_2) - (u_1 + p_2 v_1)\right] \end{aligned}$$

Note that $p_2 = p_1 = p$. The combination $(u + pv)$ occurs so frequently in thermodynamics that it is given the name *specific enthalpy* and is denoted by the symbol h.

The heat required in a constant pressure process then is $Q_{12} = m(h_2 - h_1)$. For a perfect gas, we note that the specific internal energy $u = C_v T = RT/\gamma - 1$.

The specific enthalpy then is

$$h = u + pv = \frac{RT}{\gamma - 1} + pv = \frac{RT}{\gamma - 1} + RT = \frac{\gamma}{\gamma - 1} RT$$

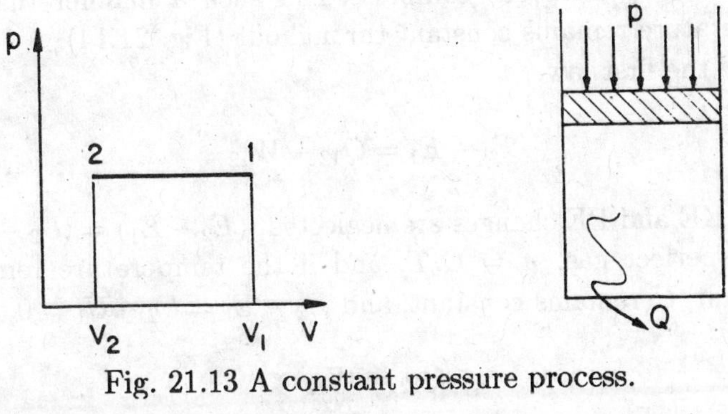

Fig. 21.13 A constant pressure process.

since $pv = RT$, for a unit mass of a perfect gas. And thus,

$$Q_{12} = m\frac{\gamma}{\gamma - 1}R(T_2 - T_1)$$

If we introduce C_p as the specific heat at constant pressure, then

$$C_p = \frac{\gamma R}{\gamma - 1}$$

and

$$Q_{12} = mC_p\Delta T$$

Also, note that, $C_p - C_v = R$. These are fundamental relations for perfect gas.

In the given example, we first determine the two temperatures T_1 and T_2. For air, the value of R, the gas constant is 284 J/kg K. Temperatures are determined from $pV = mRT$ as

$$T_1 = \frac{pV_1}{mR} = \frac{200 \times 10^3 \times 15}{20 \times 284} = 528 \text{ K}$$

and

$$T_2 = \frac{pV_2}{mR} = \frac{200 \times 10^3 \times 13}{20 \times 284} = 457 \text{ K}$$

A drop of $(528-457)=71$ K will result in release of $mC_p\Delta T = 20 \times [1.4/(1.4 - 1)] \times 284) \times 71 = 1.4 \times 10^6$ J of heat.

21.4.3 Isothermal Process

Consider a perfect gas compressed in such a manner that its temperature remains constant throughout (Fig. 21.14).

By the first law,

$$E_2 - E_1 = Q_{12} - W_{12}$$

If the KE and PE changes are neglected, $(E_2 - E_1) = (U_2 - U_1)$. For a perfect gas, $u = C_vT$, and if the temperature remains constant, U remains constant, and $E_2 - E_1 = U_2 - U_1 = 0$, or

$$Q_{12} = W_{12}$$

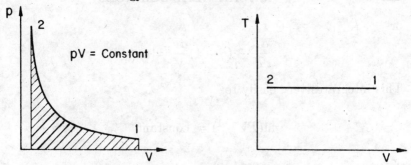

Fig. 21.14 An isothermal process.

This implies that the system must loose as much heat through conduction as the work done on it. Also,

$$W_{12} = \int p dV$$

Using the perfect gas equation of state $pV = mRT$ to replace p in terms of V gives

$$W_{12} = \int_{v_1}^{v_2} mRT \frac{dV}{V} = mRT \ln \frac{V_2}{V_1} = p_1 V_1 \ln \frac{V_2}{V_1}$$

21.4.4 Reversible Adiabatic Process

Consider a gas which is compressed quasi-statically in such a manner that no heat interaction takes place. This may occur if the compressor is insulated. Then Q_{12} is zero, and

$$dU = \bar{d}W$$

For a perfect gas

$$dU = mC_v dT = m \frac{R}{\gamma - 1} dT$$

and

$$\bar{d}W = pdV = \frac{mRT}{V} dV$$

Thus,

$$m \frac{R}{\gamma_1} dT - \frac{mRT}{V} dV = 0$$

or

$$\frac{dT}{T} - (\gamma - 1)\frac{dV}{V} = 0$$

This integrates very easily to

$$\ln(TV^{\gamma-1}) = \text{constant}$$

or

$$\frac{T_2}{T_1} = \left(\frac{V_1}{V_2}\right)^{\gamma-1}$$

Replacing T with pv/mR in this equation, gives

$$\frac{p_2}{p_1} = \left(\frac{V_1}{V_2}\right)^{\gamma} \quad \text{or} \quad pV^{\gamma} = \text{constant}$$

This is the well known equation for a reversible adiabatic process. Once we know the relation of pressure and volume, we can evaluate the work done in an adiabatic process as

$$W_{12},\text{adaiabatic} = \int_{V_1}^{V_2} p\,dV = \int_{V_1}^{V_2} \frac{C}{V^{\gamma}}\,dV = -\frac{CV^{-\gamma+1}}{\gamma-1}\bigg|_1^2$$

$$= -\frac{pV}{\gamma-1}\bigg|_1^2 = \frac{p_1V_1 - p_2V_2}{\gamma-1} = \frac{mR(T_1 - T_2)}{\gamma-1}$$

The same result is also obtained from the First Law formulation which gave the work done as equal to the change in internal energy.

Example 21.7

Figure 21.15 shows a piston-cylinder assembly which contains 2 kg of water initially at 60°C, with a volume of 2 m³ when the piston rests on stops as shown. Obtain the heat transfer required to evaporate all the water. It is given that the weight of the piston is such that it results in a total pressure of 3 bar (300 kPa) when it is off the stops.

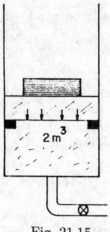

Fig. 21.15

Solution

Let us assume that the piston does not leave the stops when all the water has just evaporated giving a specific volume of saturated vapour $v_g = 1$ m³/kg. From Table A21.1, $v_g = 1.0352$ at 169.06 kPa (115°C) and 0.8905 at 198.54 kPa (120°C). This implies that our assumption is correct and the piston has not lifted off its stops. The corresponding temperature is, by interpolation

$$115 + \frac{0.8905 - 1.0}{0.8905 - 1.0352}(120 - 115) = 118.8 \text{ K}$$

and the corresponding pressure, similarly, is 191.14 kPa. The corresponding value of u is obtained by interpolation from the values of u_g at 115 K and 120 K. Since Table A21.1 gives values of h rather than u, we obtain at 115 K,

$$u_g = h_g - pv_g = 2698.7 - 169.06 \times 1.0352 = 2523.7 \text{ kJ/kg}$$

and at 120 K, $u_g = 2706.0 - 198.54 \times 0.8905 = 2529.2$ kJ/kg.

The value of u at state 2 is found by interpolation at 118.8 K,

$$u_2 = 2523.7 + \frac{118.8 - 115}{120 - 115}(2529.2 - 2523.7) = 2527.9 \text{ kJ/kg}$$

At the initial state, steam is wet, $u = 1$ m³/kg and $T = 60$°C. The corresponding $v_f = 0.0010171$ and $v_g = 7.679$ m³/kg. The quality of steam is obtained from $v = (1-x)v_f + xv_g$　as $x = 0.13$.

$$u_g = h_g - pv_g = 2609.7 - 19.92 \times 7.678 = 2456.71 \text{ kJ/kg}$$

$$u_f = h_f - pv_f = 251.09 - 19.92 \times 0.0010171 = 251.07 \text{ kJ/kg}$$

and, therefore, $u_1 = u_f(1 - 0.13) + u_g \times 0.13 = 537.8 \text{ kJ}$. The quantity of heat supplied, then, is

$$Q_{12} = m(u_2 - u_1) = 2(2527.9 - 537.8) = 3980 \text{ kJ}$$

21.5　Applications of First Law to Flow Processes

In many engineering applications, the working substance is continually flowing across a physical device which is either extracting energy from it or supplying it with energy. For example, steam flows through a turbine and delivers power, a refrigerator compressor compresses refrigerant vapour and delivers it to the condenser, and water is being continuously cooled in an automobile radiator. Such processes are termed as *flow processes*. Figure 21.16 shows some examples of flow processes.

It is difficult to use the concept of closed system in such cases. Instead, we use the alternative formulation of an *open system*. An open system is one in which matter crosses the boundaries of the system. Such an open system is also termed as a *control volume*, for in such a case we turn consider a fixed volume in space rather than a fixed mass, as we do in the case of a closed system. The identity of mass in a control volume (CV) is, thus, continually changing. Figure 21.17 shows the schematic of a CV with one inlet and the outlet ports.

Clearly, the first law formulation of Eq. (21.1) is valid for a closed system and not for a CV. We have to arrive at a new formulation of the energy principle for control volumes, i.e., open systems.

Let us look at the energy contained within CV. This energy can change in any of the following ways:

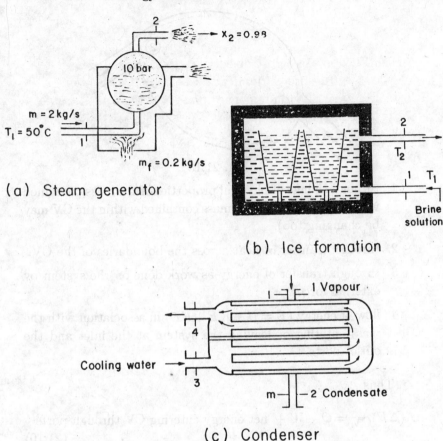

(a) Steam generator

(b) Ice formation

(c) Condenser

Fig. 21.16 Some examples of flow processes.

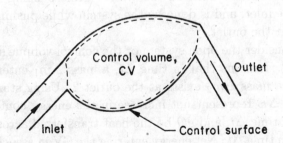

Fig. 21.17 A control volume with one inlet and one outlet.

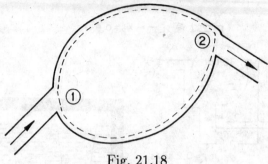

Fig. 21.18

1. through change of physical properties of the mass contained within the CV (the total mass contained within the CV may be changing too).

2. through transfer of heat across the boundaries of the CV.

3. through transfer of energy as work done on the system by external forces.

4. flow of energy $(E = U + KE + PE)$ in association with the mass entering or leaving the system at the inlet and the outlet ports.

Thus,

$$(\Delta E)_{CV} = Q - W + \text{ net energy entering CV through ports}$$
$$(21.10)$$

In Eq. (21.10), the last term is new. We also have to be careful about the work done on the system. In addition to any shaft work, the work is also done on the system while pushing in the matter at the inlet, and is done by the system while pushing the matter out at the outlet.

Let us consider the open system or the control volume shown in Fig. 21.18. Imagine that in time Δt, a mass Δm_1 enters the inlet 1, and a mass Δm_2 exists at the outlet 2. Using standard symbols, let ΔE represent net increase in the energy contained in the CV in time Δt, and $\bar{d}Q$ be the heat transferred across the boundaries in time Δt. Net energy entering the CV in association with mass Δm_1 at the inlet is $e_1 \Delta m_1$ or $(\frac{1}{2}V_1^2 + gz_1 + u_1)\Delta m_1$.

Similarly, the net energy leaving at the outlet in association with the mass Δm_2 is $e_2 \Delta m_2$ or $(\frac{1}{2}V_2^2 + gz_2 + u_2)\Delta m_2$. The work done term must be evaluated carefully. Besides $\bar{d}W$, the shaft work done by the system, the system also does work in pushing the matter out at the outlet 2. If p_2 is the pressure there, then the work done is $p_2 d\bar{V}_2$, where $d\bar{V}_2$ is the volume of mass dm_2. Clearly $d\bar{V}_2 = v_2 dm_2$ and, hence, the work done at the exit is $(p_2 v_2) dm_2$.

Similarly, the work done on the system in pushing the matter in at the inlet is $(p_1 v_1) dm_1$. Putting together all the terms in Eq. (21.10) gives

$$
\begin{aligned}
(\Delta E)_{CV} = {} & \bar{d}Q - \bar{d}W - \left(p_2 v_2 dm_2 - p_1 v_1 dm_1 \right) \\
& + \left[(\frac{1}{2}V_1^2 + gz_1 + u_1)dm_1 - (\frac{1}{2}V_2^2 + gz_2 + u_2)dm_2 \right]
\end{aligned}
$$
(21.11)

or

$$
\begin{aligned}
\bar{d}Q - \bar{d}W = {} & (\Delta E)_{CV} + (\frac{1}{2}V_2^2 + gz_2 + u_2 + p_2 v_2)dm_2 \\
& - (\frac{1}{2}V_1^2 + gz_1 + u_1 + p_1 v_1)dm_1
\end{aligned}
$$
(21.12)

This is the fundamental energy equation for a control volume.

We are usually interested in simpler cases, such as, the assumption of steady-state wherein the matter steadily flows into CV and out of it with properties that do not change with time. In addition, work is done by the system and heat is transferred to it at constant rates. Also, the properties of CV do not change with time, such that $(\Delta E)_{CV}$ in the above equation is necessarily zero.

Since the system is at steady-state, the mass contained in CV must be constant and, therefore, $dm_2 = dm_1 = \dot{m}\delta t$. Dividing across by $\dot{m}\delta t$, and taking the limit as tends to zero gives

$$
\frac{\dot{Q}}{\dot{m}} - \frac{\dot{W}}{\dot{m}} = \left[\frac{1}{2}(V_2^2 - V_1^2) + g(z_2 - z_1) + (u_2 + p_2 v_2) - (u_1 + p_1 v_1) \right]
$$
(21.13)

The property combination $(u + pv)$ occurs so often in open systems (i.e., control volumes) that it is recognized as a property in

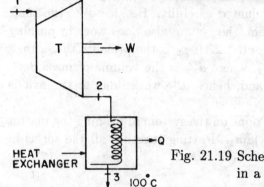

Fig. 21.19 Schematic of the processes in a steam-turbine operation

its own right and given the name *specific enthalpy*. Thus, the specific enthalpy,

$$h = u + pv \qquad (21.14)$$

Example 21.8

A steam turbine extracts energy as work from the steam entering it at high enthalpy. A common assumption is to treat the turbine as insulated, so that any heat interactions are neglected. In one application, steam enters turbine as dry saturated steam at 20 bar and with a velocity of 200 m/s. At the exit, the pressure is 2 bar and the quality of the steam is 0.86 and its velocity 100 m/s. If the flow rate of steam is 1.5 kg/s, determine the power generated. If the steam at the exit is passed through a heat exchanger cooling it to a condensate at 100°C, determine the heat transferred in the heat exchanger.

Solution

The system is sketched in Fig. 21.19. From steam tables, the enthalpy h_1 of saturated steam at 20 bar is 2797.2 kJ/kg.

At exit, at a pressure of 2 bar, $h_f = 504.70$ kJ/kg and $h_g = 2706.3$ kJ/kg.

Therefore, for $x = 0.86$,

$$h_2 = (1 - 0.86)504.7 + 0.86 \times 2706.3 = 2398 \text{ kJ/kg}$$

From the equation of first law for open systems,

$$w = \dot{m}\left[(\frac{1}{2}V_1^2 + h_1) - (\frac{1}{2}V_2^2 + h_2)\right]$$

$$= 1.5\left[\left(\frac{1}{2}(200)^2 + 2792.2 \times 10^3\right)\right.$$

$$\left. -\left(\frac{1}{2}(100)^2 + 2398.1 \times 10^3\right)\right]$$

$$= 613.6 \times 10^3 \text{ kW}$$

To determine the heat interaction in the heat exchanger, first find h_3, the enthalpy of saturated water at 100°C at the exit. From the tables, this is seen as 419.06 kJ/kg. Therefore,

$$q = -\dot{m}\left[(\frac{1}{2}V_2^2 + h_2) - (\frac{1}{2}V_3^2 + h_3)\right]$$

$$= -1.5\left[(\frac{1}{2} \times 100^2 + 2398.1) - 419.06\right] = -2976 \text{ kW}$$

Problems

21.1 A hydraulic turbine uses water stored in a reservoir 100 m above it. What is the minimum flow rate in kg/s to produce a turbine output of 5 MW?

21.2 In a process on a closed system, 150 kJ of heat is added. The system is then restored to the initial state. Heat and work transferred in the reversed process are - 50 kJ and +75 kJ, respectively. What is the work transfer in the first process?

21.3 The energy content of a closed system increases by 5 kJ while 50 kJ of work is done by it. Determine the amount of heat transferred to or from the system.

21.4 The value of u, p and v of a fluid at a given state are 2500 kJ/kg, 100 kPa and 1.8 m³/kg. Calculate the value of h at this state.

21.5 In a constant pressure process at 250 kPa, 1 kg of a fluid loses 75 kJ of heat while its volume decreases to 0.2 m³ from 0.5 m³. Determine the changes in internal energy and enthalpy of the fluid.

21.6 A piston of diameter 25 cm moves through 1 m against a pressure of 500 kPa. Determine the work done by the gas.

21.7 At a constant temperature, one mole of a gas expands from 5000 kPa to 500 kPa inside a cylinder. Determine the work done.

21.8 A gas in a vessel is initially at 100 kPa and 25°C. A total of 30 kJ of heat is added to it. If the increase in enthalpy is 40 kJ, what is the final pressure of the gas if the volume of the vessel is 0.5 m³?

21.9 A vapour is compressed in a water-cooled compressor. The enthalpy of the vapour at the inlet and the outlet are 185 kJ/kg and 210 kJ/kg, respectively. If the heat removed from the vapour is 10 percent of the work input, what is the power input for a flow rate of 75 kg/min?

21.10 Steam enters a condenser steadily with an enthalpy of 2330 kJ/kg and a velocity of 350 m/s. The condensate leaves with an enthalpy of 140 kJ/kg and a velocity of 6 m/s. What is the amount of heat picked up by the cooling fluid for each kilogram of steam condensed?

21.11 A rigid vessel of 0.5 m³ volume contains 0.8 quality steam at 200°. A total of 140 kg of water is now pumped into the vessel. If the final temperature is 80°, what is the final pressure?

21.12 One kilogram of water at 120°C and 200 kPa is converted to steam at 300°C under constant pressure. Determine the change in enthalpy of the fluid.

21.13 Steam expands from an initial state of 20 MPa and 550°C to a final pressure of 5 kPa. If the quality of steam at the final pressure is 0.9, determine the change in enthalpy and internal energy.

21.14 Determine specific volume and enthalpy for superheated steam at 1000 kPa and 520°C.

21.15 Water changes from 25°C and 100 kPa to 50°C and 100 kPa. Determine the change in enthalpy.

21.16 Dry saturated steam at 600 kPa is stored in a rigid vessel which is cooled till its pressure is reduced to 100 kPa. What is the amount of heat given up for each kg of steam?

21.17 5 kg of air is first expanded at constant pressure from 300 kPa and 50°C until its volume is doubled. It is then heated at constant volume until the pressure is doubled. Assuming air to be an ideal gas, find the heat and work transferred in the entire process.

Appendix A21

Table A21.1 Properties of Saturated Water and Saturated Steam

Temp. °C	Press. kPa	Volume. m³/kg			Enthalpy. kJ/kg			Entropy. kJ/kg · K		
		Water v_f	Evap. v_{fg}	Steam v_g	Water h_f	Evap. h_{fg}	Steam h_g	Water s_f	Evap. s_{fg}	Steam s_g
0.01	0.6112	0.0010002	206.16	206.16	0.00	2501.6	2501.6	0.0000	9.1575	9.1575
5	0.8718	0.0010000	147.16	147.16	21.01	2489.7	2510.7	0.0762	8.9507	9.0269
10	1.2270	0.0010003	106.43	106.43	41.99	2477.9	2519.9	0.1510	8.7510	8.9020
15	1.7039	0.0010008	77.98	77.98	62.94	2466.1	2529.0	0.2244	8.5583	8.7827
20	2.337	0.0010017	57.84	57.84	83.86	2454.3	2538.2	0.2963	8.3721	8.6684
25	3.166	0.0010029	43.40	43.40	104.77	2442.5	2547.2	0.3672	8.1923	8.5591
30	4.241	0.0010043	32.93	32.93	125.66	2430.7	2556.4	0.4365	8.0181	8.4546
35	5.622	0.0010060	25.25	25.25	146.56	2418.8	2565.3	0.5049	7.8494	8.3543
40	7.375	0.0010078	19.545	19.546	167.45	2406.9	2574.4	0.5721	7.6862	8.2583
45	9.582	0.0010099	15.280	15.281	188.35	2394.9	2583.3	0.6383	7.5278	8.1661
50	12.335	0.0010121	12.045	12.046	209.26	2382.9	2592.2	0.7035	7.3741	8.0776
55	15.741	0.0010145	9.578	9.579	230.17	2370.8	2601.0	0.7677	7.2249	7.9926
60	19.920	0.0010171	7.678	7.679	251.09	2358.6	2609.7	0.8310	7.0798	7.9108
65	25.009	0.0010199	6.201	6.202	272.02	2346.3	2618.4	0.8933	6.9389	7.8322
70	31.16	0.0010228	5.045	5.046	292.97	2334.0	2626.9	0.9548	6.8017	7.7565
75	38.549	0.0010259	4.133	4.134	313.94	2321.5	2635.4	1.0154	6.6681	7.6835
80	47.36	0.0010292	3.408	3.409	334.92	2308.8	2643.8	1.0753	6.5379	7.6132
85	57.81	0.0010326	2.828	2.829	355.92	2296.1	2652.0	1.1343	6.4111	7.5454
90	70.11	0.0010361	2.3603	2.3613	376.94	2283.2	2660.1	1.1925	6.2874	7.4799
95	84.52	0.0010399	1.9810	1.9820	397.99	2270.1	2668.1	1.2501	6.1665	7.4166
100	101.33	0.0010437	1.6720	1.6730	419.06	2256.9	2676.0	1.3069	6.0485	7.3554
105	120.80	0.0010477	1.4182	1.4193	440.17	2243.6	2683.7	1.3630	5.9332	7.2962
110	143.27	0.0010519	1.2089	1.2099	461.32	2230.0	2691.3	1.4185	5.8203	7.2388
115	169.06	0.0010562	1.0352	1.0363	482.50	2216.2	2698.7	1.4733	5.7099	7.1832
120	198.54	0.0010606	0.8905	0.8915	503.72	2202.3	2706.0	1.5276	5.6017	7.1293
125	232.1	0.0010652	0.7692	0.7702	524.99	2188.0	2713.0	1.5813	5.4956	7.0769
130	270.1	0.0010700	0.6671	0.6681	546.31	2173.6	2719.9	1.6344	5.3917	7.0261
135	313.1	0.0010750	0.5807	0.5818	567.68	2158.9	2726.6	1.6869	5.2897	6.9766
140	361.4	0.0010801	0.5074	0.5085	589.10	2144.0	2733.1	1.7390	5.1894	6.9284
145	415.5	0.0010853	0.4449	0.4460	610.59	2128.7	2739.3	1.7906	5.0909	6.8815
150	476.0	0.0010908	0.3914	0.3924	632.15	2113.3	2745.4	1.8416	4.9942	6.8358

155	543.3	0.0010964	0.3453	0.3464	653.77	2097.4	2751.2	1.8923	4.8988	6.7911
160	618.1	0.0011022	0.3057	0.3068	675.47	2081.2	2756.7	1.9425	4.8050	6.7475
165	700.8	0.0011082	0.2713	0.2724	697.25	2064.8	2762.0	1.9923	4.7125	6.7048
170	792.0	0.0011145	0.2414	0.2426	719.12	2048.0	2767.1	2.0416	4.6214	6.6630
175	892.4	0.0011209	0.21542	0.21654	741.07	2030.7	2771.8	2.0906	4.5315	6.6221
180	1002.7	0.0011275	0.19267	0.19380	763.12	2013.2	2776.3	2.1393	4.4426	6.5819
185	1123.3	0.0011344	0.17272	0.17386	785.26	1995.2	2780.4	2.1876	4.3548	6.5424
190	1255.1	0.0011415	0.15517	0.15632	807.52	1976.7	2784.3	2.2356	4.2680	6.5036
195	1398.7	0.0011489	0.13969	0.14084	829.88	1957.9	2787.8	2.2833	4.1821	6.4654
200	1554.9	0.0011565	0.12600	0.12716	852.37	1938.5	2790.9	2.3307	4.0971	6.4278
205	1724.3	0.0011644	0.11386	0.11503	874.99	1918.8	2793.8	2.3778	4.0128	6.3906
210	1907.7	0.0011726	0.10307	0.10424	897.73	1898.5	2796.2	2.4247	3.9292	6.3539
215	2106.0	0.0011811	0.09344	0.09463	920.63	1877.6	2798.3	2.4713	3.8463	6.3176
220	2319.8	0.0011900	0.08485	0.08604	943.67	1856.2	2799.9	2.5178	3.7639	6.2817
225	2550	0.0011992	0.07715	0.07835	966.88	1834.3	2801.2	2.5641	3.6820	6.2461
230	2798	0.0012087	0.07024	0.07145	990.27	1811.7	2802.0	2.6102	3.6005	6.2107
235	3063	0.0012187	0.06403	0.06525	1013.83	1788.5	2802.3	2.6561	3.5195	6.1756
240	3348	0.0012291	0.05843	0.05965	1037.60	1764.6	2802.2	2.7020	3.4386	6.1406
245	3652	0.0012399	0.05337	0.05461	1061.58	1740.0	2801.6	2.7478	3.3579	6.1057
250	3978	0.0012513	0.04879	0.05004	1085.78	1714.6	2800.4	2.7935	3.2773	6.0708
255	4325	0.0012632	0.04463	0.04590	1110.23	1688.5	2798.7	2.8392	3.1967	6.0359
260	4694	0.0012756	0.04086	0.04213	1134.94	1661.5	2796.4	2.8848	3.1162	6.0010
270	5506	0.0013025	0.03429	0.03559	1185.23	1604.6	2789.9	2.9763	2.9541	5.9304
280	6420	0.0013324	0.02879	0.03013	1236.84	1543.6	2780.4	3.0683	2.7903	5.8586
290	7446	0.0013659	0.02417	0.02554	1290.01	1477.6	2767.6	3.1611	2.6237	5.7848
300	8593	0.0014041	0.020245	0.021649	1345.05	1406.0	2751.0	3.2552	2.4529	5.7081
310	9870	0.0014480	0.016886	0.018334	1402.39	1327.6	2730.0	3.3512	2.2766	5.6278
320	11289	0.0014995	0.013980	0.015480	1462.60	1241.1	2703.7	3.4500	2.0923	5.5423
330	12863	0.0015615	0.011428	0.012989	1526.52	1143.7	2670.2	3.5528	1.8962	5.4490
340	14605	0.0016387	0.009142	0.010780	1595.47	1030.7	2626.2	3.6616	1.6811	5.3427
350	16535	0.0017411	0.007058	0.008799	1671.94	895.8	2567.7	3.7800	1.4377	5.2177
360	18675	0.0018959	0.005044	0.006940	1764.17	721.2	2485.4	3.9210	1.1390	5.0600
370	21054	0.0022136	0.002759	0.004973	1890.21	452.6	2342.8	4.1108	0.7036	4.8144
374.15	22120	0.00317	0.0	0.00317	2107.37	0.0	2107.37	4.4429	0.0	4.4429

Table A21.2 Properties of Superheated Steam ν in m²/kh, h in kJ/kg, s in kJ/kg.K

Abs. Press. kPa (Sat. Temp.) °C		50	100	150	200	250	300	350	400	450	500	550	600	650	700	750	800
5.0 (32.90)	ν	29.782	34.417	39.042	43.661	48.280	52.897	57.513	62.129	66.745	71.360	75.976	80.592	85.207	89.822	94.438	99.063
	h	2593.7	2688.1	2784.3	2879.6	2977.6	3076.7	3177.4	3279.7	3383.6	3489.2	3596.5	3705.6	3816.3	3928.8	4043.0	4158.7
	s	8.4981	8.7698	9.0094	9.2248	9.4211	9.6021	9.7705	9.9283	10.0772	10.2184	10.3529	10.4815	10.6049	10.7235	10.8379	10.9483
10.0 (45.63)	ν	14.869	17.195	19.512	21.825	24.136	26.445	28.754	31.062	33.370	35.679	37.987	40.295	42.603	44.910	47.218	49.535
	h	2592.7	2687.5	2783.1	2879.6	2977.5	3076.6	3177.3	3279.6	3383.5	3489.1	3596.5	3705.5	3816.3	3928.8	4042.9	4158.7
	s	8.1757	8.4486	8.6889	8.9045	9.1010	9.2820	9.4505	9.6083	9.7572	9.8984	10.0329	10.1616	10.2849	10.4036	10.5180	10.6234
20.0 (60.09)	ν		8.5847	9.748	10.907	12.064	13.210	14.374	15.529	16.684	17.838	18.992	20.146	21.300	22.455	23.609	24.762
	h		2686.3	2782.4	2879.2	2977.2	3076.4	3177.2	3279.4	3383.4	3489.0	3596.4	3705.4	3816.2	3928.7	4042.9	4158.7
	s		8.1261	8.3676	8.5839	8.7806	8.9618	9.1304	9.2882	9.4372	9.5784	9.7130	9.8416	9.9650	10.0836	10.1980	10.3065
40.0 (75.89)	ν		4.2792	4.8657	5.4478	6.0277	6.6065	7.1846	7.7625	8.3401	8.9176	9.4950	10.072	10.649	11.227	11.804	12.381
	h		2683.8	2780.9	2878.2	2976.5	3075.9	3176.8	3279.1	3383.1	3488.8	3596.2	3705.3	3816.1	3928.6	4042.8	4158.6
	s		7.8009	8.0449	8.2625	8.4598	8.6413	8.8101	8.9680	9.1171	9.2583	9.3929	9.5216	9.6450	9.7636	9.8780	9.9665
60.0 (85.95)	ν		2.8440	3.2382	3.6281	4.0157	4.4022	4.7881	5.1736	5.5590	5.9441	6.3292	6.7141	7.0991	7.4839	7.8687	8.2535
	h		2681.3	2779.4	2877.3	2975.8	3075.4	3176.4	3278.8	3382.9	3488.6	3596.0	3705.1	3816.0	3928.5	4042.7	4158.5
	s		7.6085	7.8551	8.0738	8.2718	8.4536	8.6225	8.7806	8.9297	9.0710	9.2056	9.3343	9.4577	9.5764	9.6908	9.8013
80 (93.51)	ν		2.1262	2.4245	2.7183	3.0097	3.3000	3.5898	3.8792	4.1683	4.4574	4.7463	5.0351	5.3239	5.6126	5.9013	6.1899
	h		2678.8	2777.8	2876.3	2975.2	3075.0	3176.0	3278.5	3382.6	3488.4	3595.8	3705.0	3815.8	3928.4	4042.6	4158.4
	s		7.4703	7.7195	7.9395	8.1381	8.3200	8.4893	8.6475	8.7967	8.9380	9.0727	9.2014	9.3248	9.4436	9.5580	9.6685
100.0 (99.63)	ν		1.6955	1.9362	2.1723	2.4061	2.6387	2.8708	3.1025	3.3340	3.5653	3.7966	4.0277	4.2588	4.4898	4.7208	4.9517
	h		2676.2	2776.3	2875.4	2974.5	3074.5	3175.6	3278.2	3382.4	3488.2	3595.6	3704.8	3815.7	3928.2	4042.5	4158.3
	s		7.3618	7.6137	7.8349	8.0342	8.2166	8.3861	8.5442	8.6935	8.8348	8.9696	9.0982	9.2217	9.3405	9.4549	9.5654
150.0 (111.4)	ν			1.2851	1.4444	1.6013	1.7570	1.9122	2.0669	2.2215	2.3759	2.5303	2.6845	2.8386	2.9927	3.1468	3.3008
	h			2772.5	2872.9	2973.0	3073.3	3174.7	3277.5	3381.8	3487.6	3595.3	3704.4	3815.3	3927.9	4042.2	4158.0
	s			7.4194	7.6439	7.8447	8.0280	8.1976	8.3562	8.5057	8.6472	8.7820	8.9108	9.0343	9.1531	9.2676	9.3781
200.0 (120.2)	ν			0.9595	1.0804	1.1989	1.3162	1.4328	1.5492	1.6653	1.7812	1.8971	2.0129	2.1286	2.2442	2.3598	2.4754
	h			2768.5	2870.5	2971.2	3072.1	3173.8	3276.7	3381.1	3487.0	3594.7	3704.0	3815.0	3927.6	4041.9	4157.8
	s			7.2794	7.5072	7.7096	7.8937	8.0638	8.2226	8.3722	8.5139	8.6488	8.7776	8.9012	9.0201	9.1346	9.2452
300.0 (133.5)	ν			0.6337	0.7164	0.7964	0.8733	0.9535	1.0314	1.1090	1.1865	1.2639	1.3412	1.4185	1.4957	1.5728	1.6499
	h			2760.3	2865.5	2967.9	3069.7	3171.9	3275.2	3379.8	3486.0	3593.7	3703.2	3814.2	3927.0	4041.4	4157.3
	s			7.0771	7.3119	7.5176	7.7034	7.8745	8.0338	8.1838	8.3257	8.4608	8.5898	8.7135	8.8325	8.9471	9.0577
400.0 (143.6)	ν			0.4707	0.5343	0.5952	0.6549	0.7139	0.7725	0.8309	0.8892	0.9474	1.0054	1.0634	1.1214	1.1793	1.2372
	h			2752.0	2860.4	2964.5	3067.2	3170.0	3273.6	3378.5	3484.9	3592.8	3702.3	3813.5	3926.4	4040.8	4156.9
	s			6.9285	7.1708	7.3800	7.5675	7.7395	7.8994	8.0497	8.1919	8.3271	8.4563	8.5802	8.6992	8.8139	8.9246
500 (151.8)	ν				0.4250	0.4744	0.5226	0.5701	0.6172	0.6640	0.7108	0.7574	0.8039	0.8504	0.8968	0.9432	0.9896
	h				2855.1	2961.1	3064.8	3168.1	3272.1	3377.2	3483.8	3591.8	3701.5	3812.8	3925.8	4040.3	4156.4
	s				7.0592	7.2721	7.4614	7.6343	7.7948	7.9454	8.0879	8.2233	8.3526	8.4766	8.5957	8.7105	8.8213
600 (158.8)	ν				0.3520	0.3939	0.4344	0.4742	0.5136	0.5528	0.5918	0.6308	0.6696	0.7084	0.7471	0.7858	0.8245
	h				2849.7	2957.6	3062.3	3166.3	3270.6	3376.0	3482.7	3590.9	3700.7	3812.1	3925.1	4039.8	4155.9
	s				6.9662	7.1829	7.3740	7.5479	7.7090	7.8600	8.0027	8.1383	8.2678	8.3919	8.5111	8.6259	8.7368
800 (170.4)	ν				0.2608	0.2932	0.3241	0.3543	0.3842	0.4137	0.4432	0.4725	0.5017	0.5309	0.5600	0.5891	0.6181
	h				2838.6	2950.4	3057.3	3162.4	3267.5	3373.4	3480.5	3589.0	3699.1	3810.7	3923.9	4038.7	4155.0
	s				6.8148	7.0397	7	7.4107	7.5729	7.7246	7.8678	8.0038	8.1336	8.2579	8.3773	8.4923	8.6033

P (sat T)		(1)	(2)	(3)	(4)	(5)	(6)	(7)	(8)	(9)	(10)	(11)	(12)	(13)
1000.0 (179.9)	v	0.2059	0.2327	0.2560	0.2824	0.3065	0.3303	0.3540	0.3775	0.4010	0.4244	0.4477	0.4710	0.4943
	h	2826.8	2942.9	3052.1	3158.5	3264.4	3370.8	3478.3	3587.1	3697.4	3809.3	3922.7	4037.6	4154.1
	s	6.6922	6.9259	7.1251	7.3031	7.4665	7.6190	7.7627	7.8991	8.0292	8.1537	8.2734	8.3885	8.4997
1500 (198.3)	v	0.1324	0.1520	0.1697	0.1865	0.2029	0.2191	0.2350	0.2509	0.2667	0.2824	0.2980	0.3136	0.3292
	h	2794.7	2923.3	3038.9	3146.7	3256.6	3364.4	3472.8	3582.4	3693.3	3805.7	3919.6	4034.9	4151.7
	s	6.4508	6.7099	6.9207	7.1044	7.2709	7.4253	7.5703	7.7077	7.8385	7.9636	8.0836	8.1993	8.3108
2000 (212.4)	v		0.1114	0.1255	0.1386	0.1511	0.1634	0.1756	0.1876	0.1995	0.2114	0.2232	0.2349	0.2467
	h		2902.4	3025.0	3138.6	3248.7	3357.8	3467.3	3577.7	3689.2	3802.1	3916.5	4032.2	4149.4
	s		6.5454	6.7696	6.9396	7.1296	7.2859	7.4323	7.5706	7.7022	7.8279	7.9485	8.0645	8.1763
3000 (233.8)	v		0.07050	0.08116	0.09053	0.09931	0.1078	0.1161	0.1243	0.1323	0.1404	0.1483	0.1562	0.1641
	h		2854.0	2995.1	3117.5	3232.5	3344.6	3456.2	3568.1	3681.0	3795.0	3910.3	4026.8	4144.7
	s		6.2857	6.5422	6.7471	6.9246	7.0854	7.2345	7.3748	7.5079	7.6349	7.7564	7.8733	7.9857
4000 (250.3)	v			0.05883	0.06643	0.07338	0.07996	0.08634	0.09260	0.09876	0.1049	0.1109	0.1169	0.1229
	h			2962.0	3095.1	3215.7	3331.2	3445.0	3558.6	3672.8	3787.9	3904.1	4021.4	4140.0
	s			6.3642	6.5870	6.7733	6.9388	7.0909	7.2333	7.3680	7.4961	7.6187	7.7363	7.8495
5000 (263.9)	v			0.04530	0.05193	0.05779	0.06325	0.06849	0.07360	0.07862	0.08356	0.08845	0.09329	0.09809
	h			2925.5	3071.2	3196.3	3317.5	3433.7	3549.0	3664.5	3780.7	3897.9	4016.1	4135.3
	s			6.2105	6.4545	6.6508	6.8217	6.9770	7.1215	7.2578	7.3872	7.5108	7.6292	7.7431
6000 (275.5)	v			0.03614	0.04221	0.04738	0.05210	0.05659	0.06094	0.06518	0.06936	0.07348	0.07755	0.08159
	h			2885.0	3045.8	3180.1	3303.5	3422.2	3539.3	3656.2	3773.5	3891.7	4010.7	4130.7
	s			6.0692	6.3386	6.5462	6.7230	6.8818	7.0285	7.1664	7.2971	7.4217	7.5409	7.6554
8000 (295.0)	v			0.02426	0.02995	0.03431	0.03814	0.04170	0.04510	0.04839	0.05161	0.05407	0.05788	0.06096
	h			2786.8	2989.9	3141.6	3274.3	3398.8	3519.7	3639.5	3759.2	3879.2	3999.9	4121.3
	s			5.7942	6.1349	6.3694	6.5597	6.7262	6.8778	7.0191	7.1523	7.2790	7.3999	7.5158
10000.0 (311.0)	v				0.02242	0.02641	0.02974	0.03276	0.03560	0.03832	0.04096	0.04355	0.04608	0.04858
	h				2925.8	3099.9	3243.6	3374.6	3499.8	3622.7	3744.7	3866.8	3989.1	4112.0
	s				5.9489	6.2182	6.4243	6.5994	6.7564	6.9013	7.0373	7.1660	7.2886	7.4058
15000.0 (342.1)	v				0.01147	0.01566	0.01845	0.02080	0.02291	0.02488	0.02677	0.02859	0.03036	0.03209
	h				2696.0	2979.1	3159.8	3310.6	3450.6	3579.8	3708.3	3835.4	3962.1	4088.6
	s				5.4486	5.8876	6.1469	6.3487	6.5243	6.6764	6.8195	6.9536	7.0806	7.2013
20000.0 (365.7)	v					0.009947	0.01271	0.01477	0.01655	0.01816	0.01967	0.02111	0.02250	0.02385
	h					2820.5	3064.3	3241.1	3394.1	3535.5	3671.1	3803.8	3935.0	4065.3
	s					5.5585	5.9089	6.1456	6.3374	6.5043	6.6554	6.7953	6.9267	7.0511
30000.0	v					0.002831	0.006735	0.008681	0.01017	0.01144	0.01258	0.01365	0.01465	0.01562
	h					2161.8	2825.6	3085.0	3277.4	3443.0	3595.0	3739.7	3880.3	4018.5
	s					4.4896	5.4495	5.7972	6.0386	6.2340	6.4033	6.5560	6.6970	6.8248
40000.0	v					0.001909	0.003675	0.005616	0.006982	0.008088	0.009053	0.009930	0.01075	0.01152
	h					1934.1	2515.6	2906.8	3151.6	3346.4	3517.0	3674.8	3825.5	3971.7
	s					4.1190	4.9511	5.4762	5.7835	6.0135	6.2035	6.3701	6.5210	6.6606
50000.0	v					0.001729	0.002492	0.003882	0.005113	0.006111	0.006960	0.007720	0.008420	0.009076
	h					1877.1	2293.2	2723.0	3021.1	3248.3	3438.3	3610.2	3770.9	3925.3
	s					4.0083	4.6026	5.1782	5.5525	5.8207	6.0331	6.2138	6.3749	6.5222

22. Second Law of Thermodynamics

22.1 Introduction

It is a common experience that many processes occur spontaneously, while we need to expend effort in carrying out other processes. Thus, air will rush into an evacuated tank by itself, but it takes great amount of effort, and equipment, to suck it back out. Similarly, sugar dissolves easily into water but, once dissolved, its separation is not possible without expending energy. The development of heat engines converting thermal energy into work, made it apparent that though we could easily convert mechanical energy into thermal energy, the reverse was not an easy process. In any case, not all the thermal energy available could be fully converted to mechanical energy. This lead to the development of the concept of degradation of energy. Based on the fact that it is the mechanical energy which does useful work for us, we treat it as the energy of interest, and whenever the amount of energy obtained as mechanical work decreases, we say that the energy of the system has *degraded*. The second law of thermodynamics is a formalization of this principle of degradation of energy.

On the basis of the accumulated experience with the processes involving exchange of energy, it can be stated that *all processes on an isolated system always lead to degradation of energy such that the ability to do mechanical work always decreases.*

22.2 Energy Reservoir

First, let us introduce the idea of energy reservoir. Since there are two types of energy interactions, work and heat, we introduce two types of reservoirs. A *work reservoir* is a device that interacts with a system only through work interactions. It can be visualized as a perfectly elastic spring that is compressed by any work done on it and the energy can be recovered as work done when the spring is decompressed. Similarly, a rotating shaft which raises or lowers a weight on a pulley is a work reservoir. The energy is stored as potential energy which can be made to do work as the shaft rotates to let the weight down.

A *heat reservoir*, on the other hand, is a thermodynamic system which interacts only through heat transfer. It is visualized as a heat source or sink of infinite thermal capacity, so that any heat added to it or taken from it does not change its temperature. Note that, this heat reservoir will spontaneously exchange heat with a system only in a direction governed by the temperature difference. Thus, it will gain heat from a system if its temperature (i.e., of the reservoir) is less than that of the system, and will lose heat to a system only if its temperature (i.e., of the reservoir) is more than that of the system.

22.3 Available Energy and Entropy

Since not all the energy of a system can be extracted as work, it is reasonable to introduce the concepts of available and unavailable energy. The portion of the energy content of a given system that can potentially be used to do mechanical work is termed as the *available energy* of the system. If E_{av} denotes the available energy and E_{ua}, the unavailable energy, clearly, the total energy content is given by

$$E = E_{av} + E_{ua} \qquad (22.1)$$

We introduce an extensive primitive property termed as *entropy* S, the changes in which are related to the increase in the unavailable energy. Thus, if dE is the change in energy of a system is a process, and dE_{av} represents the changes in the available part of the energy, then the change in entropy dS is defined by

$$dS = C(dE - dE_{av}) \qquad (22.2)$$

where C is a positive constant whose value depends on the standard reservoir with which the system can exchange energy.

For an isolated system, any process must result in no change in the total energy of the system, according to the first law of thermodynamics. The principle of degradation of energy then leads to

$$(dS)_{\text{isolated system}} \geq 0 \qquad (22.3)$$

Equation 22.3 is the mathematical statement of the second law of thermodynamics and states that the *entropy of an isolated system can never decrease.*

Since any one system and all its surroundings constitute an isolated system, we may write, with reference to Eq.22.3,

$$(dS)_{\text{system}} + (dS)_{\text{surroundings}} \geq 0 \qquad (22.4)$$

Let us consider a reversible process AB as shown in Fig. 22.1. The net change in entropy of the system and the surroundings

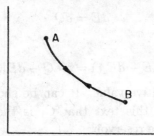

Fig. 22.1 A reversible process.

in the process AB and BA must equal zero, since the system and the surroundings return to their original states. But since the entropy change in each process must not be negative, we can conclude that *the net entropy change in any reversible process is zero.* Thus, the inequality signs in Eqs. (22.2) and (22.3) apply to an irreversible process. In other words, irreversibility of a process leads to increase in entropy of the universe. We may write

$$dS_{irr} = (dS)_{\text{system}} + (dS)_{\text{surroundings}} \qquad (22.5)$$

This signifies that the entropy of the universe is constantly being created due to irreversibilities in the various processes, and that the available energy is constantly being degraded.

22.4   Entropy Changes in Some Elementary Processes

22.4.1   Entropy Changes in a Work Reservoir

Since the energy in a work reservoir is always stored as energy available to do work (e.g., as compression energy of a perfect spring, or the potential energy of a mass raised by a shaft), a change in energy always equals the change in available work, and therefore, by Eq. (22.2), the change in entropy is zero. Thus,

$$[dS]_{\text{work reservoir}} = 0 \qquad (22.6)$$

22.4.2   Entropy Changes of a Heat Reservoir

A heat reservoir interacts with its surrounding through transfer of heat alone, with no work done. Thus, by first law

$$dE = \bar{d}Q$$

This leads to

$$dS \;=\; c(dE - dE_{av}) = c(\bar{d}Q - dE_{av}) = C_1 \bar{d}Q$$

where C_1 has a positive value. It can be shown, by arguments beyond the level of this text that C_1 is the reciprocal of the temperature T of the reservoir.
Thus,

$$[dS]_{\text{heat reservoir}} = \frac{\bar{d}Q}{T} \qquad (22.7)$$

where T is the temperature of the reservoir.

22.4.3   Entropy Change in Heat Transfer between two Heat Reservoirs

Let us consider an isolated system consisting of two heat reservoirs A and B with temperatures of T_1 and T_2, respectively (Fig. 22.2).

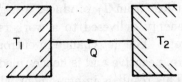

Fig. 22.2 Heat transfer between two heat reservoirs.

If a quantity Q of heat is transferred from reservoir A at temperature T_1 to B at temperature T_2, the changes in entropy are

$$(\Delta S)_A = -Q/T_1,$$
$$(\Delta S)_B = +Q/T_2$$

The two have opposite signs, since heat added to a system is positive, while heat removed from a system carries a negative sign. The net entropy increase of the isolated system then is

$$(\Delta S)_{total} = Q\left[\frac{1}{T_2} - \frac{1}{T_1}\right] \tag{22.8}$$

Since the second law postulates $(\Delta S)_{total}$ to be positive, T_2 must be less than T_1, or that heat is transferred from a higher temperature reservoir to a lower temperature one, if no other agency is involved in the process. The following formulations of the second law due to Clausius (which was arrived at independently, and is used by many authors to introduce the concept of entropy) follows directly from this:

It is impossible to operate a cyclic device whose sole effect is to transfer heat from a reservoir at lower temperature to another at a higher temperature.

22.4.4 Heat Engine

Consider a device which picks up energy Q_1 from a heat reservoir A at temperature T_1, converts it into mechanical energy W and transfers it to a work reservoir.

If this device together with the heat reservoir A and the work reservoir constitute an isolated system such that there is no interaction with any other system, the net entropy change will be

$$(\Delta S)_{total} = -\frac{Q_1}{T_1}$$

since Q_1, the heat lost from T_1 produces a negative entropy increase, and any energy delivered to a work reservoir produces no change in entropy. The net change in entropy of the isolated system is, therefore, negative and is hence ruled out by the principle of entropy. This result is in agreement with the statement of second law attributed to Kelvin and Planck:

It is impossible to construct a device which operates in a cycle and produces no effect other than taking heat from a heat source and converting it into work.

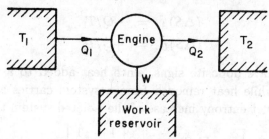

Fig. 22.3 Schematic of a heat engine working between two heat reservoirs.

If to this we add sink of heat, i.e., a reservoir B at a temperature T_2 to which Q_2 heat is lost in the process, the second law is not violated. Thus, the net entropy increase in the system depicted in Fig. 22.3 is

$$(\Delta S)_{\text{isolated system}} = \frac{-Q_1}{T_1} + \frac{Q_2}{T_2} \qquad (22.9)$$

This is possible if $Q_2/T_2 \geq Q_1/T_1$. The energy balance gives $W = Q_1 - Q_2$, so that $Q_2 = Q_1 - W$.
Therefore,

$$\frac{Q_1 - W}{T_2} \geq \frac{Q_1}{T_1} \quad \text{or} \quad W \leq Q_1\left[1 - \frac{T_2}{T_1}\right] \qquad (22.10)$$

Note that both T_1 and T_2 are absolute temperatures. Since $T_2 > 0$, we find that the work delivered by an engine is strictly less than Q_1, the energy lost by the heat source.

The equality sign in Eqs. (22.9) and (22.10) is applicable only in reversible processes which result in no increase in entropy. Thus, Eq. (22.10) establishes the limit of how much work W can be delivered per unit energy picked up as heat. We define the thermal efficiency η of a heat engine as

$$\eta = \frac{W}{Q_1} \leq \left[1 - \frac{T_2}{T_1}\right] \qquad (22.11)$$

For reversible processes, the resulting limiting efficiency is termed as the *Carnot efficiency*

$$\eta_c = \left[1 - \frac{T_2}{T_1}\right] \qquad (22.12)$$

and $\eta \leq \eta_c$. The *Carnot principle* summarizes the above as follows:

No heat engine can be more efficient than a reversible heat engine operating between the same constant temperature reservoirs.

An important corollary of the Carnot principle follows immediately, and has important implications:

All reversible engines operating between two constant temperature reservoirs have the same efficiencies irrespective of the type of working substance.

Example 22.1

Consider an engine using a perfect gas as a working substance which undergoes the following reversible cyclic operations, shown graphically in Fig. 22.4:

(a) an isothermal energy pick up $(1\rightarrow2)$ from a heat-source at temperature T_1;

(b) an adiabatic expansion $(2\rightarrow3)$ from temperature T_1 to T_2;

(c) an isothermal energy rejection $(3\rightarrow4)$ to a heat sink at temperature T_2; and

(d) an adiabatic compression resulting in an increase in temperature from T_2 back to T_1 completing the cycle.

If the two heat transfer processes are carried out quasi- statically, the cycle is reversible and is termed as the *Carnot cycle*.

Show that for this cycle the first law is satisfied, and that the thermal efficiency of the cycle is $(1 - T_2/T_1)$, in accordance with the Carnot principle.

Solution

If the mass of the working gas is m, the energy exchanges in the various processes are obtained as below:

During the isothermal heating process: Since temperature is constant, internal energy of the working substance does not change, and the heat addition Q_{12} is equal to the work done in the isothermal process form volume V_1 to V_2. Thus,

$$Q_{12} = W_{12} = \int_1^2 pdV = \int_1^2 \frac{mRT}{V}dV = mRT_1 \ln \frac{V_2}{V_1}$$

During the adiabatic expansion: $Q_{23} = 0$, and work done by the gas results in the reduction of internal energy, which is readily evaluated. Thus,

$$Q_{23} = 0$$
$$W_{23} = -\Delta U_{23} = -mC_v(T_2 T_1)$$
$$= \frac{mR}{\gamma - 1}(T_1 - T_2)$$

During the isothermal heat rejection: Again the temperature is constant, and

$$Q_{34} = W_{34} = \int_3^4 pdV = -mRT_2 \ln \frac{V_3}{V_4}$$

Since $V_3 > V_4$, Q_{34} is negative signifying that heat is lost.

During the adiabatic compression: $Q_{41} = 0$; and $W_{41} = -\Delta U_{41} = -\frac{mR}{\gamma-1}(T_1 - T_2)$. The negative sign signifies that the work is done on the system.

The net work done in the cycle

$$W = mRT_1 \ln \frac{V_2}{V_1} + \frac{mR}{\gamma - 1}(T_1 - T_2) - mRT_2 \ln \frac{V_3}{V_4} - \frac{mR}{\gamma - 1}(T_1 - T_2)$$

Since the processes 2-3, and 4-1 are adiabatic

$$\frac{V_3}{V_2} = \frac{V_4}{V_1} = \left(\frac{T_1}{T_2}\right)^{\frac{1}{\gamma-1}}$$

Then,

$$\frac{V_2}{V_1} = \frac{V_3}{V_4} = a, \text{ say,}$$

and

$$W = mR(T_1 - T_2) \ln a$$

Q_1, (heat picked up from the source reservoir) $= mRT_1 \ln a$, and Q_2 (the heat rejected at sink) $= mRT_1 \ln a$.

It is immediately verified that the net work delivered W is equal to the net heat picked up $= Q_1 - Q_2$, in accordance with the first law, and the thermal efficiency of the cycle is

$$\frac{W}{Q_1} = \frac{mR(T_1 - T_2) \ln a}{mRT_1 \ln a} = \frac{T_1 - T_2}{T_1} = 1 - \frac{T_2}{T_1}$$

which is in accordance with the Carnot principle.

Example 22.2

A Carnot engine works between 1727 °C and 27 °C. Obtain the energy picked up from the source reservoir per hour if the engine delivers 17 kW.

Solution

The Carnot efficiency

$$\eta_c = 1 - \frac{T_2}{T_1} = 1 - \frac{27 + 273}{1727 + 273} = 0.85$$

But the engine efficiency is defined as work delivered/energy supplied. Therefore, the energy supplied Q_1 is

$$\dot{Q}_1 = \frac{\dot{W}}{\eta} = \frac{17 \text{ kW}}{0.85} = 20 \text{ kW}$$

so, energy picked up in 1 hr is

$$Q_1 = 20 \text{ kW} \times 3600 \text{ sec} = 72 \times 10^6 \text{ J}$$

Example 22.3

An inventor claims to have developed a heat engine which produces 5 kW by using 400 kJ/min while operating between 1000 K and 300 K. Is the claim possibly true?

Solution

The thermal efficiency claimed by the inventor

$$\eta = \frac{5 \times 10^3}{400 \times 10^3/60} = 0.75$$

The Carnot efficiency between the two limiting temperatures is

$$\eta_c = 1 - \frac{T_2}{T_1} = 1 - \frac{300}{1000} = 0.70$$

Since the claimed efficiency is higher than the Carnot efficiency between the same temperatures, the claim cannot be true.

22.5 Heat Pump

Though the Clasius statement of the second law forbids transfer of heat from a lower temperature to a higher temperature, if this is the sole effect of the process, nothing forbids a device for transferring heat from a lower to a high temperature reservoir, if there is some other effect on the surroundings as well. In fact, reversed Carnot cycle (we have postulated the Carnot cycle to be reversible) picks up heat form the reservoir at lower temperature T_2, and delivers it to the reservoir at the higher temperature T_1 (Fig. 22.5). The principle of entropy requires that

$$(\Delta S)_{\text{isolated system}} = \frac{Q_1}{T_1} - \frac{Q_2}{T_2} \geq 0$$

From the principle of energy conservation,

$$Q_2 + W = Q_1$$

Therefore,

$$\frac{Q_2}{W} \leq \frac{T_2}{T_1 - T_2} \tag{22.13}$$

and

$$\frac{Q_1}{W} \leq \frac{T_1}{T_1 - T_2} \tag{22.14}$$

Clearly, since $Q_1 > Q_2$, we cannot expect to transfer any heat Q_2 from the cold reservoir without doing any work on the device.

Thus, while heat spontaneously moves from a hotter to a cooler reservoir, the reverse is possible only at the expenditure of work.

In a heat pump we are interested in the heat transferred to the higher temperature reservoir. Therefore, the coefficient of performance of a heat pump is defined as

$$(COP)_{\text{Heat pump}} = \frac{Q_1}{W} \le \frac{T_1}{T_1 - T_2} \qquad (22.15)$$

In a refrigerator, on the other hand, we are interested in the heat extracted from the colder reservoir. Therefore, the coefficient of performance of a refrigerator is defined as

$$(COP)_{\text{refrigerator}} = \frac{Q_2}{W} \le \frac{T_2}{T_1 - T_2} \qquad (22.16)$$

Note that COPs for both the heat pump and the refrigerator are greater than 1. That is why it is always more efficient to use a heat pump for maintaining a room warm in winters rather than using a heater which runs on electricity.

Example 22.5

The data sheet of an air conditioner gives the coefficient of performance as 12 when the ambient is 47°C and room is maintained at a cool 17°C. Is the claim possibly correct?

Solution

COP of an airconditioner (like that of a refrigerator) is given by Eq. (22.16) as

$$COP \le \frac{T_2}{T_1 - T_2} = \frac{273 + 17}{47 - 17} = \frac{290}{30} = 9.67$$

Since the advertised value is more than this, the claim cannot be correct.

Example 22.6

A heat pump adds 21 kW heat to a room which is maintained at 27°C. Find the power consumption, given that the actual coefficient of performance of the device is only 70 percent of what it can be for an ideal device.

Solution

$$(COP)_{ideal,heat\ pump} = \frac{T_1}{T_1 - T_2} = \frac{273 + 27}{27 - 7} = \frac{300}{20} = 15$$

The COP of the actual device then is $15 \times 0.70 = 10.5$, Thus,

$$\frac{Q_1}{W} = 10.5$$

or

$$W = \frac{Q_1}{10.5} = \frac{21\ kW}{10.5} = 2\ kW$$

This is the the power consumption by the device while it adds 21 kW to the room. A heating element would have used the full 21 kW!

Problems

22.1 Heat is taken up by a heat engine at a rate of 50×10^6 kJ/hr. If the power output is 5000 kW. What is the thermal efficiency of the engine?

22.2 If the power output of a heat engine is 12 MW and its thermal efficiency 40%, what must be the rate of heat rejection?

22.3 The power input to a refrigerator is 60 kW, and the heat rejection is 1000 kW. What is its coefficient of performance?

22.4 A heat pump delivers 10^5 kJ of heat per hour to a room. What is the power requirement if the coefficient of performance is 4?

22.5 A reversible heat engine operating in a cycle receives 500 kJ from a reservoir at 1800 K and rejects some heat to a reservoir at 320 K. Determine the entropy changes in the two reservoirs. What is the entropy change of the universe? Explain the significance.

22.6 An inventor claims that to have designed an engine which picks up 2000 J from a reservoir at 2000 K and rejects 300 J to a reservoir at 300 K. Can the claim be possibly correct?

22.7 A refrigerator operating on a cycle removes heat from a reservoir at 250 K, does 500 J work on it and rejects 3000 J at 300 K. Is this process reversible, irreversible, or impossible?

22.8 It is proposed to generate power from the thermal gradients in oceans. If the maximum ocean temperature is 30°C and the minimum 5°C, what is the maximum possible thermal efficiency of such an engine?

22.9 A heat pump delivers 100 kW to a room at 25°C when the outside temperature is −10°C. What is the power of the motor required to run the pump?

22.10 A reversible heat engine receives heat from a high temperature reservoir at T_1 and rejects heat at 1000 K. A second reversible engine receives this heat at 1000 K and rejects heat to another reservir at 300 K. If the two engines have the same efficiencies, what should be the value of T_1?

22.11 Repeat, problem 22.10 above, except that we want the work output of the two engines to be equal.

22.12 A scheme which may be used to produce high-temperature process heat by making use of a low-temperature heat source is to run a heat engine between the low-temperature source at 373 K (boiling water) and the ambient at 300 K. The output of this source is used in a heat pump to pump heat from the ambient to a high-temperature reservoir at 425 K. Determine the heat taken from the 373 K reservoir for each Joule of heat delivered at 425 K reservoir, if all processes are reversible.

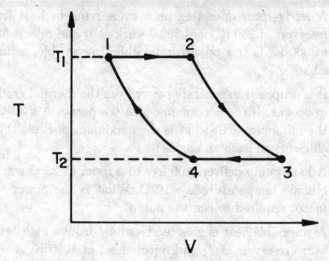

Fig. 22.4 A reversible-cycle heat engine.

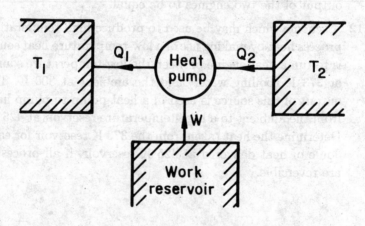

Fig. 22.5 A heat pump.

23. SOME ENERGY CONVERSION CYCLES

23.1 Otto Cycle

A German engineer, N.A. Otto, constructed an engine in 1876 which is the prototype of most of the petrol-engines used in automobiles. The engine is essentially a cylinder-piston assembly (Fig. 23.1) in which four strokes of the piston complete a cycle. In the first stroke, the piston moves down the cylinder sucking-in the fuel (petrol vapour) and air mixture across the open inlet valve. In the second stroke, valves are closed and the piston does work on the mixture compressing it to a small volume, heating it in the process. When the piston nears the top of its stroke, a spark is generated by the electrical circuitary making the gas explode in the cylinder driving the piston down in the power stroke. The fourth stroke is the exhaust stroke when the gases are driven out of the cylinder, to prepare the engine for the start of a fresh cycle. The opening of the exhaust and the inlet valves as well as the sparking is controlled by cams driven by a cam shaft which is turning at half the speed of the crank shaft, since two rotations of the crank shaft (consisting of two up and two down strokes of the piston) constitute one cycle in which each valve opens and closes only once and one spark is produced.

Figure 23.2 shows, in solid lines, a typical pressure volume curve obtained in the engine, with the major events marked with

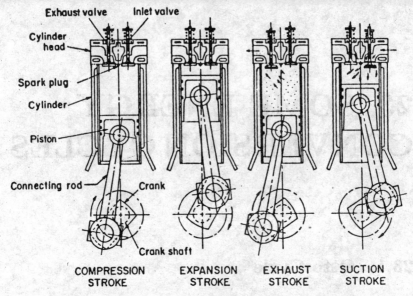

Fig. 23.1 Schematic of a spark-ignition petrol engine.

points; termed as the *indicator* diagram of the engine.

The indicator diagram is complicated enough to preclude any analysis, and therefore, we idealize it to what is termed as the *standard Otto cycle* as shown in Fig. 23.3. The four processes in the idealized standard cycle are:

(a) An adiabatic compression (1→ 2) in the compression stroke with both valves closed.

(b) A constant volume heat addition process (2→ 3) coinciding with the occurence of the spark at 2.

(c) An isentropic expansion from the high pressure p_3 in the power stroke.

(d) A constant volume heat rejection (4→ 1).

On the basis of a unit mass of gas in the Otto cycle, we can calculate the energy interactions.

Heat input Q_{in} is Q_{23}:

$$Q_{23} = C_v(T_3 - T_2) \tag{23.1}$$

Heat rejection Q_{out} is Q_{41}:

$$Q_{41} = C_v(T_4 - T_1) \tag{23.2}$$

The thermal efficiency of the engine is

$$\eta = \frac{\text{Work done}}{\text{Heat in}} = \frac{\text{Heat in} - \text{Heat out}}{\text{Heat in}}$$

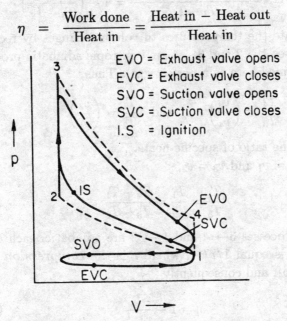

EVO = Exhaust valve opens
EVC = Exhaust valve closes
SVO = Suction valve opens
SVC = Suction valve closes
I.S = Ignition

Fig. 23.2 Indicator diagram of a petrol engine.

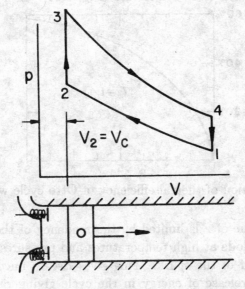

Fig. 23.3 Idealised Otto cycle.

$$= \frac{(T_3 - T_2) - (T_4 - T_1)}{T_3 - T_2} = 1 - \left(\frac{T_4 - T_1}{T_3 - T_2}\right) \quad (23.3)$$

We can relate the temperatures to volume changes by recalling that processes $1 \to 2$ and $3 \to 4$ are isentropic adiabatic processes on air (assumed here as a perfect gas). Thus,

$$\frac{T_4}{T_3} = \left(\frac{v_3}{v_4}\right)^{\gamma-1} \quad \text{and} \quad \frac{T_1}{T_2} = \left(\frac{v_2}{v_1}\right)^{\gamma-1}$$

where γ is the ratio of specific heats.

Since $v_4 = v_1$ and $v_3 = v_2$,

$$\frac{T_4}{T_3} = \frac{T_1}{T_2} = \frac{T_4 - T_1}{T_3 - T_2} \quad (23.4)$$

Since the processes $3 \to 4$ and $1 \to 2$ are adiabatic, each of the above ratio is equal $1/r^{\gamma-1}$, where r is the *compression ratio* $v_2/v_1 = v_3/v_4$, and consequently

$$\eta = 1 - \frac{1}{r^{\gamma-1}} \quad (23.5)$$

This equation indicates that the Otto cycle efficiency increases as the compression ratio r increases (Fig. 23.4). In a practical

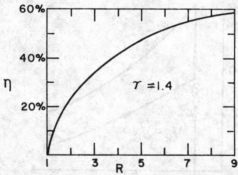

Fig. 23.4 Variation of themal efficiency of Otto cycle with compression ratio r.

engine, the value of r is limited by the tendency of the fuel-air mixture to explode at high temperatures and pressures by itself without the aid of the spark. This gives rise to uncontrolled, and untimely, release of energy in the cycle giving rise to the phenomenon of *knocking*, which leads to loss of energy.

Example 23.1

An Otto cycle starts with atmospheric air at a temperature of 300 K and a pressure of 100 kPa. The compression ratio is 8 and the amount of heat added is 2200 kJ/kg of air. Determine the thermal efficiency of the cycle, the power delivered by the engine and the maximum temperature and pressure reached by the gas. The engine is running at 3000 RPM, and the mass of air sucked-in is 5×10^{-4} kg/cycle.

Solution

We can directly find the thermal efficiency of the ideal Otto cycle from Eq. (23.5) with $r = 1.4$ as

$$\eta = 1 - \frac{1}{8}^{(1.4-1)} = 0.565$$

Since heat input is 2200 kJ/kg the work output per cycle is 2200 × 0.565 = 1243 kJ/kg of air or 0.6215 kJ/cycle. Since the engine is running at 3000 RPM or 50 revolutions/second, it completes 25 cycles/second, two revolutions of crank being equal to one complete thermodynamic cycle. Thus, the power output is 0.6215× 25= 15 kW, about 20 HP.

Also, from Eqs. (23.3) and (23.4), (Fig. 23.3)

$$\eta = 1 - \frac{T_1}{T_2}$$

Therefore, $T_2 = T_1/(1 - \eta) = 300/(1 - 0.565) = 690$ K.

The maximum temperature T_3 is obtained by Eq. (23.1) which relates heat addition. Thus,

$$Q_{23} = C_v(T_3 - T_2)$$

or

$$T_3 = T_2 + \frac{Q_{23}}{C_v} = 690 + \frac{2200}{0.7168} = 3760 \text{ K}$$

This is the maximum temperature.

To find the maximum pressure p_3, we first find p_2 which is related to p_1 by an isentropic process. Thus,

$$\frac{p_2}{p_1} = \left(\frac{v_1}{v_2}\right)^\gamma = 8^{1.4} = 18.4$$

or $p_2 = 18.4 \times 100$ kPa $= 1840$ kPa. The process $2 \to 3$ is constant volume heat addition, so that

$$p_3/T_3 = p_2/T_2$$

$$p_3 = p_2\frac{T_3}{T_2} = 1840 \text{ kPa} \left(\frac{3760 \text{ K}}{690 \text{ K}}\right) = 10024 \text{ kPa},$$

or above 100 bar.

We can compare the cycle efficiency with the efficiency of the ideal Carnot cycle operating between the maximum and the minimum temperatures. Thus,

$$\eta \text{Carnot} = 1 - \frac{\text{minimum temp.}}{\text{maximum temp.}} = 1 - \frac{300}{3760} = 0.92$$

The efficiency of the actual cycle, 56.5%, is considerably less than 92%, of the maximum possible efficiency of any engine which operates between 3760 K and 300 K.

23.2   Diesel Cycle

An engine running on the Diesel cycle differs from one running on the Otto cycle in two important ways. In a Diesel cycle, instead of the fuel and air mixture being inducted in the cylinder in the suction stroke, only the air is sucked-in. The fuel is injected near the end of the compression stroke, and it ignites without the aid of the spark. The increased compression ratio in the engine results in the temperature of the gas at the end of the compression stroke to be so high that the fuel ignites by itself. The resulting indicator diagram can be idealized as the one shown in Fig. 23.5.

The four processes are idealized as

• an isentropic compression (of air) from 1 to 2,

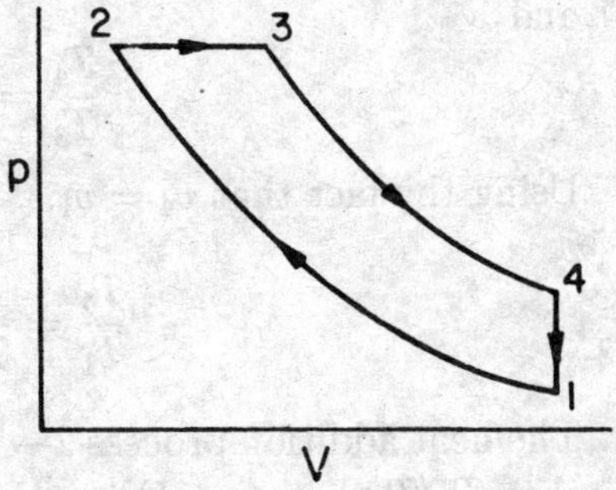

Fig. 23.5 The Diesel cycle used
in compression ignition
diesel engines.

- the constant pressure combustion of the injected fuel from
  2 to 3 during a part of the power stroke,

- an isentropic expansion from 3 to 4 during the remaining
  part of the power stroke; and

- a constant volume heat rejection from 4 to 1 completing
  the cycle.

We do the energy analysis of the cycle of the basis of a unit
mass of air.

Heat addition takes place during 2-3 at constant pressure, so
that

$$Q_{in} = C_p(T_3 - T_2) \qquad (23.6)$$

Heat rejection is at constant volume during 4 to 1, so that

$$Q_{out} = c_v(T_4 - T_1) \qquad (23.7)$$

Net work done, by first law, is

$$W = Q_{in} - Q_{out} = C_p(T_3 - T_2) - C_v(T_4 - T_1)$$

and the thermal efficiency is

$$\eta_{th} = \frac{C_p(T_3 - T_2) - C_v(T_4 - T_1)}{C_p(T_3 - T_2)}$$

$$= 1 - \frac{1}{\gamma} \frac{(T_4 - T_1)}{(T_3 - T_2)} \qquad (23.8)$$

Since the processes $1 \rightarrow 2$ and $3 \rightarrow 4$ are isentropic,

$$\frac{T_2}{T_1} = \left(\frac{v_1}{v_2}\right)^{\gamma-1} \qquad (23.9)$$

and

$$\frac{T_4}{T_3} = \left(\frac{v_3}{v_4}\right)^{\gamma-1} \tag{23.10}$$

Using the fact that $v_4 = v_1$,

$$\frac{T_4}{T_1} = \frac{T_3}{T_2}\left(\frac{v_3}{v_2}\right)^{\gamma-1}$$

The heat addition process $2 \rightarrow 3$ is a constant pressure process, so that $T_3/T_2 = v_3/v_2$. Using these temperature ratios, Eq. (23.9) gives

$$\eta_{th} = 1 - \frac{1}{r_v^{\gamma-1}}\frac{r_c^{\gamma} - 1}{\gamma(r_c - 1)} \tag{23.11}$$

where $r_v = (v_1/v_2)$ is termed as the *compression ratio* and $r_c = (v_3/v_2)$ is termed as the *cut-off ratio*. Figure 23.6 shows the

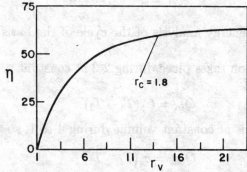

Fig. 23.6 Variation of thermal efficiency of a Diesel cycle with the compression ratio r_v for a fixed ratio r_c=1.8 cut-off.

variation of η_{th} of a Diesel cycle with the compression ratio r_v for a fixed value of the cut-off ratio $r_c = 1.8$. Diesel engines usually operate at much higher compression ratios r_v than do the petrol engines running on Otto cycles, and hence are a lot more efficient.

Example 23.2

A Diesel cycle starts with taking in air at 100 kPa and 300 K. If the compression ratio r_v is 16 and the amount of heat addition is 2200 kJ/kg of air, determine the maximum cycle pressure and temperature and the thermal efficiency. Compare it with the Carnot efficiency.

Solution

With reference to Fig. 23.5,

$$v_1 = \frac{RT_1}{p_1} = \frac{286 \times 300}{100 \times 10^3} = 0.86 \text{ m}^3/\text{kg}$$

$$v_2 = v_1/r_v = 0.86/16 = 0.054 \text{ m}^3/\text{kg}$$

$$T_2/T_1 = (v_1/v_2)^{\gamma-1} = (16)^{0.4} = 3.03$$

So that, $T_2 = 3.03 \times 300 = 909$ K. Similarly,

$$p_2 = p_1(v_1/v_2)^{\gamma} = 100 \times 10^3(16)^{1.4} = 4850 \text{ kPa}$$

This is the highest pressure.

The maximum temperature is obtained at 3.

Use $q_{in} = C_p(T_3 - T_2)$, so that $T_3 = T_2 + q_{in}/C_p$ with

$$C_p = \frac{\gamma R}{\gamma - 1} = \frac{1.4 \times 286}{1.4 - 1} = 1 \times 10^3$$

$$q_{in} = 909 + 2200 \times 10^3/1 \times 10^3 = 3109 \text{ K}$$

The cycle thermal efficiency is obtained from Eq. (23.9) as

$$\eta = 1 - \frac{1}{\gamma}\frac{(T_4 - T_1)}{(T_3 - T_2)}$$

To use this, we need to find out T_4 as everything else is known. The value of T_4 is determined by using Eq. (23.10) as

$$T_4 = T_3\left(\frac{v_3}{v_4}\right)^{\gamma-1}$$

Here, $v_3 = RT_3/p_3 = 286 \times 3109/485 \times 10^3 = 0.183 \text{ m}^3/\text{kg}$, so that $T_4 = 3109(0.183/0.86)^{1.4-1} = 1674$ K.

The value of the cycle thermal efficiency then is

$$\eta = 1 - \frac{1}{1.4}\left(\frac{1674 - 300}{3109 - 909}\right) = 0.554$$

The Carnot efficiency for an engine operating between the two extreme temperatures of 3109 K and 300 K is

$$\eta_c = 1 - \frac{300}{3109} = 0.904$$

We again see that the actual efficiency of the cycle is well below the Carnot efficiency (because the heat addition and rejection processes do not occur at constant temperatures).

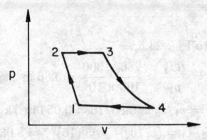

Fig. 23.7 Brayton cycle for gas turbines.

23.3   Brayton Cycle for Gas Turbines

The Brayton cycle (Fig. 23.7) is the basic cycle of the gas turbine in which a gas is first isentropically compressed along 1 to 2 in the compressor stage. The combustion and heat release process is essentially a constant pressure process $2 \to 3$. The gas expands isentropically in the turbine stages where energy is delivered as work on the shaft. The process is completed with the heat rejection at constant pressure, $4 \to 1$. For unit mass of air as the working fluid,

$$Q_{in} = Q_{23} = C_p(T_2 - T_1) \qquad (23.12)$$

$$Q_{out} = Q_{41} = C_p(T_4 - T_1) \qquad (23.13)$$

and

$$\eta_{th} = \frac{Q_{in} - Q_{out}}{Q_{in}} = 1 - \frac{T_4 - T_1}{T_3 - T_2} \qquad (23.14)$$

For the isentropic processes $1 \to 2$ and $3 \to 4$,

$$\frac{T_3}{T_4} = \left(\frac{p_3}{p_4}\right)^{(\gamma-1)/\gamma}$$

and

$$\frac{T_2}{T_1} = \left(\frac{p_2}{p_1}\right)^{(\gamma-1)/\gamma}$$

Since $p_4 = p_1$ and $p_3 = p_2$, $T_3/T_4 = T_2/T_1$. Thus,

$$\eta_{th} = 1 - \frac{T_1}{T_2} = 1 - \frac{T_4}{T_3} = 1 - \frac{1}{r_p^{(\gamma-1)/\gamma}} \qquad (23.15)$$

where r_p is the pressure ratio p_2/p_1.

Example 23.3

A gas turbine operating on Brayton cycle has the maximum and minimum pressures of 500 kPa and 100 kPa, respectively, and the maximum and minimum temperatures of 1150 K and 300 K respectively. Determine the thermal efficiency, the compressor work, the turbine work and the air- flow rate for 10 kW of net work output.

Solution

From Fig. 23.7

$$\eta_{th} = 1 - \frac{1}{r_p^{(\gamma-1)/\gamma}} = 1 - \frac{1}{5^{0.4/1.4}} = 0.37$$

The work done in the compressor is

$$W_c = C_p(T_2 - T_1)$$

where T_1 is the minimum temperature. The value of T_2 is obtained from isentropic relation

$$\frac{T_2}{T_1} = \left(\frac{p_2}{p_1}\right)^{(\gamma-1)/\gamma} = 5^{0.4/1.4} = 1.584$$

or

$$T_2 = 300 \times 1.584 = 475 \text{ K}$$

Therefore, the compressor work/kg of air is

$$W_c = C_p(T_2 - T_1) = 1.0038(475 - 300) = 176 \text{ kJ/kg}$$

The turbine work is given by $C_p(T_3 - T_4)$. The value of T_3 is the maximum cycle temperature of 1150 K, and that of T_4 is obtained as

$$\frac{T_4}{T_3} = \frac{1}{r_p^{(\gamma-1)/\gamma}} = \frac{1}{1.584}$$

so that $T_4 = 1150/1.584 = 726$ K.
The turbine work/kg of air then is

$$W_t = C_p(T_3 - T_4) = 1.0038(1150 - 726) = 426 \text{ kJ/kg}$$

The net work per cycle per kg = (426-176) = 250 kJ/kg. The flow rate required for 10 kW of net power then is 10 kW/250 (kJ/kg)=0.04 kg/s.

Problems

23.1 Calculate the thermal efficiency of an Otto cycle for the compression ratios of 2, 4, 6, 8, 10 and 12. Plot efficiency versus compression ratio.

23.2 Sketch the Otto cycle on a $T - s$ diagram.

23.3 An Otto cycle has the following specifications:

Minimum temperature: 300 K; Minimum pressure: 100 kPa; Compression ratio: 8; Maximum pressure = 8 MPa. Determine the cycle thermal efficiency and the amount of heat addition in kJ/kg.

23.4 An air standard Diesel cycle has a compression ratio of 15. Calculate and plot the thermal efficiencies for cut-off ratios of 1.5, 2, 2.5 and 3.

23.5 Sketch a Diesel cycle on a $T - s$ diagram.

23.6 An air-standard Diesel cycle has the following specifications:

Minimum pressure: 100 kPa; Minimum temperature: 300 K; Compression ratio: 15; Maximum cycle temperature: 2800 K. Determine the cycle efficiency and the amount of heat addition in kJ/kg.

23.7 Two Diesel cycles A and B have the same starting point and the same compression ratio. But cycle B has a larger cut-off ratio than cycle A. Which cycle will have a higher thermal efficiency and which will have more work output per cycle?

23.8 Calculate and plot the thermal efficiencies of a air-standard Brayton cycle for pressure ratios of 4, 6, 8, 10 and 12.

23.9 A gas turbine operates on the following specifications:

Maximum and minimum cycle temperatures of 1120 K and 300 K respectively; Pressure ratio of 8. Calculate the thermal efficiency, assuming air to be the working fluid.

23.10 A Brayton cycle operates with air at a pressure ratio of 5. The conditions at the compressor inlet are 101 kPa and 300 K. The temperature of the fluid at the turbine inlet is 1150 K. What is thermal efficiency of the cycle and that of a Carnot cycle working between the two extreme temperatures reached?

BIBLIOGRAPHY

Statics

1. Den Hartog J P, Mechanics, Dover, 1948.
2. Huang T C, Engineering Mechanics – Statics and Dynamics, Addison-Wesley, 1968.
3. Crandall S H, Dahl N C, Lardner T J, An Introduction to the Mechanics of Solids, McGraw-Hill, 1978.
4. Meriam J L, Engineering Mechanics, Vol. 1 – Statics, John Wiley, 1980.
5. Beer F P, Johnstan E R, Vector Mechanics for Engineers - Statics, McGraw-Hill, 1988.

Dynamics

1. Timoshenko S and Young D H, Engineering Mechanics, McGraw-Hill Kogakusha, 1956.
2. Meriam J L, Engineering Mechanics Vol. 2–Dynamics (S.I. Version), John Wiley, 1980.
3. Beer F P and Johnston Jr. E R, Mechanics for Engineers – Dynamics, McGraw-Hill, 1987.
4. Shames I G, Engineering Mechanics, Prentice–Hall India, 1990.

Strength of Materials

1. Timoshenko S P, Strength of Materials, Parts I and II, Van Nostrand, 1959.
2. Timoshenko S P, MacCullough G H, Elements of Strength of Materials, Van Nostrand, 1961.
3. Case J, Chilver A H, Strength of Materials and Structures, Edward Arnold, 1971.
4. Higdon A, Ohlsen E H, Stiles W B, Weese J A, Riley W F, Mechanics of Materials, John Wiley, 1978.

5. Singh P N, Jha P K, Elementary Mechanics of Solids, Wiley Eastern, 1980.

6. Srinath L S, Advanced Mechanics of Solids, Tata McGraw-Hill, 1980.

7. Beer F P, Johnston Jr. E R, Mechanics of Materials, McGraw-Hill, 1987.

8. Benham P P, Crawford R J, Mechanics of Engineering Materials, John Wiley, 1987.

9. Das B M, Hassler P C, Statics and Mechanics of Materials, Prentice—Hall, 1988.

10. Mott Robert L, Applied Strength of Materials, Prentice Hall, 1989.

Fluid Mechanics

1. Garde R J, Fluid Mechanics through Problems, Wiley Eastern, 1989.

2. Kumar K L, Engineering Fluid Mechanics, Eurasia Publishing House, 1990.

3. Gupta V, Gupta S K, Fluid Mechanics and its Application, Wiley Eastern, 1994.

Thermodynamics

1. Wylen G J V and Sonntag R E, Fundamentals of Classical Thermodynamics, John Wiley, 1976.

2. Reynolds W C, Perkins H C, Engineering Thermodynamics, McGraw-Hill, 1977.

3. Nag P K, Engineering Thermodynamics, Tata McGraw-Hill, 1981.

4. Rao Y V C, An Introduction to Thermodynamics, Wiley Eastern, 1993.

Index